Renewable Energy Technology

Renewable Energy Technology

Edited by **Benjamin Wayne**

Callisto Reference
New York

Published by Callisto Reference,
106 Park Avenue, Suite 200,
New York, NY 10016, USA
www.callistoreference.com

Renewable Energy Technology
Edited by Benjamin Wayne

International Standard Book Number: 978-1-63239-731-7 (Hardback)

Printed in China.

Contents

Preface

Every book is initially just a concept; it takes months of research and hard work to give it the final shape in which the readers receive it. In its early stages, this book also went through rigorous reviewing. The notable contributions made by experts from across the globe were first molded into patterned chapters and then arranged in a sensibly sequential manner to bring out the best results.

Renewable energy sources are the ones that are readily available and can be replenished in shorter span of time than the non-renewable sources. Wind energy, solar energy, tidal energy, bioenergy and geothermal energy are some of the main renewable sources in use and under consideration for improved technologies. Effective utilization of these sources would greatly benefit humankind as it will reduce air pollution and global warming drastically. The various studies that are constantly contributing towards advancing technologies and evolution of this field are examined in detail in this book. From theories to research to practical applications, case studies related to all contemporary topics of relevance to this field have been included in this book. It will help the readers in keeping pace with the rapid changes in this area of study.

It has been my immense pleasure to be a part of this project and to contribute my years of learning in such a meaningful form. I would like to take this opportunity to thank all the people who have been associated with the completion of this book at any step.

Editor

1

Hybrid Power System Options for Off-Grid Rural Electrification in Northern Kenya

June M. Lukuyu, Judith B. Cardell

Picker Engineering Program, Smith College, Northampton, USA
Email: junelukuyu@gmail.com, jcardell@smith.edu

Abstract

For domestic consumers in the rural areas of northern Kenya, as in other developing countries, the typical source of electrical supply is diesel generators. However, diesel generators are associated with both CO_2 emissions, which adversely affect the environment and increase diesel fuel prices, which inflate the prices of consumer goods. The Kenya government has taken steps towards addressing this issue by proposing The Hybrid Mini-Grid Project, which involves the installation of 3 MW of wind and solar energy systems in facilities with existing diesel generators. However, this project has not yet been implemented. As a contribution to this effort, this study proposes, simulates and analyzes five different configurations of hybrid energy systems incorporating wind energy, solar energy and battery storage to replace the stand-alone diesel power systems servicing six remote villages in northern Kenya. If implemented, the systems proposed here would reduce Kenya's dependency on diesel fuel, leading to reductions in its carbon footprint. This analysis confirms the feasibility of these hybrid systems with many configurations being profitable. A Multi-Attribute Trade-Off Analysis is employed to determine the best hybrid system configuration option that would reduce diesel fuel consumption and jointly minimize CO_2 emissions and net present cost. This analysis determined that a wind-diesel-battery configuration consisting of two 500 kW turbines, 1200 kW diesel capacity and 95,040 Ah battery capacity is the best option to replace a 3200 kW stand-alone diesel system providing electricity to a village with a peak demand of 839 kW. It has the potential to reduce diesel fuel consumption and CO_2 emissions by up to 98.8%.

Keywords

Hybrid Power System, Off-Grid Power System, Wind Energy, Solar Energy, Battery Storage, Multi-Attribute Trade-Off Analysis

1. Introduction

Electricity is a major contributor of global socioeconomic development. It is the foundation for technological advancement, global communication and industrialization, which results in a higher standard of living. However, numerous remote areas in developing countries are not connected to a centralized power grid. The extension of the main grid to low population areas, especially those situated in difficult terrain is associated with high capital outlays and high transmission losses. A study carried out by the World Bank on rural electrification programs placed the average cost of grid extension at between $8000 and $10,000 per km, rising to around $22,000 per km in difficult terrain [1]. For domestic consumers in the rural areas that are not served by a central power grid, the typical method of electrical supply is diesel generators. In the case of Kenya, which provides the focus of this study, there are six diesel power stations providing electricity to remote villages in the Northern part of Kenya (Mandera, Moyale, El Wak, Marsabit, Wajir and Lodwar) as shown in **Figure 1**.

The price of diesel fuel in Kenya has been escalating over the past decade and has increased by 143% since 2002 [2]. Increasing fuel prices can slow the economic growth of a country and inflate the prices of consumer goods produced in the industries that rely on fuel. This means that despite the high poverty levels in rural areas, the residents will face higher prices not only for electricity in their homes but also for consumer goods and services that are dependent on electricity.

Dependency on fuel imports for the production of electricity also affects national security. Governments in the fuel exporting countries have leverage over policies in the fuel importing countries, through which they could potentially exploit for their benefit. In addition, servicing these villages purely on diesel generation is accompanied by significant greenhouse gas emissions, which contribute to climate change.

Although Kenya's contribution to global climate change is relatively small, its greenhouse gas emissions are expected to increase rapidly as it becomes more industrialized. Since 2003, there has been a 1.7 million metric ton increase in CO_2 emissions in Kenya from electricity and heat production [3]. In the bid to ensure environmental sustainability, United Nations Energy points out the need to use cleaner fuels, and introduces renewable energy in order to reduce environmental and health hazards caused by greenhouse gas emissions [4]. This study addresses the aforementioned problems caused by the sole use of diesel fuel for electricity production by designing functional and reliable off-grid hybrid power systems, which incorporate renewable energy into the existing diesel power systems in the bid to reduce northern Kenya's dependency on diesel fuel and consequently reduce the carbon footprint. A hybrid power system is a small, often stand-alone system that uses more than one generating technology, usually consisting of one or more renewable energy sources to produce electricity. **Figure 2** shows a possible configuration of a hybrid power system.

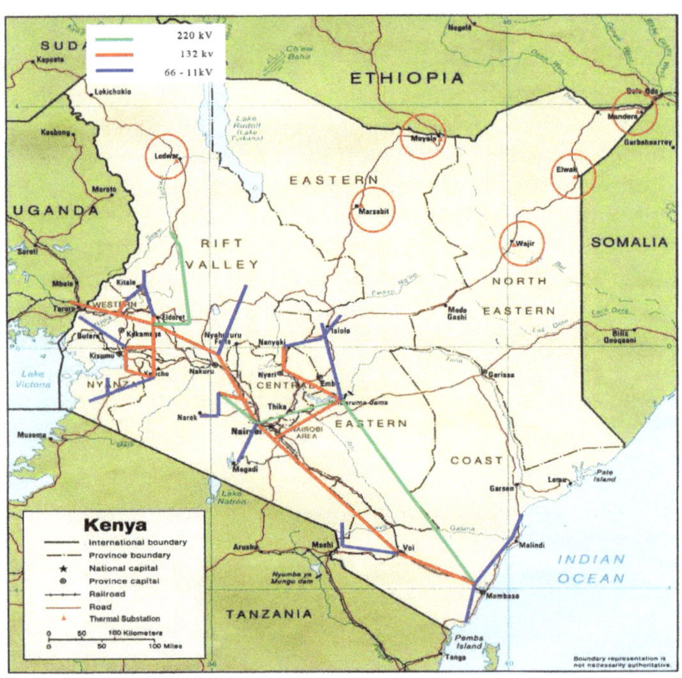

Figure 1. National electric grid system in Kenya showing the six diesel power stations in Northern Kenya [5].

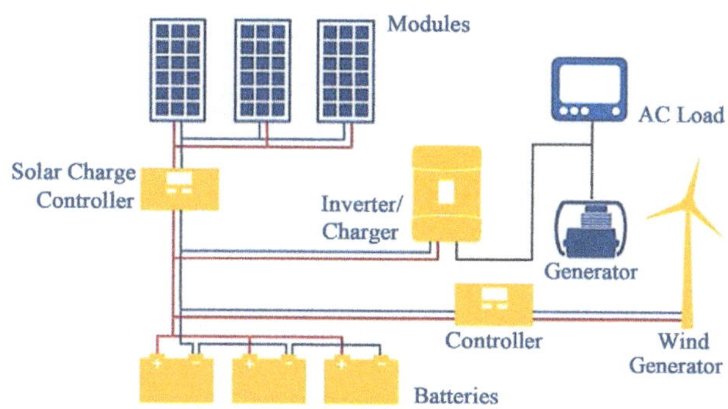

Figure 2. Typical hybrid power system incorporating wind turbines, PV panels and battery storage [6].

Section 2 presents a background of energy policy in Kenya, Section 3 discusses other hybrid power system projects in developing regions and Section 4 introduces and defines Multi-Attribute Trade-Off Analysis. Section 5 presents input data assessments for Kenyan electrical demand and the wind and solar resources. Section 6 defines the technologies and hybrid systems that are modeled in this analysis, while the results and analyses of the hybrid system options, along with sensitivity analysis on the input data uncertainties are in Section 7. Section 8 concludes and presents directions for future work.

2. Energy Policy Background

In order to achieve the long-term development strategy of the country, Vision 2030, the Kenyan Government identified energy as one of the foundations and enablers of the socio-economic transformation envisioned in the country [7]. They also recognize that climate change could hamper the ambitious development goals articulated in Vision 2030 and developed the National Climate Change Response Strategy of 2010 that integrates the principles of sustainable development into national policies and programs. A Feed-in-Tariffs (FiT) Policy was formulated in 2008, and subsequently revised in 2010, to promote the use of green (low-carbon emissions) energy for electricity generation where feasible, along with improving efficiency in the supply and the end use of electricity. The Feed-in-Tariff policy makes it mandatory for energy companies or "utilities" responsible for operating the national grid to purchase electricity from renewable energy sources at a pre-determined price that is sufficiently attractive to stimulate new investment in the renewables sector. This, in turn, ensures that those who produce electricity from identified renewable energy sources such as solar, wind and other renewable sources have a guaranteed market and an attractive return on investment for the electricity they produce [7].

In accordance with this policy, the government, in consultation with stakeholders, proposed The Hybrid Mini-grid Project, which involves the installation of 3 MW of wind and solar energy systems in facilities with existing diesel generators. It also proposes a replicable business model for installing 27 additional mini-grids, amounting to 13 MW, consequently reducing GHG emissions in rural electrification [8]. This is expected to make electricity more affordable for the poor and increase generation capacity that will enable more connections and increase access. It will also be instrumental in the advancement of environmental sustainability through the reduction of greenhouse gas emissions. One of the stakeholders and development partners, The African Development Bank, will support conversion of diesel generators into hybrid systems (wind, solar, biomass) and construction of new generators and associated mini-grids in rural areas as outlined in the SREP investment plan for Kenya [4].

3. Literature Review

Successful results have previously been obtained with hybrid systems in developing countries. Rural communities not able to connect to a national grid and lacking resources to pay the increasing fuel prices have found hybrid systems the most suitable, environmentally friendly and cost competitive solution for power delivery. A photovoltaic/diesel hybrid system installed in rural Tanzania in 2006 provides electricity to several households, community services, small workshops and technical equipment [9]. A similar hybrid system configuration installed in Algeria in 1998 provides electricity to 12 households and community services [9]. A hydro/PV/diesel hybrid system was installed in Laos in 2007 and provides electricity to 98 households and community services [9]. In China, a photovoltaic/wind/diesel hybrid system installed in 2002 provides electricity to 3 villages composed of 500 households, community services (clinic, school, postal office, and TV transferring station) and a tourist facility [9].

Studies have also been carried out to investigate the potential implementation of off-grid hybrid systems in remote villages. These studies analyzed the technical and economic feasibility of the proposed systems coupled with the resultant reduction of CO_2 emissions. A wind/PV/diesel hybrid power system was designed for a village in Saudi Arabia, which is presently powered by a diesel power plant consisting of eight diesel generator sets of 1120 kW each. The study done by [10] found that the proposed system comprising 3 wind turbines each of 600 kW, 1000 kW of PV panels and four diesel generator sets of 1120 kW each was able to meet the energy requirements (AC primary load of 17043.4 MWh/year) of the village and avoid an addition of 4976.8 metric tons of CO_2 into the local atmosphere and conserve 10,824 barrels of fossil fuel annually. A study done by [11] proposed a design of an off-grid wind-diesel hybrid system incorporating CHP technology in an island of Bangladesh to serve an energy demand of 168 MWh/day. A cost comparison based on the cost of electricity (COE) between the wind-diesel system and a PV-diesel system was done. It showed that the wind-diesel hybrid system incorporating CHP technology was the most cost effective off grid power system for the island and would result in a CO_2 emissions reduction by 33,053 metric tons per year. Reference [12] presented the economic feasibility study of adding wind energy into an isolated off-grid diesel power plant servicing a remote village in Saudi Arabia with an energy demand of 15,943 MWh/yr. Different numbers of 600 kW wind turbines were added to 6 existing diesel units of 1120 kW each and simulated in order to determine the most economically feasible option based on net present cost (NPC) and cost of electricity (COE). The study showed that the hybrid system that incorporated 5 wind turbines to 3 existing diesel generators, was the most feasible option when the wind speed of the area was 7 m/s (51% wind penetration). The system also reduced the CO_2 emissions by 3180 metric tons/yr.

Analogous to the aforementioned case studies, this study analyzes the potential implementation of off-grid hybrid power systems in northern Kenya [13]. Different hybrid system configurations consisting of the existing diesel generators plus wind turbines, PV panels and battery storage are modeled and analyzed. However, unlike the previous studies, this study employs a multi-attribute trade-off analysis to determine the best hybrid system configuration option that would reduce diesel fuel consumption, consequently minimizing the CO_2 emissions at a minimal net present cost (NPC).

4. Multi-Attribute Trade-Off Analysis

Multi-attribute trade-off analysis enables the evaluation and comparison of system design alternatives for a given planning situation [14]. This approach is based on the premise that most, if not all, system planning problems do not have a single optimum solution but rather have several alternative dominant solutions. The perceived value of these dominant, alternative solutions differs for different stakeholder groups, depending upon their preferences for different attributes, such as the level of system reliability, and different policy and societal priorities [14].

A visual representation of a trade-off analysis is shown in **Figure 3**. Each plotted point (the colored boxes and circles) represents one planning scenario and shows the performance of that scenario in terms of the attributes shown on the x- and y-axes. For the study presented in this paper, planning scenarios differ according to the hybrid system configuration and the capacity of the system components (wind turbines, PV capacity and battery capacity).

Returning to **Figure 3**, the set of dominant scenarios are shown on the Pareto Frontier/Curve, the curve closest to the lower left (origin). In this example, the Pareto frontier represents the trade-off between environmental performance and cost.

Both inferior and superior planning options are easily identified from the trade-off plot, with the Pareto frontier identifying all potentially optimal scenarios. The system design alternatives that lie on the frontier curve are considered optimal, dominant or Pareto superior. The remaining system design alternatives are therefore inferior because there is always an alternative on the Pareto frontier that is either less costly or more environmentally friendly or both [15]. For the example in **Figure 3**, the star shape identifies the scenario with jointly minimized cost and CO_2 emissions.

The results of the multi-attribute trade-off analysis performed for this study identify the best hybrid system

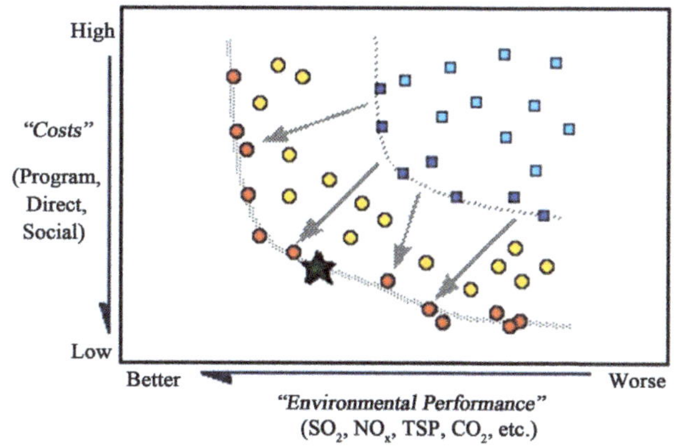

Figure 3. Typical pareto frontier [14].

designs for northern Kenya, which achieve the optimal reduction in CO_2 emissions at a reasonable cost. The results of the hybrid system design analyses for this project are presented in Section 7.

5. Energy Demand & Renewable Resources

Of the six towns in Northern Kenya identified for this study, only Marsabit had the required electrical demand (load), wind speed and solar insolation data readily available. Marsabit was therefore used as the case study area and a sensitivity analysis was carried out to determine the impact of average wind speed variations, average load variations and diesel fuel price variations on the hybrid system design for Marsabit. This analysis was then used to determine how the best hybrid system configuration option for Marsabit would change for the other five locations.

The town of Marsabit is located 170 km east of the center of the East African Rift in the Eastern Province of Kenya at 37°58'E, 2°19'N and is almost surrounded by the Marsabit National Park and Reserve. According to the 2009 Kenya population census carried out by the Kenya Bureau of Statistics, Marsabit has a population density of 5 people per square km [16]. The district has limited coverage of classified roads, 62.7 km gravel surface and 37 km earth surface roads, which are impassable during the rainy season [17]. A wind-diesel hybrid system is currently operational in Marsabit comprising of 3 diesel generators with a total capacity of 3.2 MW and a 275 kW wind turbine. **Figure 4** shows an overhead satellite map of the study area to give an idea of the terrain of the region.

5.1. Load/Demand Assessment

In many cases, actual load data for rural populations is not readily available and this was the case with the remote regions in Northern Kenya. However, a 24-hour load profile for a typical month in Marsabit was obtained, see **Figure 5** [18], and was used to develop the electrical demand data required for this study. As of June 2010, maximum demand in Marsabit was reported to be 740 kW [19]. According to Kenya Power Company, formerly known as Kenya Power and Lighting Company, power demand in Kenya is projected to grow at an average rate of 6.5% per annum during 2003 to 2023 [20] with the hourly load profile assumed to remain the same. Based on this prediction, the peak load for the year 2012 was adjusted to 839 kW. Electrical consumption does not vary much from month to month since Kenya does not experience winter and summer except for the months of April, August and December when children are on school holidays. During these three months, demand is estimated to increase by 10% [21]. Therefore, the peak load during the other seven months of the year is 755 kW.

5.2. Wind Resource Assessment

In order to estimate the expected power output of the wind turbines in a hybrid system, data pertaining to the wind resource available at the location is required. The annual average wind speeds in northern Kenya collected at an anemometer height of 50 m typically range between 3 m/s and 9 m/s as shown in **Figure 6** [22]. Marsabit typically records twice the annual average wind speed range (8.8 - 9.2 m/s) compared to the other five locations, which record an annual average wind speed range of 4.0 - 4.8 m/s. Kenya's lack of sufficient wind data is a ma-

Figure 4. Satellite map of Marsabit.

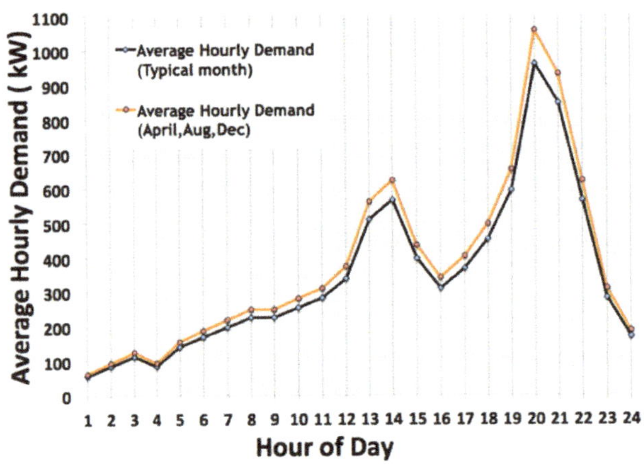

Figure 5. 24-Hour load profile, Marsabit.

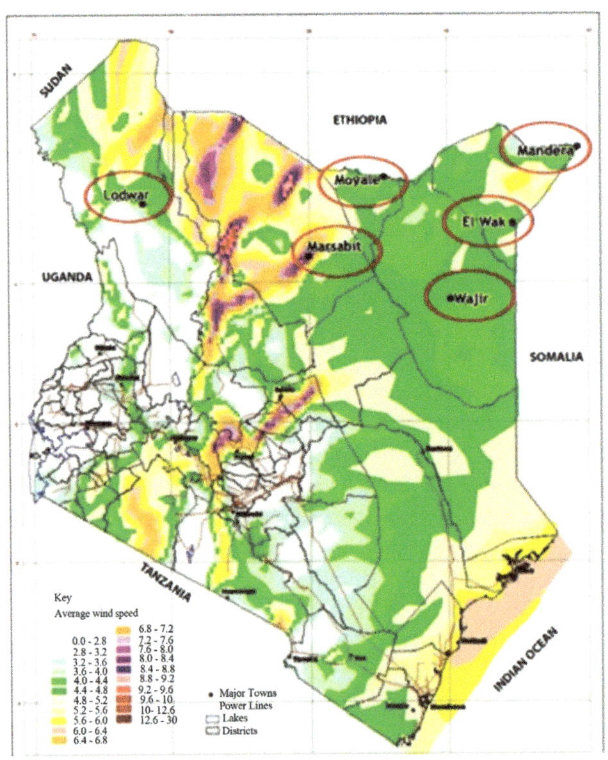

Figure 6. Annual average wind speeds (m/s) over Kenya at an anemometer height of 50 m.

jor barrier affecting the exploitation of wind energy. In particular, multi-year 10 minutes or hourly time series wind data, ideal for this type of analysis, is not available at these locations. Monthly and daily averaged wind speed data was however obtained for the Marsabit region [23]. It is therefore used for the subsequent wind resource analysis and hybrid system.

The monthly average wind speed data for Marsabit was collected and averaged over a 6-year period from 2001-2006 at an anemometer height of 10 m and is shown in **Figure 7** [23]. The standard anemometer height is 10 m, however wind power output is determined using wind speeds at the hub height, which is the distance from the turbine platform to the rotor of an installed wind turbine. Therefore, the wind speed at the hub height is predicted in terms of the measured speed at the anemometer height using Equation (1) [23].

$$\frac{v_2}{v_1} = \left(\frac{z_2}{z_1}\right)^{a} \tag{1}$$

where v_2 is the extrapolated wind speed at height z_2 (hub height) and v_1 is the measured speed at z_1 (anemometer height). The exponent α is the wind shear exponent, which determines the shape of the vertical wind profile in the lower stratum of the atmospheric surface layer (ASL), and depends on surface roughness. With a common approximation, adopted for this analysis, the $^1/_7{}^{\text{th}}$ power law assumes a constant factor of 0.143 for the shear exponent (which does not depend on geographical features of the monitoring station and ambient atmospheric condition) [24].

The hourly average wind speed data for Marsabit collected in 2006 is shown in **Figure 8** [18]. High wind speeds are prevalent during nighttime hours and they peak above 12 m/s during the early morning hours. The effect of the month-to-month averaged variation on the diurnal wind speed shape is not modeled owing to the unattainability of hourly time series wind speed data for a full year.

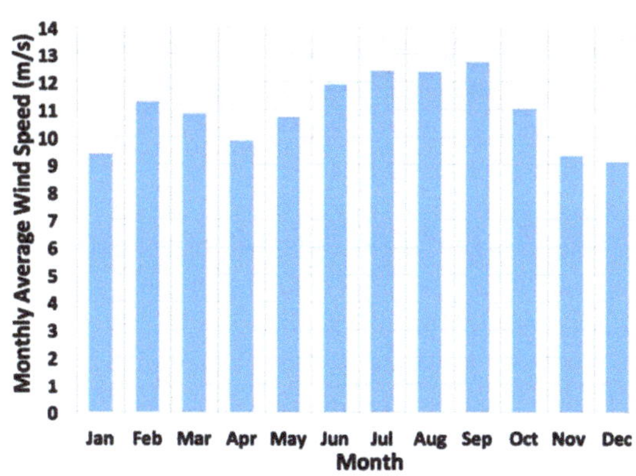

Figure 7. Monthly average wind speeds in Marsabit at an anemometer height of 10 m.

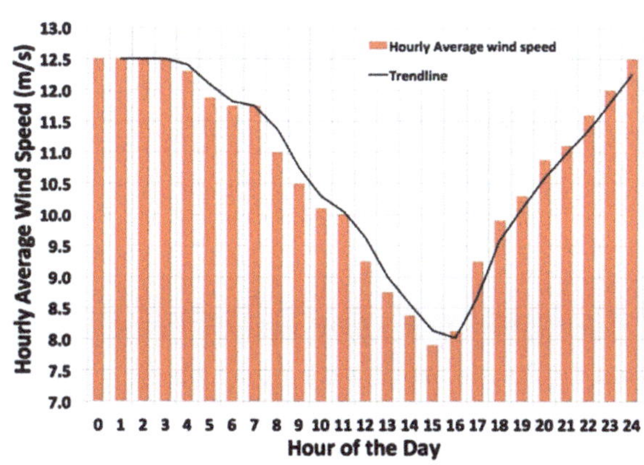

Figure 8. Hourly average wind speeds in Marsabit in 2006 at anemometer height of 10 m.

5.3. Solar Resource Assessment

Kenya is situated on the equator therefore it generally receives consistently high solar radiation. Solar insolation is a measure of solar radiation recorded during a given time, usually an hour on a unit surface area. It is measured in kWh/m^2/day. Solar insolation data is used to assess the site-specific solar availability and these values are used to determine the amount of photovoltaic energy output. The system designer must specify the monthly average solar insolation data (kWh/m^2/day) and the clearness index values for each month. The clearness index metric is used to approximate the effect of cloud cover on the amount of solar energy that can be harvested at a particular location. It is a measure of the fraction of extraterrestrial radiation, which reaches a given location [25].

Terrestrial global horizontal solar insolation values in northern Kenya typically range between 5 and 6.5 kWh/m^2/day [26]. The monthly average daily total solar insolation (**H**) on the horizontal for Marsabit was obtained from the NASA Atmospheric Science Data Center for the latitude and longitude representative of the Marsabit site over a 22-year period from 1983-2005 for the months beginning in January and ending in December [27]. The data collected on the horizontal was then elevated to the plane of array at optimal tilt angle to assess the actual plane of array solar insolation available ($\mathbf{H_T}$) and is shown in **Figure 9**. Average values for daily total solar insolation are commonly lower between May and August than between January and March.

Hourly averaged solar radiation data on a horizontal surface for Marsabit town at 3-hour intervals was obtained from the NASA Atmospheric Science Data Center over a 22-year period from 1983-2005 for the months beginning in January and ending in December [27]. These values, averaged for each of the four seasons are shown in **Figure 10**. Peak average solar radiation is reported at around solar noon.

6. Hybrid System Technologies

As discussed in the introduction, the goal of this research project is to design and evaluate a variety of off-grid hybrid electric power systems for use in rural, Northern Kenya. The hybrid system configurations proposed in this study incorporate wind energy and solar energy into the existing diesel power systems. These configurations are modeled and analyzed to determine the best configurations considering the net present cost (NPC) of the system over its lifetime, diesel fuel consumption and CO_2 emissions.

The software package used for the analysis is HOMER, which simulates the operation of various hybrid power system configurations and ranks them on the basis of the net present cost, NPC (the total cost of installing and operating the system over its lifetime). Extensive documentation and examples for use can be found at [28]. HOMER assists the system designer in assessing the technical feasibility of each system design (whether the system can adequately serve the load and other constraints imposed by the user) and then estimates the net present cost (NPC) of the system [29].

For the study presented here, five hybrid system configurations, in addition to existing stand-alone diesel system, are modeled, simulated and analyzed. These configurations include:

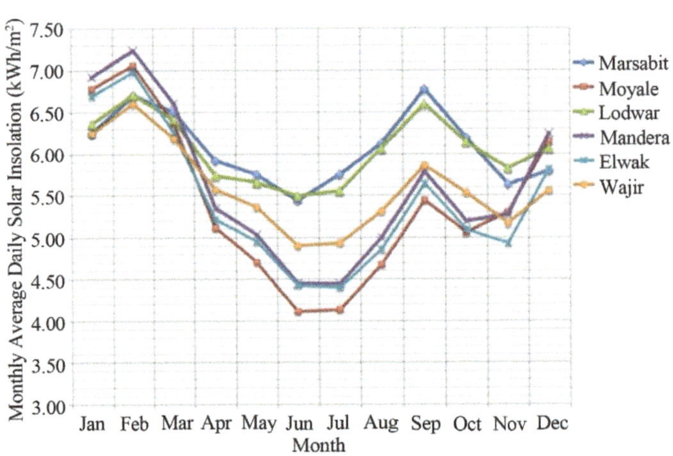

Figure 9. Monthly average daily total solar insolation elevated to plane of array at optimal tilt angle for the six locations in kWh/m^2.

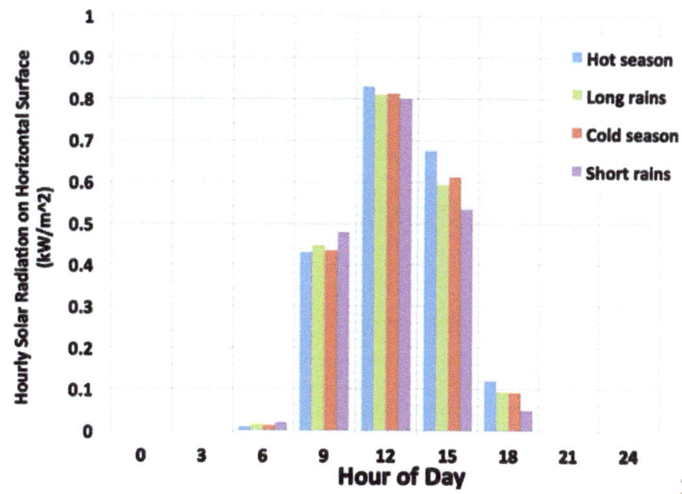

Figure 10. Hourly averaged solar insolation data on a horizontal surface for Marsabit town at 3-hour intervals.

1) Wind-diesel hybrid system
2) Wind-diesel-battery hybrid system
3) PV-diesel hybrid system
4) PV-diesel-battery hybrid system
5) Wind-PV-diesel-battery hybrid system

Multi-attribute trade-off analysis is performed for each design option, with representative trade-off graphs included for design options 2 and 5, along with the final multi-attribute trade-off graph comparing all design options.

Sensitivity analyses are conducted to evaluate the economic and technical feasibility of a large number of-technology options in order to account for variations in technology costs and energy resource availability.

Hybrid System Technology Selections

Table 1 identifies the technologies selected for the hybrid system designs. A detailed discussion and analysis justifying the selection of each technology option is provided in [13].

Possible system configurations within each design option vary the number and capacity of each technology. The configurations investigated for the wind-diesel design option, for example, include the six possible permutations of the technology mix including one or both wind turbines and one, two or three diesel generators. The number and capacity of units considered is summarized in **Table 2**. These technology options are discussed in more detail below, within the analysis sections for each design option.

7. Results and Discussion

7.1. Existing System Design

For comparison to the existing electric power supply, the design option using only diesel generators is investigated first. In this stand-alone diesel system, the diesel generators are the only source of power and run throughout the year to meet the entire electrical demand. The stand-alone diesel system currently servicing Marsabit town is comprised of three diesel generators: one CAT® 1200 kW diesel generator and two CAT® 1056 kW diesel generators (see **Table 1**).

7.2. Option 1: Wind-Diesel System

In this first proposed hybrid system configuration, wind turbines are added to the stand-alone diesel system. The diesel generators operate only during hours when the wind turbines cannot meet the electrical demand. This system is designed to show significant fuel and emissions savings in accordance with the project statement of Vision 2030 [30], and at the same time ensuring system stability. In order to adhere to this design criteria, a

Table 1. Technologies used in hybrid system designs.

Category	Technology Selection
Diesel Generators	Caterpillar 3512B-HD TA 1200 ekW, Caterpillar 3512B TA 1056 ekW
Wind Turbines	Vestas v47 RRB Energy PS500
PV Panels	24-volt Polycrystalline SP 195 PB (Sollatek East Africa)
Battery System	Trojan Deep Cycle VRLA AGM 12 Volt
Inverter	LTI Power Systems 48 VDC 10kVA

Table 2. Component input values considered in HOMER simulations.

Technology	Number of Units	Sizes Considered
Diesel Generators	1, 2 or 3	1200 kW, 1056 kW
Wind Turbines	1 or 2	500 kW
PV Panels	1 or 2	493 kW, 839 kW
Battery System	(See Section 7.3)	2, 4 and 6 autonomy hours

maximum of two 500 kW Vestas v47 wind turbines are added to the system. The capital cost of the wind turbines is $2438 per kW. In this hybrid design analysis, the system is varied to include one, or both, wind turbines, in addition to 1, 2 or all 3 diesel generators, in different combinations.

The system configuration that jointly minimizes NPC and CO_2 emissions is found to consist of one 1056 kW diesel generator and both 500 kW turbines. It is associated with an NPC of $8,349,152 and 429.4 metric tons of CO_2emissions per year.

The term *renewable fraction* defines the ratio of the total renewable energy generated to the amount used to serve primary load. This system option has a renewable fraction of 0.8 indicating that 80% of the wind energy produced serves the load. The amount of electricity produced by the wind turbines is 7,864,188 kWh/yr while the AC primary load being met is 2,417,384 kWh/yr. Therefore 71% of the electricity produced by wind is excess (5,924,869 kWh/yr). A battery system is therefore recommended to store the excess electricity produced.

7.3. Option 2: Wind-Diesel-Battery System

In this proposed system configuration, batteries are added to the wind-diesel hybrid system for improved stability of the system given the intermittent nature of the wind, and also to store the excess electricity produced by the wind turbines. If power output from the wind turbines exceeds the demand, the surplus power is stored in the batteries. If demand is higher than the wind power output, the batteries cover the energy deficit until the battery maximum depth of discharge, DOD, is reached. If this limit is reached, the diesel generators are turned on. An inverter is also added to the system to convert DC current from the battery to AC current.

In order to determine the size of the batteries needed, the autonomy hours of the system need to be specified. Thus the simulations for this hybrid design option vary the battery capacity to allow for 2, 4 and 6 autonomy hours (requiring 660, 1320 and 1976 batteries respectively), in addition to including one or both wind turbines and 1, 2 or all 3 diesel generators.

Marsabit has an average daily load of 6.6 MWh/day. Together with a DC system voltage of 48 VDC, inverter efficiency of 0.97 and a battery DOD of 0.5, the daily average capacity of the battery system was calculated to be 142,192.35 Ah/day. The cost of each VRLA AGM 12 Volt battery is $372.59. The typical life cycle of a VRLA battery is 2 years or less depending on the frequency of discharge. For this system, the batteries are assumed to be replaced every two years.

The use of the multi-attribute trade-off analysis for this design option is demonstrated in **Figure 11**. The relationship between the NPC and CO_2 emissions of each alternative system configuration is shown in this figure, with the Pareto optimal frontier drawn in as the dashed curve in the left side of the figure. The system configurations not on the frontier curve are sub-optimal to those on this curve. The use of multi-attribute trade-off analysis

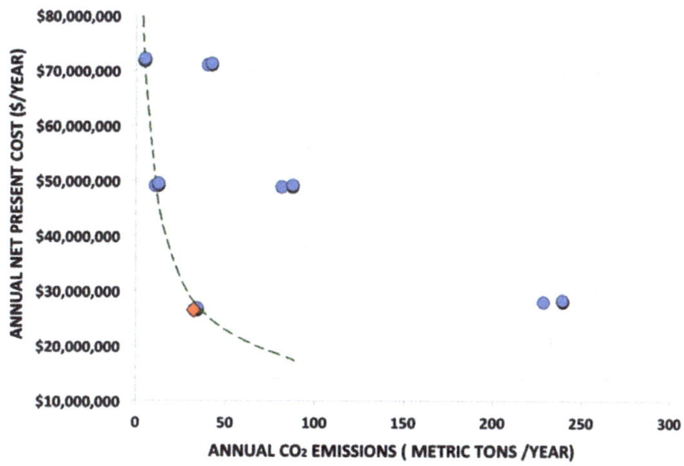

Figure 11. Pareto frontier for wind-diesel-battery hybrid system design configuration.

assists policy development and decision makers by effectively removing all system configurations from further consideration unless they are on the Pareto frontier curve.

For the analysis shown in **Figure 11**, the clusters of system configuration options along the curve vary in the number of battery autonomy hours, which determines the total capacity of the battery system. The NPC increases as autonomy hours increase (moving up the curve).

The red diamond shape toward the bottom of the curve represents the wind-diesel-battery hybrid system configuration that is associated with jointly minimal NPC and CO_2 emissions. The system configuration consists of one 1200 kW generator, two 500 kW turbines and 66,012 V batteries. It is associated with an NPC of $26,548,492 and 32.5 metric tons of CO_2 emissions per year. It has a renewable fraction of 0.98. The wind turbines produce 7,864,188 kWh/yr of electricity.

Even with the addition of a battery system, there is still an excess electricity production of 5,924,869 kWh/yr. However, based on these results the capacity factor of the wind turbines was calculated to be 89.8%, which is significantly higher than the typical range of wind capacity factor values (20% - 40%). This is likely due to the unavailability of time step series wind speed data. Instead, average wind speed values are being used to run the simulations. Based on a capacity factor of 40%, the wind turbines would actually produce 3,504,000 kWh of electricity per year, which would still be adequate to meet demand and the surplus electricity produced would be substantially less.

7.4. Option 3: PV-Diesel Hybrid System

In this proposed system, PV panels and an inverter are added to the original, stand-alone diesel system. This system design is varied for a PV capacity of 493 kW and 839 kW, representing the demand reported at the time of maximum solar insolation and the maximum system demand respectively. The design configurations analyzed are also varied by including 1, 2 or all 3 diesel generators.

The system configuration options that incorporate the lower PV capacity of 493 kW are intriguingly more costly than those with a higher PV capacity. This is likely due to the high diesel fuel prices. The higher fuel consumption (116,873 L/yr) in the low PV configuration outweighs the cost of adding 454 kW of PV capacity.

The dominant system configuration for this hybrid design option consists of one 1200 kW diesel generator, one 1056 kW diesel generator and 839 kW PV rated capacity. It is associated with an NPC of $37,568,924 and 2446.3 metric tons of CO_2 emissions per year. The PV array produces 1,602,735 kWh/yr, which is 21.8% of its rated capacity. Photovoltaic array capacity factors are typically under 25% [31]. Less than 10% of this energy goes into meeting demand. The surplus energy produced from this system is 2,031,636 kWh/yr. A battery system is therefore recommended to store the surplus energy.

7.5. Option 4: PV-Diesel-Battery Hybrid System

In this proposed system, battery storage is added to the PV-diesel hybrid option to improve the stability of the system. Excess energy from the PV system (energy above the hourly demand) is stored in the battery until full capacity of the storage is reached. The diesel generators are operated only when the PV system cannot generate enough electricity to meet demand and when the battery storage is depleted.

The battery options are the same as those for the wind-diesel hybrid system. In addition to varying the system for 1, 2 or all 3 diesel generator, 493 kW and 839 kW PV capacity, this system is also varied for 2, 4 and 6 autonomy hours for the battery system. The relationship between NPC and CO_2 emissions of each alternative system design is shown in **Figure 12** with the Pareto optimal frontier curve drawn in on the left side of the figure.

The system configuration options along the curve vary in the number of autonomy hours, which subsequently determines the total capacity of the battery system. However, they all incorporate one 1200 kW diesel generator and 839 kW PV capacity. Of these three options, the red diamond shape represents the dominant system configuration as associated with a jointly minimal NPC and CO_2 emissions for this design option. One of the remaining two options on the curve is associated with 123.7 metric tons/yr. increase in CO_2 emissions at a $21,293,988/yr decrease in cost and the other option is associated with an 81.7 metric tons/yr decrease in CO_2 emissions at a $21,903,896/yr increase in cost.

The optimal system identified here consists of one 1200 kW diesel generator, 132,012 V batteries and 839 kW PV capacity. It is associated with an NPC of $64,140,568 and 983.3 metric tons of CO_2 emissions per year. The PV array produces 1,602,643 kWh/yr, which is 21.8% of its rated capacity. 48% of this energy goes into meeting demand. The surplus energy produced from this system reduced to 290,284 kWh/yr after adding the battery system.

7.6. Option 5: PV-Wind-Diesel-Battery System

In this proposed system configuration, a PV array is added to the wind-diesel-battery hybrid system. If demand is less than power output of the wind turbines, the energy surplus is stored in the batteries along with power output from the PV array. If the battery is fully charged, the residual energy is dumped. If demand is greater than the power output of the wind turbines but less than the sum of powers of the PV array and wind turbines, the extra demand is covered by the PV array power output. Any surplus power from the PV array is stored in the batteries. If the power demand is greater than the combined power output from the two renewable sources, the batteries cover the energy deficit. In the event that the batteries have reached the maximum DOD, the diesel generators are switched on. A sophisticated control system is required to operate this system configuration, the cost of which is included in the analysis.

The system is varied for 1, 2 or all 3 diesel generators, one or both wind turbines, a PV capacity of 493 kW and 839 kW and for 2, 4 and 6 autonomy hours for the battery system. The dominant system configuration for this design option consists of one 1056 kW diesel generator, 66,012 V batteries, 2500 kW wind turbines and 839 kW PV rated capacity. It is associated with an NPC of $ 30,158,636 and 6.558 metric tons of CO_2 emissions per

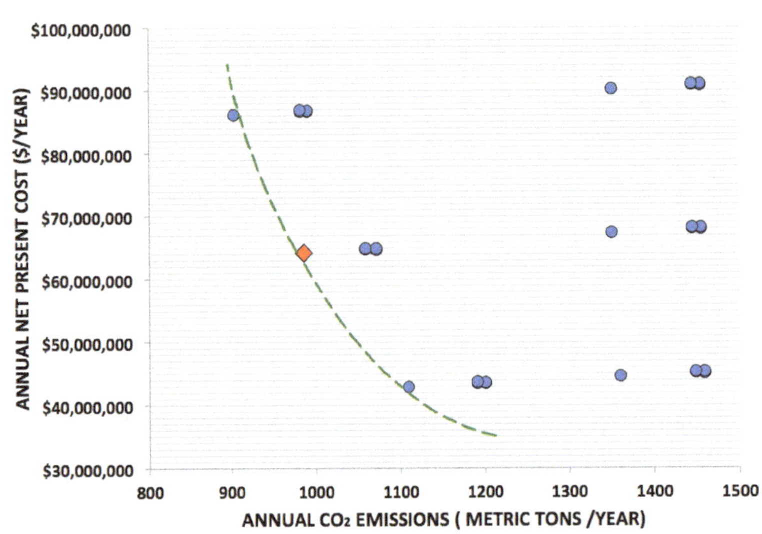

Figure 12. Pareto frontier for PV-diesel-battery hybrid design option.

year. 83% of the renewable energy produced by wind and 17% of the renewable energy produced by PV goes into meeting demand. The electricity produced by the wind turbines is 7,864,188 kWh/yr and that produced by the PV array is 1,604,643 kWh/yr. The surplus electricity produced from the system is 7,055,707 kWh/yr, which is quite high. Taking into account the unrealistically high capacity factor of the wind in the system, the electricity production from the wind turbines may be in actuality significantly less than the reported amount. Consequently, the surplus electricity produced is likely to be much less.

7.7. Hybrid System Optimization

The stand-alone diesel system and the five proposed hybrid system designs were analyzed and compared based the NPC and CO_2. The multi-attribute tradeoff analysis determines the set of dominant system configurations along with identifying the best hybrid system in terms of jointly minimizing CO_2 emissions and NPC. The system configurations for all six system design options are plotted and the resultant Pareto frontier is shown in **Figure 13**.

There are four distinct clusters of system configuration options on the curve. They vary in the number of battery autonomy hours, which subsequently determines the total capacity of the battery system. The lowest cluster of system configuration options (least overall cost) has no battery storage incorporated in the systems. The NPC of the remaining clusters of system configuration options increases with the increasing number of autonomy hours (2, 4 and 6).

The curve represents the Pareto optimal frontier of all the dominant system designs, while the red diamond shape represents the dominant hybrid system configuration that jointly minimizes annual NPC and annual CO_2 emissions.

The optimal system design identified by the red diamond symbol in **Figure 13** is a wind-diesel-battery system, with a configuration of one 1056 kW diesel generator, two 500 kW wind turbines, and 66,012 V batteries. It is associated with an NPC of $26,527,390 and 34.3 metric tons of CO_2 emissions per year. This leads to a reduction in CO_2 emissions of 2755.1 metric tons per year, which is a 98.8 % decrease with respect to the existing diesel system. There is also a 30% decrease in NPC.

The system configuration option cluster on the curve that is lower than the optimal (red-diamond) system configuration is associated with a 233.1 metric tons/yr increase in CO_2 emissions at a $15,823,188 decrease in cost. An addition of 66,012 V batteries to the optimal hybrid system configuration reduces CO_2 emissions by 19.9 metric tons/yr yet at a significant increase in cost of $22,573,040/yr. Consistent with multi-attribute tradeoff analysis in general, any of the options along the curve are preferred to all of the sub-dominant (interior points in **Figure 13**) system configurations.

7.8. Sensitivity Analysis

The sensitivity of the selection of the dominant hybrid system design to uncertainty in the inputs is investigated

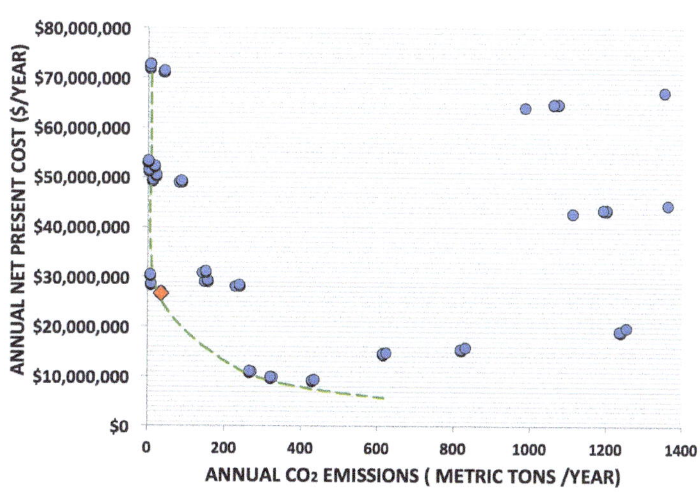

Figure 13. Pareto frontier for all system design configurations.

next. The elements analyzed include the effect of uncertainty in the diesel fuel price ($/L), scaled average wind speeds (m/s) and scaled annual average demand (kWh/d).The effect of best and worst case pairing of wind speed and load level on the NPC and CO_2 emissions is also analyzed.

7.8.1. Extreme Scenarios: Wind Speed and Demand

Due to the intermittent nature of wind speeds, three scenarios are defined, simulated and analyzed in order to determine the impact of average wind speed variations on the net present cost of the system and in turn on the amount of CO_2 emitted from the hybrid system as outlined below. These scenarios are based on the different levels of wind speeds at different levels of demand.

1) Scenario 1—Medium wind speeds serve an average demand level throughout the year. This scenario is a reference for comparison of the best and worst case scenarios.
2) Scenario 2—Worst-case: Maximum demand with wind speeds below the cut in speed of the wind turbines throughout the year.
3) Scenario 3—Best-case: Low demand with high wind speeds throughout the year.

In Scenario 1, the average demand level was determined by summing hourly average demand values and dividing by the number of hours. The average demand in the months of April, August and December was calculated to be 298 kW, with 267 kW demand during the remaining months. The medium wind speed was determined to be 6.0 m/s for the months of October and November, 8.0 m/s for the months of December to February, 6.0 m/s for the months of March to May and 5.5 m/s for the months of June to September [13].

The results of this analysis find that with medium wind speeds throughout the year and average demand, the annual NPC of the system would be $26,273,338, with 12.2 metric tons of CO_2 emissions being added to the atmosphere every year.4539 liters of diesel fuel would be consumed per year.

In Scenario 2, the maximum demand in Marsabit of 839 kW for the months of April, August and December and 755 kW during the rest of the months is used. If there was no wind throughout the day or if the wind speeds were below 4 m/s (the cut in speed of wind turbine), the annual NPC of the system would be $87,282,072, with 4927.9 metric tons of CO_2 emissions being added to the atmosphere every year. This is a 70.0% increase in NPC and a 99.7% increase in CO_2 emissions. The amount of diesel fuel consumed each year increases to 1,838,753 L/yr.

In Scenario 3, the low demand in Marsabit was determined to be 49.4 kW in the months of April, August and December and 44.4 kW during the rest of the months. The high wind speed [13] was determined to be 17 m/s for the months of October and November, 20 m/s for the months of December to February, 18 m/s for the months of March to May and 18 m/s for the months of June to September. If there were high wind speeds throughout the year and low demand, the annual NPC of the system would be $26,116,260 with 0.579 metric tons of CO_2 emissions being added to the atmosphere every year. This is a 0.599% decrease in NPC and a 95.3% decrease CO_2 emissions. The amount of diesel fuel consumed each year decreases to 216 L/yr.

This negative impact of very low or no wind speeds on the NPC of the hybrid system is much larger than the positive impact of high winds. However, the negative impact of low or no wind speeds and the positive impact of high wind speeds on CO_2 emissions are both very substantial.

7.8.2. Diesel Fuel Price Forecast

In 2009, the diesel fuel price projections in €/MWh from the year 2012 to 2034 were determined using a regression analysis [32]. A typical 3320 kW diesel generator consuming No. 2 diesel fuel oil, consumes fuel at a rate of 55.4 L/h at full load [32]. The price per liter of diesel fuel was therefore determined by multiplying the price per MWh by the capacity and fuel rate of the generator. The fuel price projection forecasts an approximately linear increase in price from $1.55/L in 2012 to $2.80/L in 2034.

7.8.3. Scaled annual Average Demand Projection

The power demand in Kenya is projected to grow at an average rate of 6.5% per annum during 2003 to 2023 [20]. The scaled annual average demand in 2012 based on the demand data available is 6623 kWh/day. The projected growth from the 2012 level is a demand of 13,300 kWh/day in 2023 [13].

7.8.4. Scaled average Wind Speed Projection

A fundamental element in the assessment of wind energy production is the prediction of the long-term wind resource at the site. If the data is not available, a statistical method known as "measure correlate predict" (MCP) is commonly used. This method uses modeling techniques to relate short-term measurements at the potential wind farm to long-term measurements of wind speeds at a nearby reference station. The on-site wind speeds are compared with concurrently measured wind speeds at the reference station. Once a correlation has been established, it is applied to long-term historic data from the reference station to estimate the long-term wind resource at the potential wind farm [33].

Since wind speed forecast data for Marsabit was not available, the sensitivity analysis for the scaled average wind speed is based on the individual scaled average wind speeds for the other five locations in Northern Kenya. This analysis is also carried out to examine the impact of average wind speed variations on the hybrid system design for Marsabit so as to determine how the hybrid system configuration would change for the other five locations.

Monthly average wind speed data for these locations was obtained from the NASA Atmospheric Science Data Center for the latitudes and longitudes representative of the locations. It was collected and averaged over a 10-year period from 1983-1993 for the months beginning in January and ending in December at an anemometer height of 50 m. Since the monthly average wind speed data for Marsabit was later re-collected and averaged over a 6 year period from 2001-2006 at an anemometer height of 10 m, the monthly average wind speed data for the other five locations was adjusted based on the ratio of the 1983-93 to the 2001-06 Marsabit wind speed data.

7.8.5. Results of Sensitivity Analysis

For sensitivity to the scaled average, the analysis finds that when there is low average demand, a hybrid system with wind and a 1056 kW diesel generator is the best design. As average demand increases, it is ideal to add PV and upon further increase in demand a 1200 kW diesel generator should be added.

For variations in diesel fuel, the sensitivity analysis determined that as fuel price increases, it is ideal to increase the amount of renewable energy in the system, *i.e.*, add a PV array to the wind turbines to meet demand. This decreases the amount of time the diesel generators run and consequently decreases the amount of diesel fuel being consumed.

As discussed above, the system configuration options on the curve in the cumulative Pareto frontier vary in the number of autonomy hours, which subsequently determines the total number of the batteries (total battery system capacity). The dominant system design option was determined to be Wind/1056 kW diesel gen set/ 66,012 V batteries. The implication of increasing diesel fuel prices with growth in demand on the dominant hybrid system design option is discussed next.

The analysis determined that an increase in diesel fuel at scaled average load levels of less than 11.7 MWh/day does not change the dominant hybrid system design option. However, at scaled average load levels of greater than 11.7 MWh/day, the dominant hybrid system design option changes to incorporate a larger capacity of diesel generation. In addition, when both the diesel fuel price is greater than $1.70 per liter and the scaled average load levels were greater than 11.7 MWh/day, the dominant hybrid system design option changes to also incorporate PV.

Next, the implication of variations in average wind speeds, as scaled average load increases, on the dominant hybrid system design option is discussed. The analysis found that at scaled average load levels of less than 11.7 MWh/day, the dominant hybrid system design option does not change for the given values of scaled average wind speed. Furthermore, the variations in average wind speed in the other five locations do not have an impact on the dominant hybrid system design option determined for Marsabit. This hybrid system design option can therefore be assumed to represent the other five locations as average demand increases over time. However, at scaled average load levels of greater than 11.7 MWh/day, the dominant hybrid system design option changes at different levels of scaled average wind speed and this assumption can no longer hold.

Finally, the implications of variations in average wind speeds, as diesel fuel price increases, on the dominant hybrid system design option is analyzed. At a diesel fuel price of less than $2.40 per liter, the analysis determined that the dominant hybrid system design option does not change for the given values of scaled average wind speed. At a diesel fuel price of greater than $2.40 per liter, the dominant hybrid system design option changes to incorporate larger diesel capacity at scaled average wind speed values of greater than 12.2 m/s. This hybrid system design option can therefore be assumed to represent the other five locations as average diesel fuel price increases over time up to $2.40 per liter.

8. Conclusions

This study analyzes the incorporation of renewable energy and storage technologies into stand-alone diesel power systems servicing six remote villages in northern Kenya, in the bid to reduce Kenya's dependency on diesel fuel and consequently reduce the carbon footprint. Successful results have already been obtained with hybrid systems in rural communities in developing countries worldwide. Owing to the availability of abundant wind and solar resources in northern Kenya, six different configurations of hybrid energy systems incorporating wind energy, solar energy and battery storage were proposed, simulated and analyzed. The study confirms the feasibility of these hybrid systems with many configurations being profitable. A multi-attribute trade-off analysis was employed to determine the best hybrid system configuration option that would reduce diesel fuel consumption, and jointly minimize the CO_2 emissions and net present cost (NPC).

The key findings of this study are:

A. The simulations and analysis results find that the wind-diesel-battery configuration consisting of two 500 kW turbines and 95,040 Ah battery capacity is the best option to replace the stand-alone diesel system providing electricity to a village with a peak demand of 839 kW. This system significantly reduces fuel consumption and CO_2 emissions at minimal net present cost. It consumes 12,114 liters of diesel fuel and emits 32.5 metric tons of CO_2 gas every year, which is a 98.8% reduction from that of the stand-alone diesel system at a 30% lower net present cost of $ 26,548,492.

B. The sensitivity analysis finds that very low or no wind speeds have a large negative impact on the net present cost (70% increase) of the hybrid system compared to the positive impact of high winds (1% decrease). However, impact on CO_2 emissions in both cases is very substantial (>95% increase and decrease respectively). Very low or no wind speeds have a 1.98% probability of occurrence, compared to a 1.25% probability of high wind speeds occurring.

C. An increase in diesel fuel prices and the demand for electricity, based on future projections, result in a substantial increase in both the net present cost of the system (75% increase) and the CO_2 emissions(57% increase).

D. As the price of diesel fuel and the electricity demand increase over the years, it would be more economical and environmentally friendly to increase the amount of renewable energy capacity in the system by adding PV panels.

E. At scaled average load levels of greater than 11.7 MWh/day, the dominant hybrid system design option changes to incorporate a larger capacity of diesel generation. In addition, when the diesel fuel price is greater than $1.70 per liter and scaled average load levels are greater than 11.7 MWh/day, the dominant hybrid system design option also changes to incorporate PV.

F. The hybrid system design option for the town of Marsabit can be assumed to represent the other five locations (Moyale, Mandera, Wajir, Lodwar and El Wak) for scaled average load levels of less than 11.7 MWh/day and diesel fuel prices less than $2.40 per liter. This is because that the variations in average wind speed in the other five locations do not have an impact on the dominant hybrid system option, which was designed for Marsabit. However, at scaled average load levels of greater than 11.7 MWh/day and diesel fuel prices greater than $2.40 per liter, the dominant hybrid system design option changes at different levels of scaled average wind speed and this assumption can no longer hold.

The key findings in this study suggest that the Hybrid Mini-grid Project could be a solid investment and an example of a successful Feed-in-Tariffs (FiT) Policy. The proposed hybrid system could potentially reshape the Kenyan energy industry, indicating that future work should explore the country's policies concerning electrical power production and distribution to ensure that these policies are met with the implementation of the proposed hybrid system project.

The key findings of this study also indicate that in order to achieve policy strategies and development targets with regards to the incorporation of renewable energy, especially wind energy to the power grids, future work should focus on a detailed analysis of the variability of the wind resource. The integration of a substantial amount of wind power in isolated electrical systems needs careful consideration, so as to maintain a high degree of reliability and stability of the system operation. Due to the high production forecasting uncertainties, there may be problems with operational scheduling (e.g., unit commitment) as well as steady state and dynamic system stability concerns. These problems may considerably limit the amount of wind generation that can be connected to the power systems, increasing the complexity of their operations [15]. Ongoing analysis should therefore include simulation of the proposed hybrid system dynamic performance in order to analyze system voltage and frequency stability and to develop guidelines for the optimal operation of the system.

The results obtained from the simulations show that there would be significant excess electricity produced from the hybrid system design options. Based on the arid nature of Marsabit, future work that explores the possibility of incorporating a water-pumping unit to the system will benefit the community. Water pumping can make use of the excess electricity produced to provide water for household use and also for agriculture.

The results presented here provide a solid foundation for this future research into the efficacy of renewable resources, in the reduction of greenhouse gas emissions. This study provides a framework upon which Kenya can transit into a more environmentally friendly and sustainable country through the energy sector.

References

[1] Moner-Girona, M. (2008) A New Scheme for the Promotion of Renewable Energies in Developing Countries: The Renewable Energy Regulated Purchase Tariff. http://www.energy.eu/publications/LDNA23284ENC_002.pdf

[2] Trading Economics (2012) Pump Price for Diesel Fuel (US Dollar Per Liter) in Kenya. http://www.tradingeconomics.com/kenya/pump-price-for-diesel-fuel-us-dollar-perliter-wb-data.html

[3] The World Bank (2012) Data Kenya: World Development Indicators. http://data.worldbank.org/country/kenya

[4] International Energy Agency (2011) Energy for All: Financing Access for the Poor. *Proceedings of the Energy for All Conference*, Oslo, 10-11 October 2011, 1-52.

[5] Kianji, C.K. (2013) Kenya's Energy Demand and the Role of Nuclear Energy in Future Energy Generation Mix. *Presented at the Joint JAPAN-IAEA Nuclear Energy Management School*, Tokai-Mura, Japan, 18 June 2012.

[6] Phaesun. (2013) Stand-Alone Solar Solutions, Hybrid Systems. http://www.phaesun.com/systems/references/rural-electrification/hybrid-systems.html

[7] The Republic of Kenya (2011) Scaling-Up Renewable Energy Program (SREP): Investment Plan for Kenya (Unpublished). http://www.renewableenergy.go.ke/downloads/policy-docs/Updated_SREP_Draft_Investment_Plan_May_2011.pdf

[8] Scaling-Up Renewable Energy Program Country Investment Plan, Kenya. *Lecture Presented at SREP Sub-Committee Meeting*, Cape Town, 21 June 2011.

[9] Alliance for Rural Electrification (2012) Hybrid Power Systems Based on Renewable Energies: A Suitable and Cost-Competitive Solution for Rural Electrification. http://www.ruralelec.org/fileadmin/DATA/Documents/06_Publications/Position_papers/ARE-WG_Technological_Solutions_-_Brochure_Hybrid_Systems.pdf

[10] Rehman, S., Al-Hadhrami, M.L. and Mahbub, A.M. (2012) Feasibility Study of a Wind-PV-Diesel Hybrid Power System for a Village. *Elsevier*: *Renewable Energy*, **38**, 258-268. http://dx.doi.org/10.1016/j.renene.2011.06.028

[11] Nahid-Al-Masood, R., Mirza, J., Ahmed, A.H., Abedin, S.R., Deeba, F., Begum, H.M. and Chowdhury, A.H. (2011) Design of a Cost Effective Off-Grid Wind-Diesel Hybrid Power System in an Island of Bangladesh. *World Academy of Science, Engineering and Technology*, **60**, 312-315.

[12] Rehman, S., Ahmad, F., Shaahid, S.M., Shash, A., El-Amin, I.M., Al-Shehri, A.M. and Bakhashwain, J.M. (2011) Feasibility Study of Hybrid Retrofits to an Isolated Off-Grid Diesel Power Plant. *Renewable and Sustainable Energy Reviews*, **11**, 635-653. http://dx.doi.org/10.1016/j.rser.2005.05.003

[13] Lukuyu, J.M. (2013) Design and Analysis of Hybrid Power System Options for Off-Grid Rural Electrification in Northern Kenya. B.Sc. Thesis, Smith College, Northampton.

[14] Tabors, R.D. and Hornby, R. (2005) The Use of Multi-Attribute Trade-Off Analysis in Strategic Planning for an Electric Distribution Utility: An Analysis of Abu Dhabi Distribution Company. *Proceedings of the 38th Hawaii International Conference on System Sciences*, Big Island, 3-6 January 2005, 61b. http://dx.doi.org/10.1109/HICSS.2005.621

[15] Karapidakis, E.S. (2010) Wind Power Impact on Power System Dynamic Performance. In: Muyeen, S.M., Ed., *Wind Power*, INTECH, Croatia, 558.

[16] Kenya National Bureau of Statistics (2013) Population and Housing Census. Constituency Population by Sex, Number of Households, Area and Density. http://www.knbs.or.ke/index.php?option=com_content&view=article&id=165:population-distribution-by-sex-number-of-households-area-and-density-by-administrativeunits&catid=114&Itemid=639

[17] Elliot, F. and Roth, E.A. (2005) As Pastoralists Settle: Social, Health, and Economic Consequences of the Pastoral Sedentarization in Marsabit District. Kluwer Academic, New York.

[18] Mbogho, M.S., Wekesa, C., Murage, D. and Toyoda, J. (1996) Economic Analysis of Supplying an Electro Village

Using Alternative Electrical Power Sources. *IEEE 4th AFRICON*, **2**, 968-971. http://dx.doi.org/10.1109/AFRCON.1996.563026

[19] Mwangangi, S., Koech, W., Akinala, J. and Omwega, T. (2010) Environmental and Social Impact Assessment Project Report for Proposed Two 250 KW Wind Turbines at Marsabit. http://www.kplc.co.ke/fileadmin/user_upload/Documents/ESIA/ESIA%20of%20Two%20250%20KW%20Wind%20Turbines%20at%20Marsabit%20District.pdf

[20] Mwangi, D. (2003) Kenya's Electricity Demand Projections and Role of Geothermal Resource in National Power Development Plan. *Eastern Africa Geothermal Market Acceleration Conference*, Nairobi, 9-11 April 2003, 23-43.

[21] Ouma, C. (2011) Hybrid Energy System for Off-Grid Rural Electrification (Case Study Kenya). Ph.D. Thesis, Gotland University, Sweden.

[22] Oludhe, C. (2008) Assessment and Utilization of Wind Power in Kenya—A Review. *Journal of Kenya Meteorological Society*, **2**, 39-52.

[23] Kamau, J.N., Kinyua, R. and Gathua, J.K. (2010) 6 Years of Wind Data for Marsabit, Kenya Average over 14 m/s at 100 m Hub Height; An Analysis of the Wind Energy Potential. *Renewable Energy*, **35**, 1298-1302. http://dx.doi.org/10.1016/j.renene.2009.10.008

[24] Apratim, R. (2012) Derivation of Surface Roughness and Capacity Factor from Wind Shear Characteristics. *International Journal of Renewable Energy Research*, **2**, 348-355.

[25] McKahn, D. (2012) Bringing Extraterrestrial Radiation to the Plane of Array. Lecture presented at Smith College, Northampton, 11 September 2012.

[26] National Renewable Energy Laboratories (2005) Kenya Global Horizontal Solar Radiation [Image]. GIS Database.

[27] NASA (2012) Surface Meteorology and Solar Energy. Atmospheric Science Data Center Database. https://eosweb.larc.nasa.gov/order-data

[28] HOMER Energy (2013) Optimizing Clean Power Everywhere. Energy Modeling Software for Hybrid Renewable Energy Systems. http://homerenergy.com/index.html

[29] Rehman, S. and Al-Hadhrami, L.M. (2010) Study of a Solar PV/Diesel/Battery Hybrid Power System for a Remotely Located Population near Rafha, Saudi Arabia. *Renewable and Sustainable Energy Reviews*, **35**, 4986-4995. http://dx.doi.org/10.1016/j.energy.2010.08.025

[30] The Kenya Government (2007) Kenya Vision 2030, a Globally Competitive and Prosperous Kenya. http://www.kilimo.go.ke/kilimo_docs/pdf/Kenya_VISION_2030-final.pdf

[31] Wikipedia (2013) Photovoltaics. http://en.wikipedia.org/wiki/Photovoltaics

[32] Vuorinen, A. (2007) Planning of National Power Systems. http://www.optimalpowersystems.com/stuff/planning_of_national_power_systems.pdf

[33] Wind Energy: The Facts (2013) Analytical Methods for the Prediction of the Long-Term Wind Regime at a Site. http://www.wind-energy-the-facts.org/analytical-methods-for-the-prediction-of-the-long-term-wind-regime-at-a-site.html

2

Solar Energy Generation Potential Estimation in India and Gujarat, Andhra, Telangana States

T. Harinarayana[1], K. Jaya Kashyap[2]

[1]Gujarat Energy Research & Management Institute (GERMI), Gandhinagar, India
[2]Malla Reddy Engineering College, Hyderabad, India
Email: harinarayana@germi.res.in, kashyap.thewall@gmail.com

Abstract

It is well known that the rampant increase for the demand of electricity and rapid depletion of the fossil fuels has called for immediate response in the direction of energy sufficiency. To accomplish this, one of the important tasks is to identify the locations of high potential for renewable energy generation. It is a well-established fact that solar energy proved to be the most sought after source for energy generation. Although, solar energy potential maps of India have been prepared based on solar irradiation maps in the earlier studies, the present research study has been carried out with a focused attention directly on solar energy generation considering various parameters. In this work it is shown that solar energy generation does not depend on solar radiation alone at a location. Instead, there are various other factors that influence the energy generation. Some of them are ambient temperature, wind velocity and other parameters like weather and topographic conditions. In this study the locations with high and low solar energy generation potential in India have been identified through systematic analysis by computing the solar energy parameters at every grid point (1° × 1°). The work has been extended with more detailed study for Gujarat, Andhra Pradesh and the newly formed Telangana states. The data points considered for the states are 0.25° × 0.25° having resulted in adding more number of locations. Our results indicate that the total annual energy generation in India varies from 510,000 KWH to 800,000 KWH per acre of land. The least energy generation location pertains to the eastern parts of Arunachal Pradesh and eastern part of Assam and the highest annual solar energy generation has been identified in the eastern parts of Jammu & Kashmir and eastern part of Uttarakhand.

Keywords

Solar Energy, Estimation, Financial Benefits, Gujarat, Andhra Pradesh, Telangana, India

1. Introduction

Solar energy incident on the Earth's surface also known as insolation, is dependent on many parameters such as geographic location, earth-sun movements, tilt of earth's rotational axis, atmospheric parameters, suspended particles etc. Quantification of solar energy potential in any region depends mainly on the intensity of insolation [1]. The grid-tied renewable energy from the installed PV capacity is around 6% in India [2]. The solar insolation of India is very much encouraging. The daily global radiation on an average is around 5 KWH per square meter per day and in a year the sunshine ranges from 2300 to 3200 hours [2]. Another advantage is that India lies in an ideal geographical location between the equator and tropic of cancer and has an average annual temperature ranging from 25˚C to 27.5˚C. All these advantages can be effectively and strategically used if we can identify the locations of high potential solar energy generation. This helps to achieve the target of Jawaharlal Nehru Solar Mission which is 20 GW power by the year 2022 [3].

Among all possible alternative energy options, for example, wave energy, geothermal energy, solar energy, wind energy, and hydroenergy, solar energy is becoming more popular in India. This is mainly due to 1) the availability of plenty of sunlight in all the seasons and also at all the locations of India and 2) the initiation of solar mission by the government of India with attractive incentives to the developers [4]. If we look at the total renewable energy generation of the world, which is around 5×10^{20} J per year, solar thermal contributes to 0.5%, wind 0.3%, geothermal 0.2%, bio-fuel 0.2%, and solar photovoltaic (PV) is only about 0.04% as per statistical review of world energy during 2007 [5]. In recent years, the technology upgradation has made solar photovoltaic technology as a viable technology with competitive price. It is projected that by the year 2030, the solar PV electricity will dominate compared to other sources of energy [6]. From the study growth of photovoltaic, an average about 45% annual increase is noticed during the years 2000 to 2009 [6]. From the study of cost economics of a solar photovoltaic power plant, PV modules cost about 40% - 45% and the other 55% - 60% is due to its components [7]. Additionally, cost of the power plant also depends on the land value. If the solar power plant is close to the substation near the populated area, the transmission of energy losses will be minimum, but the cost of the land will be high [8]. If the power plant is at a remote location, the cost of the land is low but the energy losses will be high. On the other hand, with less population and in a remote location, the use of energy is limited to the local community. Ideally, the solar power plant needs to be located at a place where the energy generation from the plant can be connected directly to the power grid at an optimum distance from the plant. Apart from the government of India's national solar mission program, the recent initiation by the government of Gujarat to establish the solar photovoltaic plants is commendable. During 2012, while Gujarat alone crossed 600 MW power through solar, the rest of the country was far behind with only about 200 MW. The Gandhinagar Photovoltaic Rooftop Programme for solar energy generation using PV modules has set an example by government of Gujarat to save the land cost (see http://www.gpclindia.com/gpcl_rsg/index.html). Another way to save the land cost is to adopt a new methodology to get maximum output from the solar power plant in a limited area (Ref: 2 layer panel). In India, the cost of the land has grown up five to ten times for the last 10 years. This is true in all the urban and semi-urban regions of India.

There are many places in the world still suffering to use the electricity for their needs. It is reported that many islands around the world are facing acute shortage of electricity. For instance, Papua New Guinea, Fiji islands, Ports of Andaman & Nicobar and Lakshadweep islands in India are some of the examples. Major electricity generation in these islands is based on using diesel generators, for which diesel oil needs to be transported from nearby main land to these islands [9]. For effective use of renewable energy, storage and conservation are the important factors. In fact, many nations are now looking at various renewable resources available with them and plan to utilize them optimally. In this direction nations like Nigeria have taken several steps on the use of energy utilization for equipment, lighting, machines and engines in the industry and also new designs for buildings to make them energy-efficient [10]. In the same direction we have suggested earlier to use the national highways for solar energy generation by putting the solar panels at a reasonable height of 5.7 meters above the road. This helps to save the land use [11] [12]. Realizing the importance of climate change and global warming, use of renewable energy is on the rise in many countries, especially the grid-connected PV solar installations [13] [14]. For example there is a 55% increase in solar PV with a cumulative installed capacity of around 3.1 GW during the year 2005 as compared to 2 GW during the year 2004 in the world [15]. India's solar energy scenario is also following a similar growth rate for the past few years. This is mainly due to new initiatives taken by the government, for example subsidies to the developers with tax exemption. The other necessity factors are the large gap between the energy generation and demand. It is very much an evident fact that since India has entered into a phase of rapid development, the energy and in particular the electricity demand would always be on the rise as illustrated in the **Figure 1** [16]. As seen from **Figure 1**, the present (2014) electricity generation in India is close to 248 GW. If India plans to increase its rate of growth by 7% or 8% or 9%, the target energy generation also needs to be increased. By 2034 (after 20 years) our target will be 960, 1156 and 1390 GW respectively. This means the target is close to 4 to 6 times the present value. All these challenges would require meticulous planning and start with the identification of strategic locations with high energy generation potential. The present study is aimed towards this direction. Thus the main objective of this research work is to estimate the solar energy potential at different locations of India with high and low energy regions.

2. Methodology

The methodology followed in our study is through modelling. The modelling has been carried out using PVsyst, Google Earth and Surfer software programs. A brief description of these software programs is presented in the following.

PVsyst: PVsyst simulation software is a popular tool among solar energy engineering community. It is being used extensively to analyze the detailed performance of the solar plant in field conditions. It can be used in many ways, for example, to investigate different loads on the system, to estimate the size of the system, to determine the optimal size of the system, to assess the energy production in the system etc. One of the most important features of PVsyst is that the economical evaluation of the solar photo-voltaic installation can be accomplished with reasonable accuracy. Other capabilities and options available in the PVsyst software can be seen in [17]. In the present study the 6.25 version of the PVsyst software [17] has been used.

Surfer: It is graphical software that can be used for contouring the values, 2D and 3D visualization of the data. The software runs on Windows operating system. It can also be used for bathymetric, terrain modelling studies, visualization of the landscape, watershed analysis, interpolation, extrapolation, gridding, digitization etc. [18]. In the present study, this software has been used to digitize the map boundaries, creation and plotting the contour maps. In addition to these features numerous other features mostly related to extrapolation, interpolation, and other statistical algorithm do exist.

Google Earth: It is also software that can be used to visualize the imagery for the satellite maps, 3D buildings, terrains etc. [19]. It also superimposes the various images, derived from aerial photography records, satellite imagery maps having the same coordinating system. Additionally, digital elevation map can also be prepared from the data derived from SRTM (Shuttle Radar Topography Mission) of NASA [20]. One of the key features is to obtain an altitude of a location for a given geographic parameter such as latitude and longitude. Some of these parameters are used in our study.

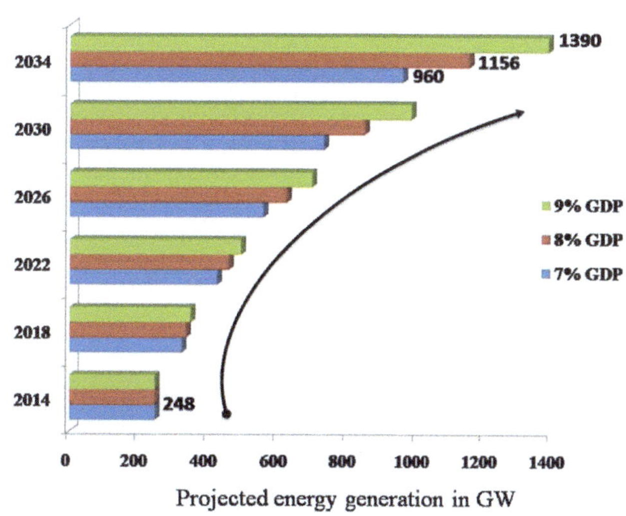

Figure 1. Projected energy generation for the next 20 years from the year 2014 until 2034. Initial energy generation is assumed as 248 GW of electricity.

3. Site Selection

Since the main objective of the study is to estimate the total solar energy potential of India, the site selection should be carried out in such a way that it effectively and uniformly covers the entire land mass of India. For this purpose, 1° × 1° grid by varying the latitude and longitude has been considered. This has rendered a total of 286 geographical locations covering the entire land mass of India. Similar method has been extended to the states of Gujarat, Andhra Pradesh and Telangana. In order to acquire high density values for individual states, the latitude and longitude has been reduced to 1/4° (15 arc minutes). This has been done to obtain more geographical locations that would subsequently aid in obtaining better and accurate results. Detailed computation has been carried for three states owing to the following reasons:

Rajasthan and Gujarat are the two states in India which lie in the sunny belt (>2000 KWH per Km) having bright sunlight weather for more than 250 to 300 days in a year. This way these two states receive maximum amount of annual global radiation [4] [7] [12] [21]. The newly formed states of Andhra Pradesh and Telangana are badly in need of electricity and require well directed plans aiming at energy independence.

4. The Data

In the present study grid-connected system is chosen. The basic input parameters required are PV component database, grid inverter database, geographical site information and monthly meteorological data for horizontal global irradiance and temperature. More details of the parameters are provided in **Table 1**.

The PV module chosen in the modelling is DelSolar (D6P 230 A3E). Each module has a maximum power output of 230 W. The module dimensions are 1.62 m in length and 0.99 m in width. Although the module measures 1.62 m in length an additional space of 0.05 m (5 cm approximately) has been given for easy operations, by taking mechanical aspects of the panel etc into consideration. Among the various inverters, Siemens' inverter has been chosen.

The meteorological data has been acquired from meteonorm version 6.1 (Fabrikstrasse, Bern, Switzerland) [21]. We have a choice of using meteorological data acquired from meteonorm or NASA-SSE data [22].

The meteonorm software has inbuilt database with a monthly irradiation data for 1200 stations with an average values for 20 years from 1981 to 2000. Data related to Indian region is clearly documented in meteonorm. Additionally, different European countries' data are also available. Apart from this, data for a new site can be easily derived through interpolation of values, usually carried out between three nearest stations [18]. The NASA data contains monthly average data in a region of 1° × 1° for the whole world for the years 1983 to 1993. Accordingly, the data from meteonorm data available with us is more accurate and recent as compared to NASA data. For this purpose in our study we have used meteonorm data.

Table 1. Details of the PV module and the corresponding grid inverter.

Area considered	1 acre (~=4047 Sq·m) (63 × 63 m)
Name of the manufacturer (PV module)	Del Solar
Technology (PV module)	Silicon-Polycrystalline
Pmpp	230 Wp
Impp	7.8 A
Vmpp	29.49 V
Name of the manufacturer (grid inverter)	Siemens
Nominal AC power (grid inverter)	465 kW
Total No. of PV modules in 1 acre	1944
Dimension of the PV module	L = 1.67 m (1.62 m + 0.05 m) W = 0.99 m
Active (sensitive) area of the PV modules	3129 m^2
Source of data	Meteonorm V6.1

5. Solar Panel Configuration

The area and solar panel orientation have been carried out on similar lines mentioned in the previous papers [12] [23]. The solar panels have been arranged in an area of 1 acre (4047 m^2). This can be conveniently approximated to a square with dimensions of 63 m × 63 m. The solar panels have been placed horizontally on the ground. This facilitates the accommodation of maximum number of panels. The thumb rule of solar panel placement in general states that the azimuth of the solar panel should be equal to the latitude of the place [23]. But this has been disregarded in the present work and all the panels have been placed in horizontal direction. This has been done to provide a standard and a uniform platform for energy computation throughout the present study covering the whole country. One such typical configuration is presented in **Figure 2**.

In the current study, the solar panels pertaining to the above mentioned configuration and orientation have been placed at numerous geographical locations of India considering 1 × 1 degree. For individual states, with more focused attention given to the states of Gujarat, Andhra Pradesh and Telangana considering smaller area covering 0.25 × 0.25 degree at each location. Grid locations used in our study in all the four cases, namely, India, Gujarat, Andhra and Telangana regions are shown in **Figure 3**. A total of 286 locations in India, 266 in Gujarat, 231 in Andhra Pradesh and 165 in Telangana state have been obtained. As can be seen at each location, we have computed solar energy generation for all the months and also for the annual generation. This has generated large data base of 948 locations for analysis. All these results are described in the following for each case.

6. Results

Results derived in our present study are solar energy generation at every location for each month and also the total energy for the year. All these values at each location have been considered and contouring has been carried out using gridding method. The total annual energy generation for India is presented in **Figure 4(a)** varies from 510,000 KWH to 800,000 KWH per acre of land. In **Figure 4(b)** [24], different states of India are shown. The least energy generation location pertains to the eastern parts of Arunachal Pradesh and eastern part of Assam. The highest annual solar energy generation (750,000 - 800,000 KWH/acre) potential has been identified in the eastern parts of Jammu & Kashmir and eastern part of Uttarakhand. Major part of Indian landmass has the solar energy generation potential ranging from 680,000 KWH - 730,000 KWH, among which southern parts of Saurashtra, eastern part of Rajasthan, southern part of Karnataka and western part of Tamil Nadu region have shown the average annual energy generation close to 730,000 KWH. Northern part of Rajasthan and Punjab, Haryana states and northern part of Jammu & Kashmir, West Bengal, Bihar, Nagaland, Manipur, Meghalaya and western

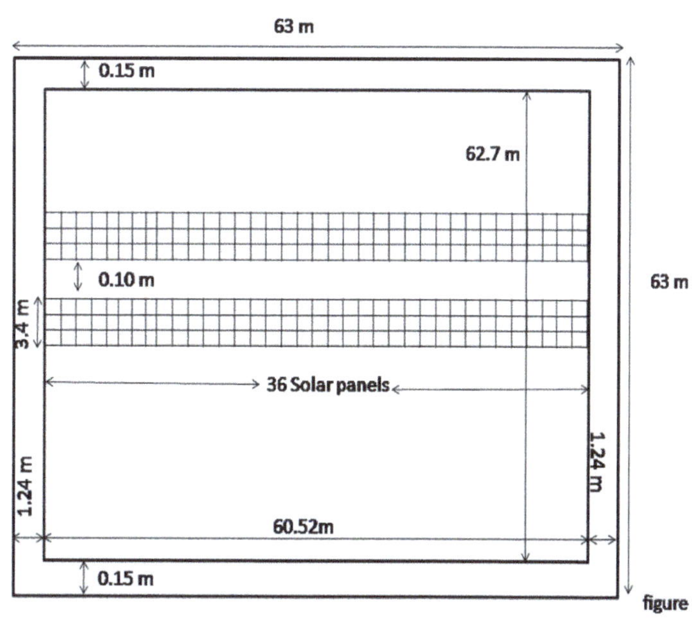

Figure 2. Typical field design of solar panels considered in one acre of land. In this design, all the panels are symmetrically arranged in a horizontal direction and giving a small area all around the land.

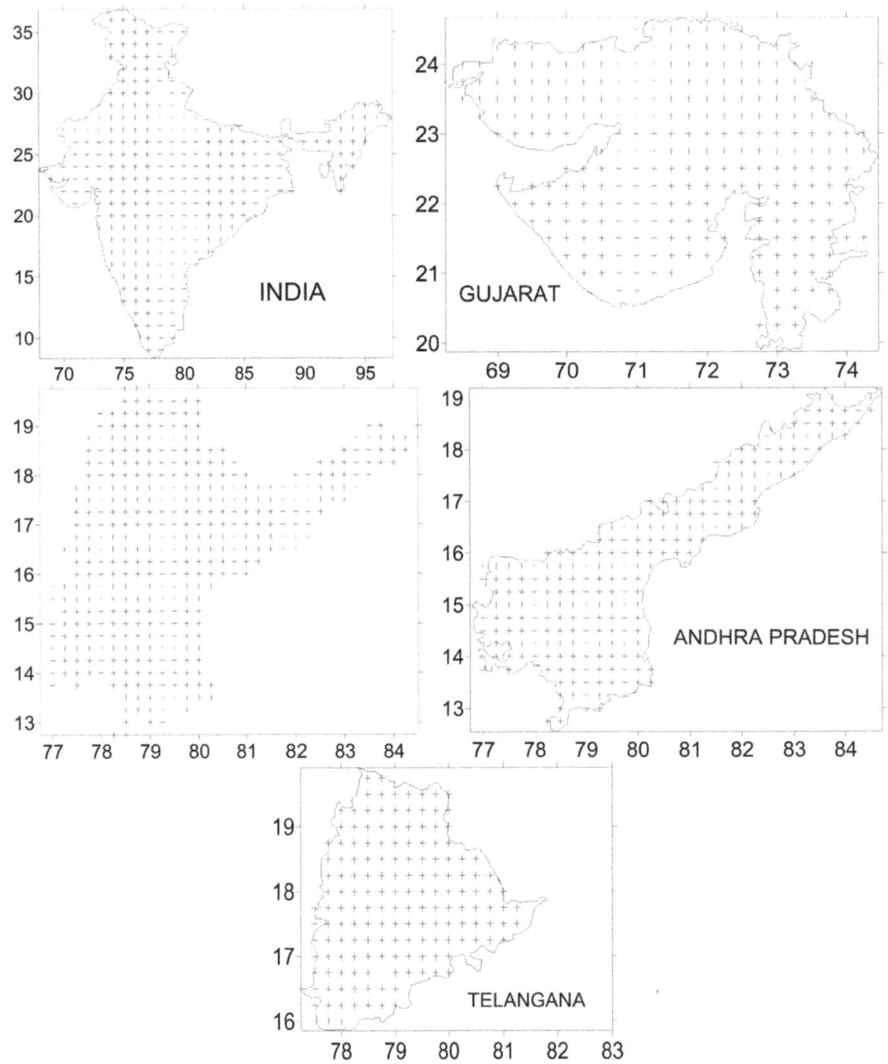

Figure 3. Data locations considered in our study. The grid locations used in our study for the entire country India (1° × 1°), Gujarat, Andhra Pradesh and Telangana states (0.25° × 0.25°).

part of Assam have shown average annual energy generation values ranging from 620,000 KWH - 650,000 KWH. The results derived from the present study gave a clear evidence that major portion of Indian sub-continent has substantial solar energy potential except in the area of Arunachal Pradesh & eastern part of Assam states. **Figure 5** gives the solar energy generation averaged for each month. Annual seasonal variation of weather parameters is normal in India. It is almost cyclic in each year. June to August months are considered to be monsoon months. Due to the effect of monsoon, solar radiation on the surface of the Earth decreases. This affects the solar energy generation. This can be seen clearly from **Figure 5**. The monthly energy generation picks up from the month of September. Similarly, March to May is considered as summer months. During this period, solar energy is maxima as compared to other months.

Similarly, in the state of Gujarat, solar annual energy generation ranges from 630,000 KWH to as much as 800,000 KWH (**Figure 6(a)**). Different districts of Gujarat can be seen in **Figure 6(b)** [25]. **Figure 7** shows monthly energy generation for Gujarat. The maximum solar energy generation (800,000 KWH) is limited to small area in the district of Junagarh. Interestingly, this small area has shown a near-circular shape. The reason might be attributed to the higher solar radiation received in this small area. This can be understood from the observation of the elevation map of Gujarat. This small area is located in the well-known Gir forest region where the elevation of Girnar hills is close to 1000 m. The energy generation is least in the northern part of Rann of

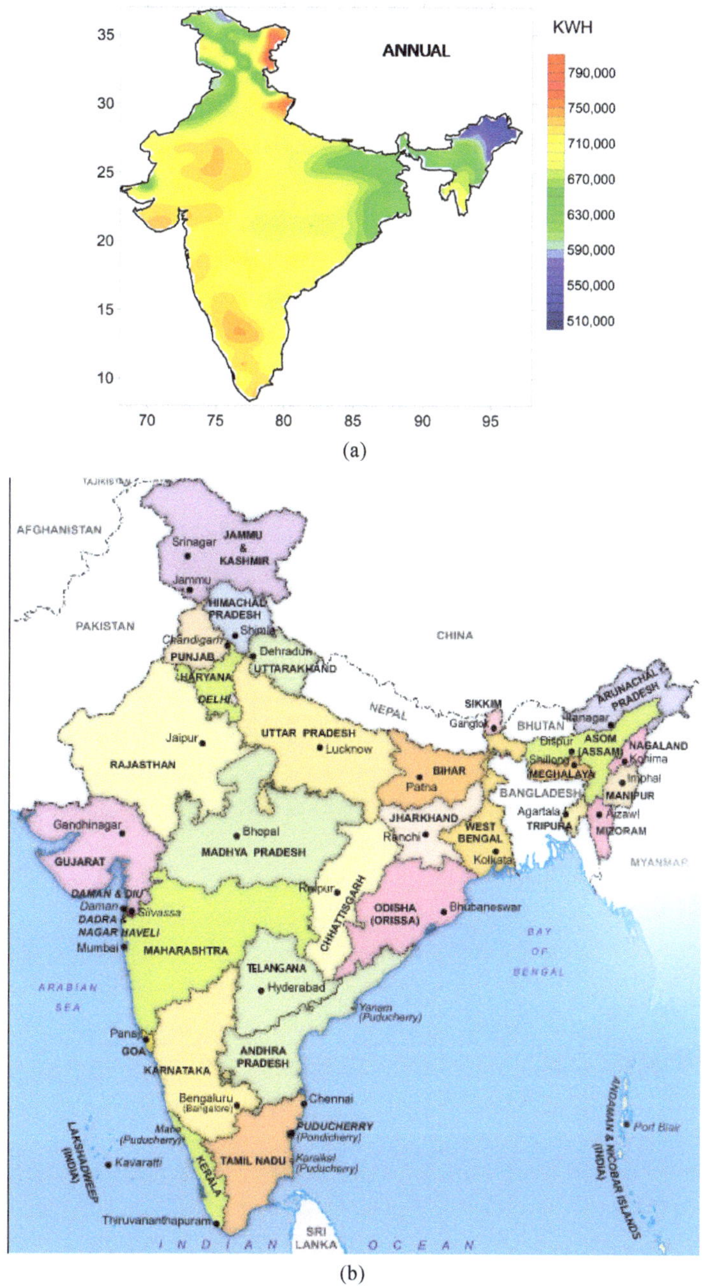

Figure 4. (a) Annual solar energy generation potential map for India estimated from the present study. Blue colour shows the low energy generation and Red indicates the high energy generation as shown in the Index; (b) India map showing different states.

Kutch region and northern part Kutch district. Majority of the region in Gujarat has the potential to generate energy with magnitude ranging from 700,000 KWH - 730,000 KWH. Due to low monsoon effect as compared to the rest of India, Gujarat solar energy is affected only during parts of July to August.

In the case of Andhra Pradesh state, the annual solar energy generation map, has shown a variation from 670,000 KWH - 740,000 KWH (**Figure 8(a)**). Different districts are shown in **Figure 8(b)** [26]. The least annual generation has been identified in the area of eastern part of Kurnool, northern part of Kadapa, major portion of Ongole and Guntur districts. The maximum energy potential (730,000 KWH - 740,000 KWH) is limited to the western part of Ananthapur district. The districts of Srikakulam, Vizianagarm, Vizag, east and west Godavari

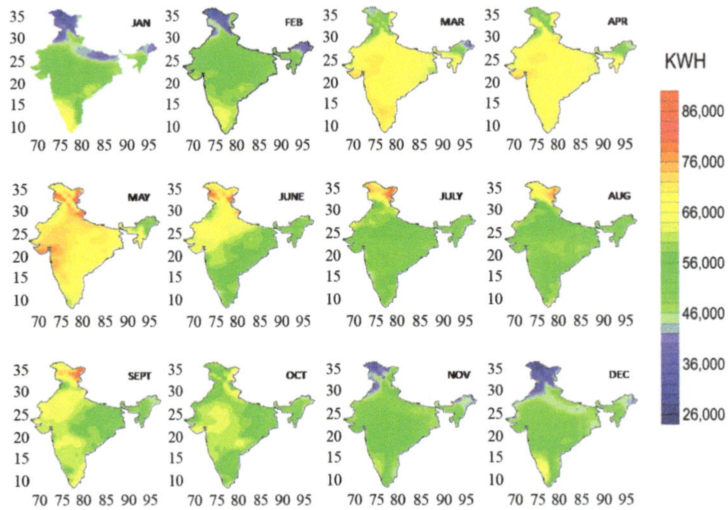

Figure 5. Solar energy generation for India, showing the variation of energy in different months. Lowest is about 26,000 KWH and the highest is close to 86,000 KWH.

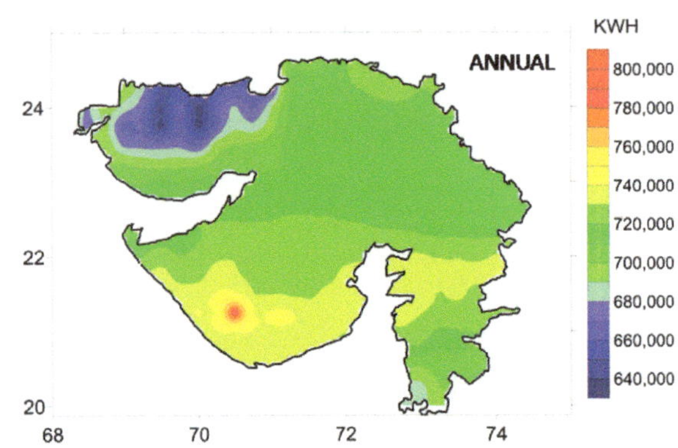

(a)

(b)

Figure 6. (a) Annual solar energy generation for the Gujarat state. Southern part of Saurashtra shows high potential compared to northern part; (b) Gujarat map showing distribution of different districts.

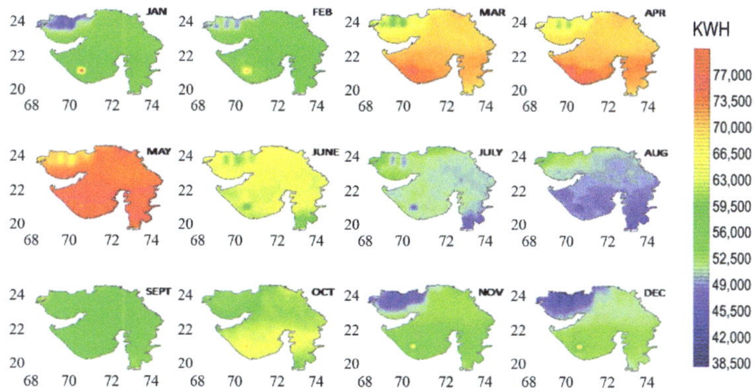

Figure 7. Monthly solar energy generation for Gujarat state, showing the variation in different months.

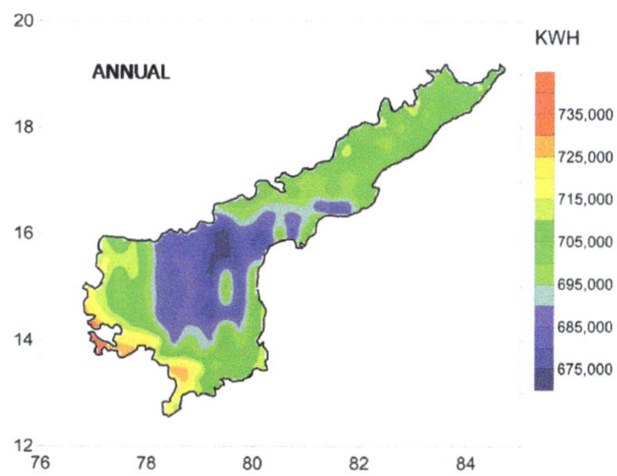

(a)

(b)

Figure 8. (a) Annual solar energy generation potential map for Andhra Pradesh state. Highest being about 740,000 KWH and the lowest being 670,000 KWH; (b) Andhra Pradesh map showing different districts of the two states, Andhra and Telangana.

and northern part of Krishna district have shown a variation ranging from 695,000 KWH - 715,000 KWH. Monthly variation can be seen in **Figure 9**. Due to frequent cyclonic effects, coastal districts show low energy generation during the months of October-November-December.

In the newly formed Telangana state, northern part of Adilabad district and southern part of Mahboobnagar district have the least energy generation potential ranging from 670,000 KWH - 690,000 KWH (**Figure 10(a)**). The district map of Telangana region is shown in **Figure 8(b)** [27]. The districts of Khammam and western parts of Medak seem to have the highest energy potential ranging from 715,000 KWH - 725,000 KWH. In all other districts, solar energy generation varies from 695,000 KWH - 710,000 KWH. Monthly variation can be seen in **Figure 11**. Similar to other states, Telangana region showed low energy generation during July and Aug compared to other months.

The results generated from our study are close to the results reported in earlier studies [12] [23]. For instance, in Ahmedabad district, we have shown the energy generation value close to 710,000 KWH, whereas in earlier studies, 678,574 KWH. This small variation is due to data considered in our study being meteonorm data, whereas in other studies NASA-SSE data is considered. Thus our study is close to the previous studies.

7. Discussion

An attempt has been made in our study, to estimate the solar energy generation for the whole country of India in a regional scale and more focused detailed study for three different states—Gujarat, Andhra and Telangana. It is observed that solar energy generation at a particular place does not solely depend on the solar irradiation or solar insolation. Earlier studies have made attempts in estimating the solar potential of India using solar irradiation data alone. Although the amount of solar radiation plays a significant role in energy generation, it is observed that there are other parameters which tend to influence the energy generation to a great deal [28]. Some of these parameters are ambient temperature, altitude of the place, wind velocity, weather conditions etc. The fact that weather conditions influence the energy generation can be seen from the monthly energy generation plots of India and the individual states of Gujarat, Andhra Pradesh and Telangana. It is a well-conceived fact that higher the altitude at a particular place lower is the ambient temperature. This combination of high altitude and low ambient temperature plays a crucial role in the efficient performance of the PV modules. The optimum wind velocity is also a deciding factor in the energy generation process from the view point of heat transfer.

Examination of the yearly energy generation map of India (**Figure 4(a)**) would reveal that eastern part of the states of Jammu & Kashmir and Uttarakhand which house the Himalayas have shown great potential of solar energy generation. One noteworthy remark to be made in this context is that although the radiation in these regions is low when compared to the rest of the country, the energy generation potential is high. The reason for this might be attributed to the ideal combination of solar radiation, ambient temperature and wind velocity. In **Figure 12**, the results presented in earlier sections are shown as 3D perspective diagram indicating elevated hills and valleys. Higher the elevation, greater the solar energy generation and vice versa.

From comparison of solar radiation maps of India in **Figure 13(a)** [29], and that of Germany in **Figure 13(b)**

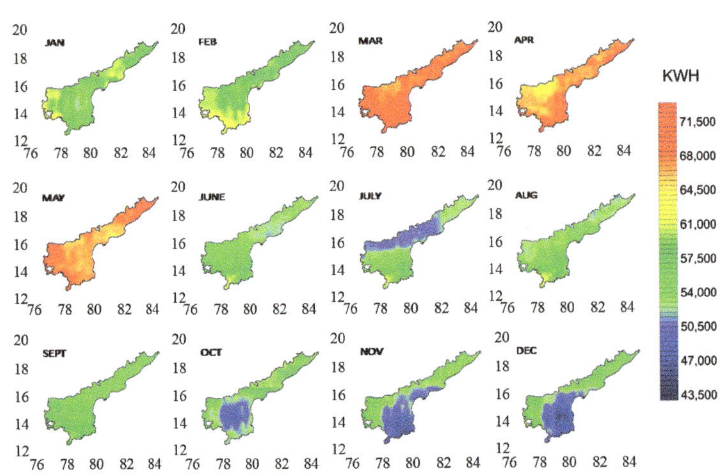

Figure 9. Monthly solar energy generation for Andhra Pradesh state, by showing the variation in different months.

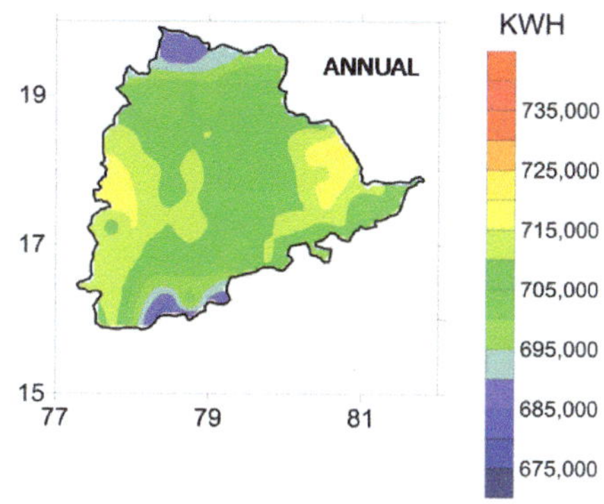

(a)

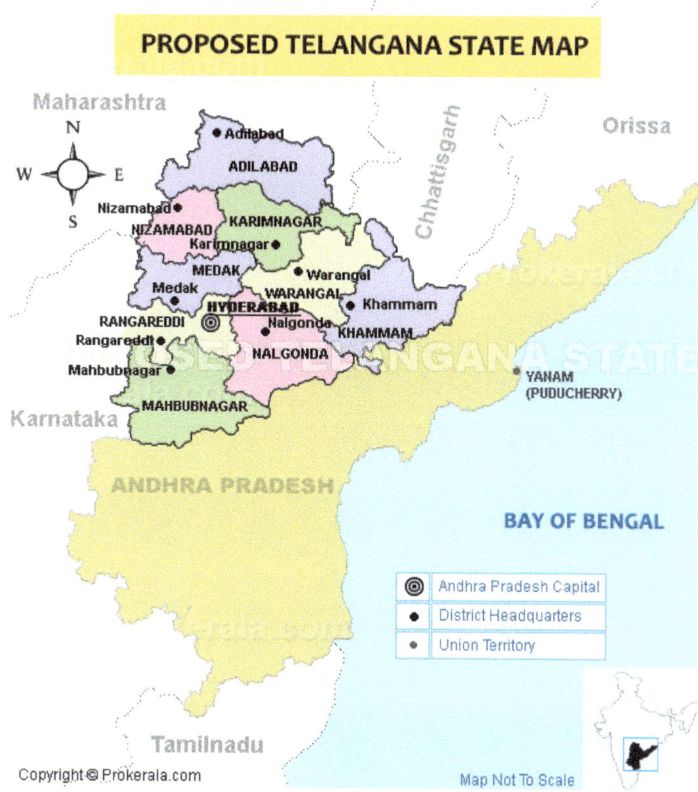

(b)

Figure 10. (a) Annual Solar energy generation for Telangana state; (b) Different districts of Telangana state.

[29] offers a platform for some interesting observations. It is a well established fact that Germany is heading towards solar super power with the aid of its research and development capabilities. The solar radiation map of Germany reveals that it is making tremendous strides in solar energy generation with minimal amount of radiation it receives as compared to India. Examination of maps reveal that maximum amount of solar radiation received by Germany is approximately equal to the minimum amount of solar radiation received by India. This fact clearly lays emphasis on the abundance of resources India possesses owing to its ideal geographical location. The present day reported solar energy generation in India is close to 2.6 GW [30], whereas Germany's solar energy generation is close to 22 GW as of 25th May, 2012 [30]. From the comparison of land area between Germany and India, Indian land mass is approximately 9.2 times greater than Germany. This is another good reason to indicate that India needs to give priority in solar energy generation and formulate policies more aggressively.

Another useful result that can be drawn from our study for solar developers in India is the number of solar panels required for a fixed amount of energy generation and the cost factor involved. These results are summarized in **Table 2**. For example, in India one can generate a maximum annual energy generation ranges from

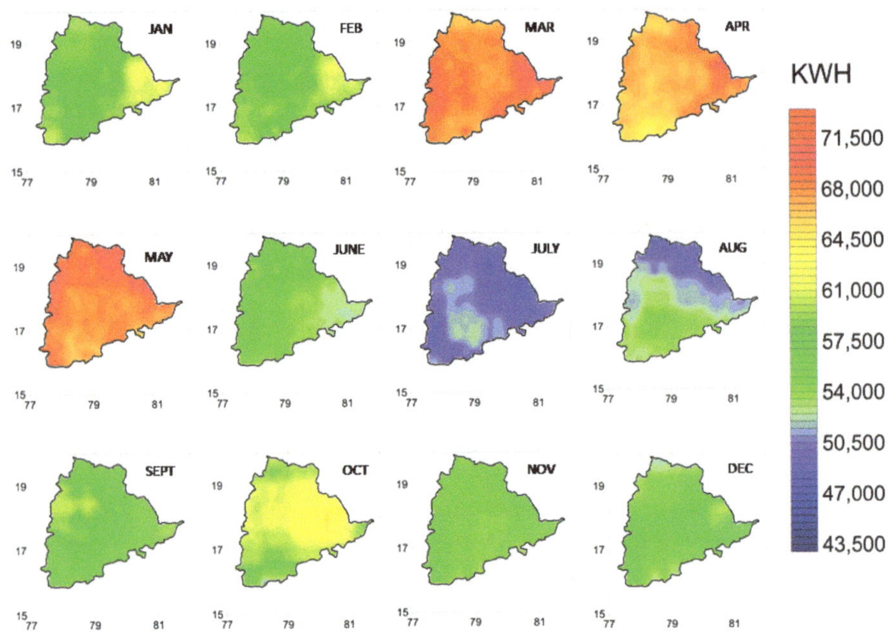

Figure 11. Monthly solar energy generation for Telangana state, showing the variation in different months.

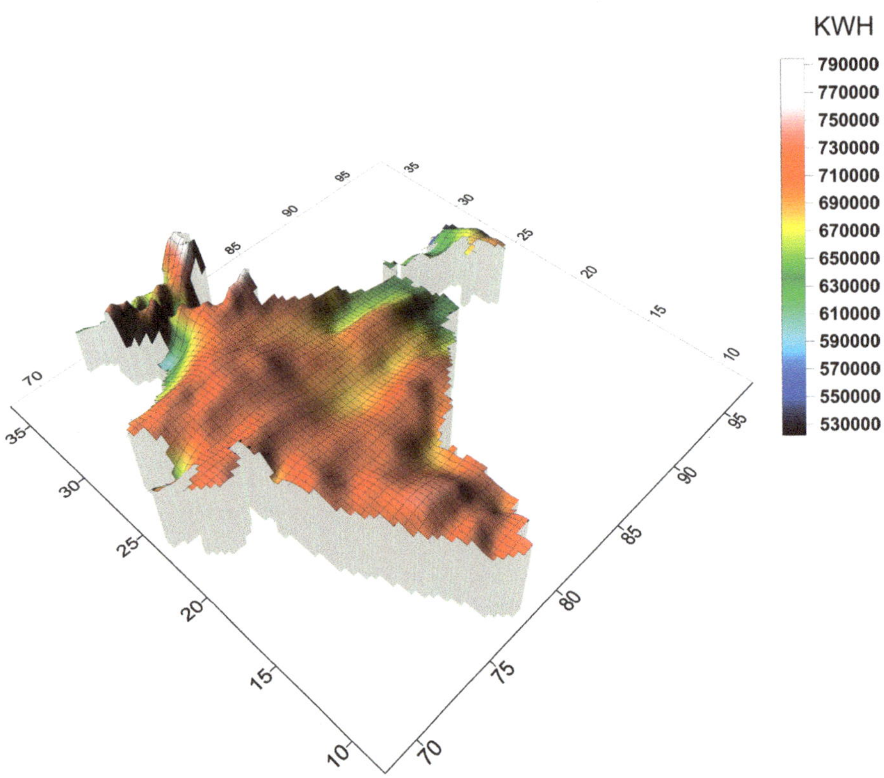

Figure 12. 3D representation of annual solar energy potential. Elevated area indicates high energy potential and the less elevated area shows low energy potential.

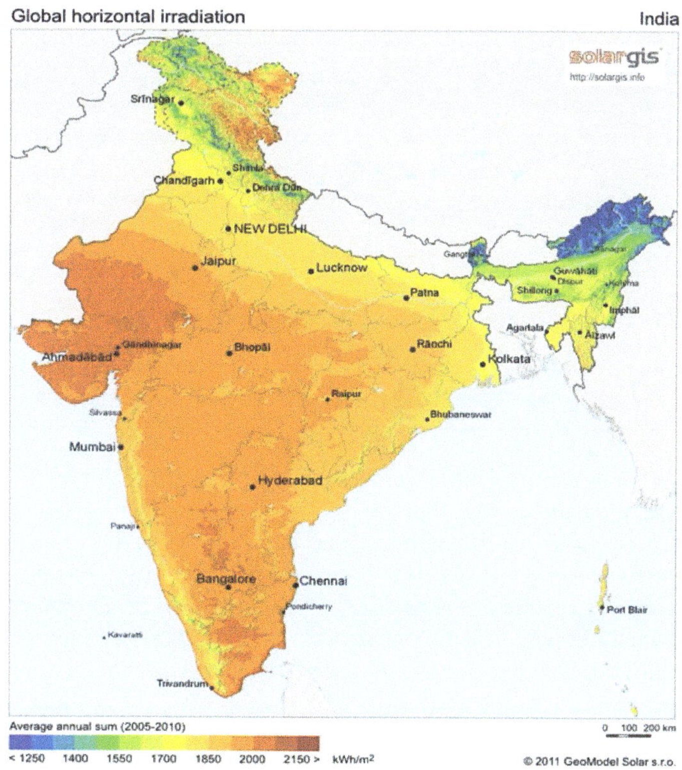

(a)

(b)

Figure 13. (a) Global horizontal irradiation map for India; (b) Global horizontal irradiation map for Germany (Courtesy: Geomodel Solar).

Table 2. Financial analysis keeping the target solar energy generation.

Annual energy generation in an area KWH	Annual target energy	Area of land (acres)	Number of panels	Cost of solar panels ($)	Indian rupees
800,000	800,000	1.00	1944.00	291,600	17,496,000
750,000	800,000	1.06	2060.64	309,096	18,545,760
700,000	800,000	1.43	2779.92	416,988	25,019,280
650,000	800,000	1.23	2391.12	358,668	21,520,080
600,000	800,000	1.33	2585.52	387,828	23,269,680
550,000	800,000	1.45	2818.80	422,820	25,369,200
500,000	800,000	1.60	3110.40	466,560	27,993,600

500,000 - 800,000 KWH per acre of land. In general we keep a wide gap between the rows of panels. For the sake of uniform computation throughout the country and also to estimate the maximum energy possible, we have kept the panels, as described earlier, in horizontal direction. Thus the values provided in our study are the maximum possible energy at any given location. For this type of energy generation, at a high energy (800,000 KWH) potential location, one requires 1944 panels. To achieve the same amount of energy generation in an area of 500,000 KWH potential per acre of land, one need to have 1.6 acres of land and nearly 3110 number of panels. This can be looked from the view point of financial terms as well. For example, with the assumption of 230 W solar panel cost equal to 150 $, one need 291,600 $ (nearly Rs. 1.74 crores @ 1 $ = Rs60/-) for 1944 solar panels. To achieve the same amount (800,000 KWH) of energy generation, in the low energy potential area one requires 466,460 $ (nearly 2.8 crores). This thumb rule hopefully helps the solar developers to estimate the energy generation correctly for better planning before deciding on the solar plant development in an area.

8. Conclusion

Thus, the present research study has been carried out with a focused attention directly on solar energy generation considering various parameters, such as ambient temperature, wind velocity and also other parameters like weather and topographic conditions. In future, we plan to carry out a detailed analysis on the affecting factors and determine their priority.

Acknowledgements

We would like thank the Vice Chairman, GERMI, Sri Tapan Ray for his continuous encouragement and support to this work. Discussions with Dr. Sagar and Dr. Jani and the entire solar team are very helpful at various stages. We thank Ms. KSV Vasavi for her help in many ways in finalizing the manuscript. Coordination of the summer Intern programme by Sri Ankur Bhatnagar, Dharmik Lay, Prashant Gopiyani and Ms. Sandhya are gratefully acknowledged.

References

[1] Ramachandra, T.V., Jain, R. and Krishnadasa, G. (2011) Hotspots of Solar Potential in India. *Renewable and Sustainable Energy Reviews*, **15**, 3178-3186. http://dx.doi.org/10.1016/j.rser.2011.04.007

[2] Khare, V., Nerma, S. and Baredar, P. (2013) Status of Solar Wind Renewable Energy in India. *Renewable and Sustainable Energy Reviews*, **27**, 1-10. http://dx.doi.org/10.1016/j.rser.2013.06.018

[3] Ministry of New and Renewable Energy, Government of India (2009) Jawaharlal Nehru National Solar Mission towards Building SOLAR INDIA. http://www.mnre.gov.in/file-manager/UserFiles/mission_document_JNNSM.pdf

[4] Goswami, D. (2012) India's Solar Sunrise. *Renewable Energy Focus*, **13**, 28-30. http://dx.doi.org/10.1016/S1755-0084(12)70037-6

[5] BP Statistical Review of World Energy (2014). www.bp.com

[6] European Photovoltaic Industry Association (2014) Global Market Outlook for Photovoltaics. http://www.epia.org/news/publications/

[7] Sharma, A. (2011) A Comprehensive Study of Solar Power in India and World. *Renewable and Sustainable Energy Reviews*, **15**, 1767-1776. http://dx.doi.org/10.1016/j.rser.2010.12.017

[8] Raman, P., Murali, J., Sakthivadivel, D. and Vigneswaran, V.S. (2012) Opportunities and Challenges in Setting Up Solar Photovoltaic Based Micro Grids for Electrification in Rural Areas of India. *Renewable and Sustainable Energy Reviews*, **16**, 3320-3325. http://dx.doi.org/10.1016/j.rser.2012.02.065

[9] Lal, S. and Raturi, A. (2012) Techno-Economic Analysis of a Hybrid Mini-Grid System for Fiji Islands. *International Journal of Energy and Environmental Engineering*, **3**, 10. http://dx.doi.org/10.1186/2251-6832-3-10

[10] Oyedepo, S. (2012) Efficient Energy Utilization as a Tool for Sustainable Development in Nigeria. *International Journal of Energy and Environmental Engineering*, **3**, 1-12.

[11] Zweibel, K. (2010) Should Solar Photovoltaics Be Deployed Sooner Because of Long Operating Life at Low, Predictable Cost? *Energy Policy*, **38**, 7519-7530. http://dx.doi.org/10.1016/j.enpol.2010.07.040

[12] Sharma, P. and Harinarayana, T. (2013) Solar Energy Generation Potential along National Highways. *International Journal of Energy and Environmental Engineering*, **4**, 16. http://link.springer.com/article/10.1186%2F2251-6832-4-16

[13] Dinçer, F. (2011) Overview of the Photovoltaic Technology Status and Perspective in Turkey. *Renewable and Sustainable Energy Reviews*, **15**, 3768-3779. http://dx.doi.org/10.1016/j.rser.2011.06.005

[14] Bhutto, A.W., Bazmi, A.A. and Zahedi, G. (2012) Greener Energy: Issues and Challenges for Pakistan—Solar Energy Prospective. *Renewable and Sustainable Energy Reviews*, **16**, 2762-2780. http://dx.doi.org/10.1016/j.rser.2012.02.043

[15] Renewable Energy Policy Network, Renewables Global Status Report. http://www.ren21.net/Portals/0/documents/activities/gsr/RE_GSR_2006_Update.pdf

[16] Electricity Sector in India. http://en.wikipedia.org/wiki/Electricity_sector_in_India

[17] PVSYST. http://www.pvsyst.com/en/

[18] Surfer v11. http://downloads.goldensoftware.com/guides/Surfer11TrainingGuide.pdf

[19] Google Earth v7.1.2.2041. http://www.google.com/earth/explore/products/index.html

[20] Google Earth. http://en.wikipedia.org/wiki/Google_Earth

[21] Global Meteorological Database, Meteonorm Switzerland Version 6.1.0.23.

[22] NASA, Surface Meteorology and Solar Energy (SSE). https://eosweb.larc.nasa.gov/sse/

[23] Harinarayana, T. and Vasavi, K.S.V. (2014) Solar Energy Generation Using Agriculture Cultivated Lands. *Smart Grid and Renewable Energy*, **5**, 31-42. http://dx.doi.org/10.4236/sgre.2014.52004

[24] Maps of India. http://www.mapsofindia.com/

[25] District Map of Gujarat, File. http://commons.wikimedia.org/wiki/File:Map_Gujarat_state_and_districts.png

[26] District Map of Andhra Pradesh. http://www.mapsofindia.com/maps/andhrapradesh/andhrapradesh-district.htm

[27] District Map of Telangana. http://www.prokerala.com/maps/telangana-state-map.html

[28] Solanki, C.S. (2011) Solar Photo-Voltaics: Fundamentals, Technologies and Applications. 2nd Edition.

[29] SolarGIS. http://solargis.info/doc/postermaps

[30] Renewable Energy. http://en.wikipedia.org/wiki/Renewable_energy

3

A Study on Eco-Tourism and Sustainable Development of Economic Underdevelopment Areas—An Example from Kanas Nature Reserve, Xingjiang Province, Northwest China

Jianxiong Qin[1], Pei Zhang[1], Guiping Deng[2], Lu Chen[3]

[1]College of Tourism Historical Culture, Southwest University for Nationalities, Chengdu, China
[2]Science Department, Jiuzhaigou National Nature Reserve Administration, Jiuzhai, China
[3]Institude of Geographic Sciences and Natural Resources Research, CAS, Beijing, China
Email: jx-qin@vip.163.com, 283428785@qq.com, gdeng@umich.edu, chenlu@igsnrr.ac.cn

Abstract

Based on the assessment of eco-tourism resource and its location advantage, the paper has discussed the principle, projects, and facilities of eco-tourism, in the Kanas Nature Reserve in Xinjiang as well as the direction of related industries. To solve the problems occurred at present in the Kanas Nature Reserve, strategic counter measures of sustainable development of eco-tourism have also been put forward, including: 1) The establishment of eco-tourism planning legislative system of eco-tourism; 2) The establishment of environmental protection planning system of eco-tourism; 3) The establishment of environmental audit system of eco-tourism; 4) The establishment environmental legislation and law enforcement system of eco-tourism; 5) The establishment ethic education and management system of eco-tourism.

Keywords

Resources Characteristics, Eco-Tourism, Sustainable Development, Economic Underdeveloped Areas

1. Introduction

Kanas National Nature Reserve is located at the west side of Aletai Mountain and the north banks of Burqin

River and Geba River. It borders on Kazakhstan and Russia in the northwest. Kanas Lake is its center. Kanas National Nature Reserve is 66 km in length from south to north and 74 km in width from west to east with an area of about 250,000 hectares, accounting to around 23.72% of Burqin's total area. It mainly protects the natural landscapes including the cold temperate zone mixed coniferous and broad-leaved forests eco-system and the alpine lake Kanas (**Figure 1**).

Kanas enjoys advantageous geographical position and natural condition. Its biological combination is also extraordinary. The Taiga forest in this area is the only southern Siberia zone of flora and fauna in China, especially the Xinjiang five-leaved pine and fir can only be found here. Thus Kanas is of great value for nature protection and scientific research. Here it can be seen that the forests alternate with the grassland. The communities keep their original state. It has clear vegetation vertical band and the contemporary glaciers are well preserved. Therefore, Kanas possesses unique eco-system and beautiful natural scenery. Many kinds of flowers, trees, birds and animals in this zone are unique in Xinjiang and even whole China. It is also the only area where the Tuvans live in a compact community.

As a high level tourism pattern demonstrating the concept of sustainable development [1]-[4], eco-tourism's core objective is to promote the coordination between natural resources protection and local economic development [5]-[7]. Developing eco-tourism in Kanas Nature Reserve not only provides new opportunities for the construction and development of the nature reserve, but also is of vital significance in driving local economic development, enhancing the public's awareness to protect the nature and expanding the social influence of the nature reserve.

2. The Characteristics of the Resources and the Comment

2.1. The Characteristics of the Eco-Tourism Resources

1) Lake landscape. The Kanas lake surface is 1374 m in altitude. The lake is 24 km in length from south to north and its widest part reaches 9 km from west to east. The depth of the lake is at most 188.5 m and its area is 45 km^2. Kanas Lake is the deepest alpine freshwater lake in China. Its characteristic landscapes include lake view, discoloration lake, water monster, sea of clouds and glory of the Buddha. The towering snow mountains all around, green slopes and thick forests, gorgeous flowers and colorful butterflies and the natural beauty of lakes and mountains present a fantastic view.

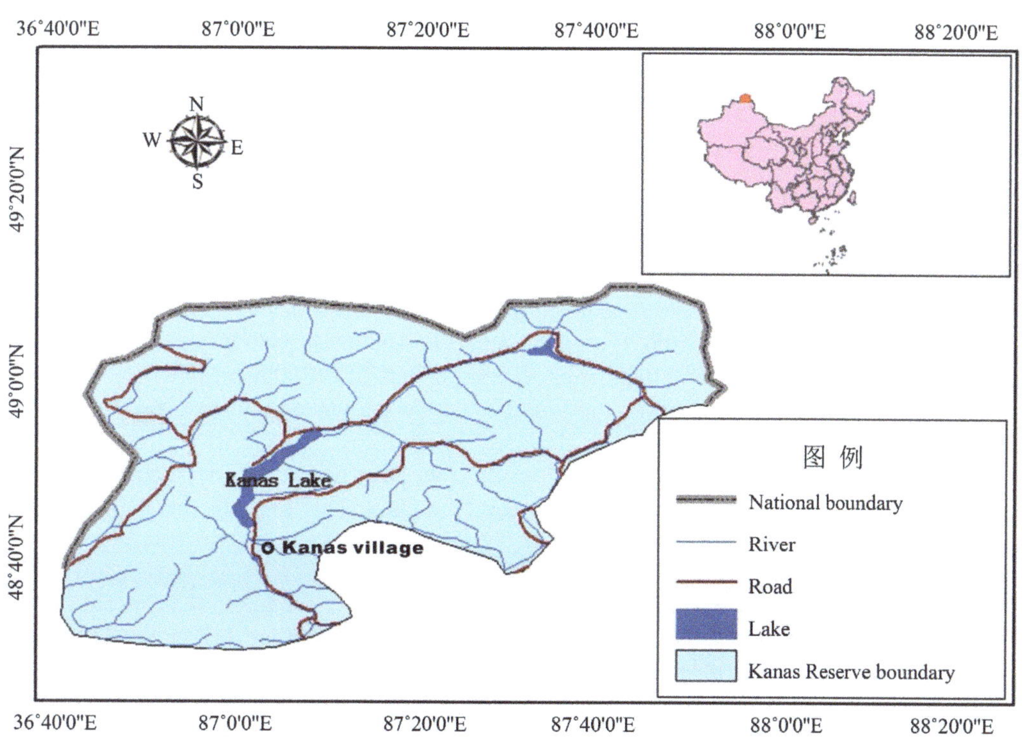

Figure 1. Location map of Kanas National Nature Reserve.

Akekule Lake is located at the foot of the Mountain Friendship in the northeast of Kanas. It is 38 km from the white lake to the head of Kanas Lake and 60 km to the Hemu village. The path is treacherous and rarely visited by people from outside world. The altitude of the lake surface is 1954 km and its area is 9.5 km^2. It is formed by glaciations.

2) River landscape. Kanas River is the major river in this zone. It flows through the whole zone from the northeast to the southwest. It's the largest branch of the downstream Burqin River, which is the largest branch of Irtysh River—the only river belonging to the Arctic Ocean water system. The Kanas River enjoys sound water quality and abundant hydropower. The major landscapes include riverside scenery, Wolong Bay, Moon Bay and Sanctuary Bay.

3) Mountain landscape. The Kanas Nature Reserve locates at the western part of Altay Mountain. The terrain slopes from the northeast to the southwest. It's hilly and very complicated with continuous snow mountains and ravines criss-cross. The altitude is between 1300 km and 4374 km. The highest mountain is the Mountain Friendship in the northeast, at the top of which exist well preserved Quaternary glaciers.

4) Canyon landscape. The Kanas river canyon with turbulence, high mountains and deep valleys, forests and landforms provides a mysterious, tranquil and spectacular view.

5) Rare plant and fungus. The rare tree species in this area is the Siberian pine. There are more than 540 higher class plants. Because the moist soil and humid air are fit for the spread and grow of different kinds of fungus, there are various kinds of fungus and this type of resource is quite rich, most of which are great to eat and materials for medicine.

6) Wild animals. The complete natural and ecological system in Kanas has created a sound environment for the survival of the wild animals and an ideal place for their rest and reproduction. The animals known include 39 kinds of beasts, 4 kinds of reptiles, 117 kinds of birds and 10 kinds of fish. Among them, snow leopard, sable, glutton, goat and black stork are the first-grade national protection animals. The second-grade national protection animals include snow hare, brown bear, otter, lynx, rabbit manul, red deer, original musk deer, bonasa, goshawk, snow pigeon, rock ptarmigan, pana altaica, viviparous lizard, vipera berus and common grouse.

7) Forest landscap. Kanas Nature Reserve is the only distribution area of European/Siberian flora and fauna in China. It mainly consists of Siberian mountain conifers. The advantageous tree is unique in Siberia and basically stays at its original state. The nearly mature and over mature area and storage volume all exceed 97%. The forest coverage rate is 19.4% and the rate reaches 82% in forest land. This area has the most kinds of plants in temperate grassland and has the only Taiga landscape, thus it is of great value for protection and scientific research.

8) Ice and snow landscape. Because of the impact of the Quaternary glaciers and the arctic climate, the Altai Mountain within this zone is one of the major distribution areas of contemporary glaciers. The area and reserve of glaciers in Kanas Nature Reserve account to 71.46% and 70.08% respectively of the total of Altai Mountain area. Looking into the distance at the side of Kanas Lake, one can see very clearly the magnificent ice and snow world at the Peak of Altai Mountain. Especially the Mountain Friendship with snow and glaciers is quite a grand vision. The quaternary glaciers here are well preserved and the glacial landforms are in full growth.

9) Meadow landscape. There is a fascinating meadow landscape with lush herbage and blooming flowers, which offers a perfect natural pasture. We can see different scenery at different altitude. The alpine meadow mainly locates between 2500 m and 3200 m, subalpine meadow between 1800 m and 2400 m, mountain brown coniferous forest between 130 0m and 2300 m and cyperaceae between 1300 m and 1500 m.

10) Clinate land scape. Kanas Nature Reserve is situated at the highest part of the Altai Mountain. The special climatic conditions, the location in the hinterland of the Eurasian continent, far from the sea, the high latitude, distinct altitude difference and complicated terrain result in the divergence in climate compared with neighboring regions and the obvious vertical zoning and horizontal zoning. This leads to the staggered distribution of vegetation vertical band and the forest and pasture on the Yin and Yang slopes, which is a typical representative of the cold temperate climate.

11) Folk custom. Kanas is the only habitation of the Tuvans. They were called dubo people in Sui and Tang dynasty and tuba people in Yuan dynasty. They speak Turki, which is similar to Kazakh. They live on hunting and grazing. In the last 400 years, they have settled down here and left historic and cultural relics including abundant cliff carvings and paintings. They have also maintained the ancient tribal concepts and religious beliefs. The Tuvans build their houses with wood and live on hunting. They are good at archery, horse riding and skiing. They are industrious, simple, brave and strong. Every year they will perform some religious activities such as

sacrificing to heaven, to lake, to mountain, to fish and to fire. The Aoba Festival is the most important festival for them.

The Kazak is a nomadic nationality and they migrate to wherever water and grass are available. Therefore, they are also called the people on the horseback. The Kazak people are hospitable and good at singing, dancing and horse riding. Their traditional ethnic activities include buzkashi, horse racing and girls chasing. The major festivals are Kurban Festival and Roza Festival. The Aken Concert is a civil activity with the largest scale.

12) Scenery of the four seasons. Spring in Kanas is blazing with color and the mountain flowers are in full bloom. All kinds of wild flowers are blossoming, forming a sea of flowers. The white yurts, flocks of sheep and cloud clusters are radiating with the colorful grassland. In spring, the Kanas Lake is rippling with green waves and is as clear as crystal.

Summer in Kanas is so green as if the earth is covered with a green carpet without any stain. The smog is like holy white gauze floating and winding in the valleys. Take a breath you will find the air is so refreshing. Thanks to the white water from the white lake in the upper stream, Kanas Lake is dyed into milky white, which is a mysterious and spectacular view.

When it is autumn in Kanas we can see all the forests are dyed and the colorful hilltop has already begun to store snow. The original green slopes and forests turn into golden yellow, crimson and rosy. The whole wilderness is becoming increasingly attractive with autumn flavor.

Winter in Kanas is covered by snow. The perky land is almost sealed up by snow, presenting a sacred and silent atmosphere. The snow and fog hang on the pine branches quietly. Only the Tuvans' sleighs pulled by horses are still active on the vast snow land.

2.2. Comment on the Eco-Tourism Zone

1) The index of the regional ranks. The nature reserve is divided into 5 levels including world relics, national, provincial, regional and county level. According to the standard of the National Environment Protection Administration, Kanas Nature Reserve is a national level reserve, indicating a relatively high regional rank, especially under the special geographical conditions of Xinjiang where deserts take a large part.

2) Biological diversity. Biological diversity includes ecological diversity, diversity of species and genetic diversity. The visibility of biological diversity makes up the major characteristic of the Kanas eco-tourism zone. The complexity and good taste of the biodiversity are very attractive to tourists. Preliminary investigation has showed that there are over 1500 kinds of plants and animals, suggesting Kanas Nature Reserve is rich in species.

3) The specificity of the elements of the eco-system structure. The specificity can be divided into 5 levels including unusual in the world, rare in China, unique in the province, exceptional in the region and common. According to relevant materials, Kanas Nature Reserve is the only distribution area of European/Siberian flora and fauna and the only Taiga landscape. The obvious specificity demonstrates the good taste of the eco-tourism zone.

4) The primitiveness of the ecological environment. Primitiveness represents the extent of human's interference on the regional ecological environment. The more primitive the region is, the more attractive it is to the tourists. The level of primitiveness can be divided into primitive, relatively primitive, light interfered, damaged and not primitive. Researches reflect that Kanas Nature Reserve remains primitive and undeveloped, which is quite an attraction for the tourists.

5) The stability of the eco-system. The stability of the eco-system consists of the stability of the structure elements and the normality of the functioning. The stability can be divided into stable, relatively stable, ordinary, relatively bad and bad. Kanas Nature Reserve is situated at the west side of the Altai Mountain, where earthquakes are rarely seen and the ecological environment is quite stable. Although people once chopped down forests, the vegetation recovered soon thanks to the superior climatic conditions. The eco-system, as the mountain composite belt, is extremely stable.

6) The diversity and specificity of the scenic spots. The more sightseeing tourism resources and scenic spots within the region, the more valuable it is to travel here and the more attractive it is to the tourists. While among the numerous sightseeing tourism resources, the more the special tourism resources and the higher quality they possess, the more attractive they are. Kanas Nature Reserve is a major eco-tourism zone in our country, enjoying at least 12 types of characteristic tourism resources and scenic spots. It is thus clear that the scenic spots within the region have evident diversity and specificity.

2.3. Comment on the Eco-Tourism Zone

Kanas Nature Reserve borders on Kazakhstan and Russia in the north. To the south, it is 170 km away from the gold city Altay, 210 km away from Burqin county, 80 km away from Habahe county, 120 km away from Beitun county, 140 km away from Fuhai county and 300 km from the oil city Karamay. There are high class roads connecting all of them. The Heiliutan airport in Burqin county was finished in 2007, providing convenience for other regions in Xinjiang and tourists from other provinces. Therefore, Kanas Nature Reserve enjoys sound location and market conditions.

To sum up, the tourism resources, regional comment results and location and market conditions of Kanas Nature Reserve have proved that it has high development value and sound prospect.

3. The Development of the Eco-Tourism Resources

3.1. The Principles for Development

The eco-tourism development in the nature reserve must be guided by the concepts of sustainable development and environment ecology. The following principles should be abided by:

- The principle of function division. All nature reserves should have function divisions. The development of eco-tourism must be carried within the scope of the function divisions. According to the distribution and characteristics of the tourism resources, the function divisions of Kanas Nature Reserve can be divided into core zone (only for scientific research), buffer zone (transitional zone for some eco-tourism projects such as teaching and sightseeing), pilot zone (peripheral zone operating eco-tourism under the precondition of protecting the nature).
- The principle of environment protection theme. Developing tourism in nature reserve should strengthen the theme of environment protection. From the setting of the tourism projects and the composition of the guide words to the selection of the building materials, we should follow the principle of adjusting measures to local conditions and advocate the protection of the nature. For example, we should promote ecological buildings and pollution-free materials and exhibit the use of chemical coating and paint. We should also make the most of solar power and wind power. People can appreciate natural scenery and experience balanced environment through travelling. Thus they can receive ecological and scientific education and are inspired to love nature and think about life.
- The principle of controlling the capacity. The ecological balance in the nature reserve depends on the influence of the human's method and intensity towards the environment and resources, as well as the nature's self cleaning capacity against this kind of influence. When the number of tourists reaches a level that the environment and resources are damaged and cannot recover by nature's force, this number is called the critical capacity of a certain scope. Before developing, constructing and operating the scenic spot, we should have a planning of the tourist capacity according to the characteristics of the tourism resources in order to control the number of tourists effectively.
- The principle of coordination and integration. Necessary facilities in the nature reserve must keep coordinated and integrated with the natural landscapes. We should adopt the methods of making things done with whatever available, using local materials, fitting to the scenery and situation and adjusting measures to local conditions. The layout of the buildings should be small in size, private and separated.
- The principle of non-urbanization. We should try to limit the facilities of accommodation, catering, transportation, communication and management and arrange them in the periphery zones. Namely, traveling within the region and living outside the region. Blindly improving the grade of the reception facilities must be strictly prohibited. We should ensure the non-urbanization of the nature reserve.

3.2. The Activity Projects of Eco-Tourism

As the eco-system of the eco-tourism zone is fragile, the setting of the eco-tourism projects should uphold the following principles: 1) causing least damage to the ecological environment; 2) enabling the tourists to participate in the experimental activities and easy to control and manage; and 3) returning to and understanding nature while strengthening the awareness to protect the environment [8]. According to the type, structure, distribution and location features of the tourism resources in Kanas Nature Reserve and taking into consideration of the principles above, the optional projects are as follows:

1) Ecological sightseeing. Ecological sightseeing means that the tourists can integrate with the nature and have visual perception of the environment by walking or other simple tools. For instance, appreciate the scenery of the four seasons; admire precious plants and animals; enjoy meteorological views including snow, sunrise and glory of the Buddha; observe mountains and rivers. This category of projects is one of the most common in various eco-tourism zones and provides a complete feeling.

2) Drifting. Drifting is a relatively exciting eco-tourism activity when one floats downstream in rivers and trenches aided by certain tools such as rubber boat, bamboo raft and rubber life belt. Drifting includes professional and leisure drifting. Ecological drifting is often for leisure and less demanding for tourists' skills. With simple tools like raft and boat, it is not as exciting as professional drifting but it is less dangerous, so one can feel excited floating downstream in Kanas River.

3) Horse riding. Horse riding is a kind of eco-tourism activity which tourists ride domesticated horses or other livestock along a certain path and enjoy the scenery on the way. Generally, the horses are tamed for the tourists, thus they are not dangerous but can excite the tourists. Horse riding should be accompanied and guided by specially-assigned person and follow fixed line in order to reduce danger and the damage to the environment.

4) Grass skiing. Grass skiing is a fresh eco-tourism activity. It offers a brand new feeling to ski along the slope with certain tools. The grassland should have high quality and specialized slide way under the management of specialized personnel.

5) Snow skiing. Skiing is a popular activity in many mountainous tourist attractions. Skiing with tools on gentle slope in mountainous area covered by snow requires thick snow and paths fit for skiing. Tourists can participate in this activity after simple training.

6) Gliding. Gliding is an activity which the tourists glide in the sky from high place to lower place with certain tools. This is a high demanding activity thus participants need specialized training and guidance.

7) Cliff climbing. Cliff climbing means climbing on some steep and firm cliffs with certain tools. Cliff climbing calls for high security and the participants should be guided. This activity gives people a strong sense of achievement and is most attractive to young people.

8) Mountain cycling. Riding bicycle in mountainous area offers a stronger sense than on flat ground. So it's also attractive to ordinary tourists.

9) Camping. Camping is staying in the wild overnight within tents or other facilities. Surrounding the bonfire, putting on performances, listening to the nature's various sounds or talking to the nature are all charming activities for people who live in the city for a long time.

10) Scientific expedition. It has become a popular activity in recent years to explore creatures and geological sites in primitive environment. It can provide more direct knowledge. The expedition includes investigating the diversity of creatures, collecting specimen, inspecting geologic section, studying the origin of the Tuvans, exploring precious plants and animals and geographical expedition in the cold temperate zone.

11) Spending the holiday in the cabin. Spending the holiday in the cabin means living in the specially built cabin for a relatively long time in a good environment. The tourists can take forest bath, drink mountain spring and breathe fresh air as well as take part in a series of activities such as sightseeing and body-building. When building the cabin, we should make sure it is simple, clean and tidy. It should also have little impact on the environment.

12) Exploration. Exploration consists of exploring the caves, mountain climbing, drifting, exploring the glaciers, cliff climbing and discovering precious plants and animals. All these explorations are challenging and exciting.

13) Leisure agriculture. We should carry out popularity education on ecological agriculture. The tourists can visit ecological orchard and sample ecological agriculture zone. They also have the opportunity to appreciate flowers, taste fruits, visit the gardens, taste green food, relax and visit ecological folk habitat. Through participating in agricultural labor, they can also enjoy the pleasure and happiness of the countryside.

14) Mountain spring bath. Taking a bath in the clear mountain stream, lake, hot spring or pond not only offers a cool feeling but also makes one feel like returning to his original nature. With the development of hot springs, hot spring vocational tourist attractions emerge in some mountainous regions. As many hot springs have a role to play in people's recuperation, these attractions are becoming increasingly popular.

3.3. The Construction Projects of Eco-Tourism

1) Visitors village. The village should be constructed at the flat and broad Tumu town, which is located at the

entrance of the reserve and outside the pilot zone. The projects include interpretive center, environment protection education center, reception station, information agency, administrative office, shopping mall and ecological park. The interpretive center should offer multi-media exhibition and consulting service so as to introduce the natural resources and cultural value of the eco-tourism zone, code of conduct and security guideline of camping to the tourists. The village should keep harmony with surrounding environment. Tall buildings must be strictly prohibited.

2) Accommodation facilities. The construction of accommodation facilities should pay attention to the following points: a) they should be non-permanent facilities; b) it is not appropriate to pursue the grade of the facilities and the facilities should not have more than three floors; c) they should be consistent with the natural characteristics and ethnic customs; d) the number of bed needed should be predicted according to the demand of tourism; e) supporting drainage system should be built; f) green toilets should be constructed. The eco-tourism accommodation facilities are mainly cabins, youth hotels and tents.

3) Catering facilities. We should supply green restaurants and set up pubs, bars, cafés, buffet restaurants and tea houses. We should also provide green food, fuel, power supply, source of water, lighting, purchasing, catering and management. Supporting sewage treatment plants must be built to eliminate pollution sources.

4) Green shopping and entertainment facilities. Associated facilities should not be assigned in the eco-tourism zone. Precious plants and animals are banned to be used as materials of souvenirs. We should promote entertainment arena and activities projects advocating environment protection and ecological theme.

5) Denoter. We should set various kinds of pegs and scutcheons in the scenic spots and along the paths according to terrain and actual demand. The four kinds of denoters are perimeter pegs, boundary pegs, interpretive scutcheons and restrictive scutcheons. Thus the tourists can avoid getting lost or entering non-tourism zone and dangerous area.

6) Communication and security. Electricity and communication circuits as well as green rescue system should be constructed to ensure the smooth operation of eco-tourism.

3.4. Industrial Development of Eco-Tourism

1) Sparing no efforts to popularize ecological sightseeing agriculture

We should carry demonstration experiments such as the introduction and domestication of precious plants and animals and other creatures, research on biodiversity, construction of natural products base and ecological reconstruction between buffer zone and periphery zone. We should utilize protection belt, woodland and hilly land to develop ecological orchard, sightseeing medical materials, ecological agriculture, sightseeing animal husbandry and scientific agriculture, thus the tourists can appreciate them and receive science education.

2) Striving to develop garden economy

The development of garden economy should be based on the peasant household and under the united training and guidance of the administrative department. They can carry out reception duty and run household hotels. They can establish bases for planting and raising pollution free Chinese medicine, growing characteristic fruits, cultivating high quality ornamental plants and nursery stocks, planting tea, growing and processing wild vegetables. These will help to drive the planting and breeding work of the farmers and increase their income.

4. Countermeasure for the Sustainable Development of Eco-Tourism

4.1. Current Problems to Be Solved Immediately

According to relevant regulations of the local government, Kanas eco-tourism development zone is generally divided into two parts. The scenic region management committee is in charge of the development, construction and management of this region, while the investors are responsible for the service, infrastructure, operation and management outside the scenic area. As the administrative department and investors lack eco-tourism concept and environment protection awareness, the management of the nature reserve is in a mess. It mainly shows in the following aspects:

1) Lacking government control and increasingly prominent anarchy

a) Tourism planning, development and construction disjoint from operation, even the situation of detailed planning before overall planning, development before planning, construction before development and operation before construction has emerged. Under the circumstance of developing and constructing without overall planning, the scenic zone often suffers from overburden.

b) The development of the scenic zone, the construction of the projects and investment are in anarchy state. On one hand, the government doesn't play its dominant role in the region's planning, development, construction and management. On the other hand, the government doesn't do well in administrating the investors, examining the investment programs and controlling the number of investors and development projects.

c) The investors do business in their own way, conduct vicious competition and rush into mass action. Without analyzing the guest market and assessing the ecological capacity of the environment, they enhance the grade of the reception facilities blindly and increase beds arbitrarily.

d) Unauthorized constructions haven't been removed in time. They continue to exist with destructive development and illegal operation.

e) The general layout of the eco-tourism zone is unreasonable. For instance, there are no basic function divisions according to the requirement of the nature reserve. As the core of the nature reserve, there are flooding unauthorized constructions around Kanas Lake.

2) The damage of the tourism resources and the pollution of the ecological environment are becoming increasingly serious

a) Because of the impact of the above factors, the sustainable development of the Kanas National Nature Reserve is severely restricted. This mainly shows in the following aspects:

b) The damage of the tourism resources is becoming increasingly serious, which can be demonstrated by the following phenomenon: a large number of construction projects are going on in the core zone; stocks are used to build flats and pave roads; constructing high class cement roads in the scenic zone; damaging grassland and meadow; using wood as fuel; using gasoline and diesel to power pleasure boat; occupying forest arbitrarily; operating the scenic zone exceeding its capacity. The natural resources and environment in Kanas Nature Reserve are suffering severe damage because of human factors.

The pollution of the ecological environment is getting more grievous. This mainly shows in the following aspects:

- Tail gas from the vehicles entering the scenic area without permission, exhaust gas from the factories and mines outside the scenic area, smoke from burning coal by hotels, restaurants and households are all polluting the air in the scenic zone;
- The number of the unauthorized touring ship powered by gasoline and diesel exceeds normal level. Waste water from the hotels and domestic sewage are polluting the water resources in the scenic region.
- Solid wastes pollution from local residents and tourists.
- After the text edit has been completed, the paper is ready for the template. Duplicate the template file by using the Save As command, and use the naming convention prescribed by your journal for the name of your paper. In this newly created file, highlight all of the contents and import your prepared text file. You are now ready to style your paper.

4.2. Strategic Resolution for Sustainable Development

The following strategies are put forward to cope with the problems above:

1) Establishing a legislation system for the eco-tourism programming

Whether the eco-tourism programming is perfect is of profound significance for the eco-tourism and sustainable development of the Kanas Nature Reserve. First, we should formulate and perfect the basic law of tourism programming and direct it into the legalized track; second, we should explicit the legal nature of the behaviors related to tourism programming; third, we should resolve bull management in the process of tourism programming; fourth, we should establish and improve the comment system about the influence on the environment in tourism programming; fifth, improve the accreditation regime through setting up qualification system for the judges in order to ensure the scientific of tourism programming.

2) Establishing an eco-tourism environment protection planning system

Through this system we can have scientific assumption, design and prediction about the prospect of the eco-tourism development in Kanas Nature Reserve thus we can ensure its sustainable development [9] [10]. The following aspects deserve our attention: a) the matching development of eco-tourism environment protection industry and other related industries; b) the coordination and integration of the ecological environment system and social economic system; c) setting up a regime about the influence on the environment; d) dividing the eco-tourism zone according to its function and protecting it; e) attaching importance to the overall environment construction when developing eco-tourism

3) Establishing an eco-tourism environment protection auditing system

Eco-tourism environment protection auditing is a process in which we predict the organization and operation of the enterprises of the developers. This system helps us to find out whether the enterprises adhere to the designated regulations, standards and policies. Eco-tourism environment protection auditing system operates through environment influence assessment (ELA), environment influence statement (EIS) and environment management system (EMS). Environment protection auditing system emphasizes the following points: a) the developers should work out a systematic comment and scheme about the influence on natural resources, economic conditions, aesthetic appreciation, social environment and cultural field during the process of eco-tourism development; b) the developers must employ qualified professional institution to formulate a specified environment influence document which consists of the possibility, probability and inevitability evaluation about the influence on the environment as well as a resolution on the EMS's quality, scope and depth; c) the government should supervise and control the environmental protection, influence, detection and recovery of the developers involved projects.

4) Establishing an eco-tourism environment protection legislative and law enforcement system

The most serious problem the eco-tourism environment protection of Kanas Nature Reserve is faced with is lack of strict legislation and powerful law enforcement. In terms of legislation, we should strengthen our efforts in formulating administrative laws and regulations, departmental rules and local laws and regulations and improving various implementing regulations. Except for the environment protection departments, different levels of tourism management sections should set up specialized institutions to strengthen law enforcement efforts. We must have strict legislation, powerful law enforcement and punishment on those who break the law [11].

5) Establishing an eco-tourism ethical education and management system

We should set up a virtuous circle mechanism of the coordinated development between eco-tourism education and eco-tourism industry. The targets of education and training include administrative staff, investors, operators, tour guides, tourists and community residents.

Acknowledgements

The research was supported by Southwest Universities of Nationalities, with the project "academic degrees construction subject of College of History and Culture, project code 2012XWD-S1202". The research was also financed by The Asian Development Bank with the project "Study on Suggestions and Implementation of Reconstruction of Urban and Transport Infrastructure in the Sichuan Earthquake-Stricken Areas".

References

[1] Nabin, B. and Anal, D. (2014) Diversifying Finance Mechanisms for Protected Areas Capitalizing on Untapped Revenues. *Forest Policy and Economics*, **41**, 60-67. http://dx.doi.org/10.1016/j.forpol.2014.01.002

[2] Tran, L. and Walter, P. (2014) Ecotourism, Gender and Development in Northern Vietnam. *Annals of Tourism Research*, **44**, 116-130. http://dx.doi.org/10.1016/j.annals.2013.09.005

[3] Yin, K.P. and Yan, H.L. (2012) Eco-Tourism and Sustainable Development. Sichuan University Press, Chengdu.

[4] Chakrabarty, A. (2011) Ecotourism Development and Security Restructuring: A GI Based Planning for Peaceful Dissuasion of Anarchism in Forest Provinces of India. *Procedia Social and Behavioral Sciences*, **21**, 108-115. http://dx.doi.org/10.1016/j.sbspro.2011.07.006

[5] Qin, J.X. (2003) Humble Opinion about Sustainable Development of the Tourism Industry in Chengdu City. *China Sustainable Development*, **10**, 10-20. http://sd.thnet.gov.cn/zgkcxfzzz/

[6] Qin, D.H. (1999) China's Population Source Environment and Sustainable Development. Xinhua Press, Beijing.

[7] Li, J.Q. (2001) Discussion on the Eco-Tourism Development of Labagou Nature Reserve in Beijing. *China's Sustainable Development*, **10**, 10-20. http://sd.thnet.gov.cn/zgkcxfzzz/

[8] Yoder, A. and Bowler, P.A. (1998) Sustainable Development, Ecotourism, and Globalization: Are They Compatible? Interdisciplinary Minor in Global Sustainability. University of California, Irvine.

[9] Cleveland, C. (1999) Sustainable Tourism, Globalization, Regionalization and Third World Development. *Regional Studie*, **343**, 34-56.

[10] Milton (1999) Sustainable Tourism: A Local Authority Perspective. *Proceedings of the 7th Tourism and Sustainable*

Development commission, New York, 19-30 April 1999, 19-30.

[11] Pearce, D. (2001) Cultural Tourism and Sustainable Development in the Arab Region. *Journal of Environment Management*, **45**, 1-10. http://link.springer.com/search?query

4

Integrating Renewable Energy and Smart Grid Technology into the Nigerian Electricity Grid System

Emodi Nnaemeka Vincent, Samson D. Yusuf

Department of Technology Management Economics and Policy Program, College of Engineering, Seoul National University, Seoul, South Korea
Email: icecube4ever2000@gmail.com

Abstract

The electricity situation in Nigeria can be described as epileptic with no sign in view of improvement. This epileptic power situation affects the manufacturing, service and residential sectors of the economy which in turn affects the country's economic growth. Even with the recent reforms in the power sector, more than half of the country's population still lack access to electricity. The epileptic condition of the power sector can be attributed to the inadequate and inefficient power plants, poor transmission and distribution facilities, and outdated metering system used by electricity consumers. This paper attempts to present the way forward for the Nigerian poor electricity situation by reviewing the power sector as a whole and the renewable energy potentials. We identified the problems in the national grid and then proposed a smart grid model for the Nigerian power sector which will include renewable energy source. We believe that the content of this review paper will solve the poor epileptic condition of the power sector in Nigeria and also enable the proper integration of smart grid technology into the national grid.

Keywords

Epileptic Power, Power Sector, Smart Grid, Renewable Energy, National Grid

1. Introduction

Electricity is an important facet of any nation's development. In Nigeria, electricity is the pillar of its growth and development with roles in the nation's production of goods and services in the industrial sector as well as agriculture, health and education [1]. Nigeria, a county known as "the giant of Africa" is blessed with an abundant amount of fossil fuel and renewable energy resources, but the country is battling with an acute epileptic power situation. According to the World Bank data; only about 50.9% of Nigerians have access to electricity [2].

The just concluded World Economic Forum on Africa which took place in Abuja, Nigeria (ended on 9th May, 2014) saw an investment boost to the tune of $68 billion from foreign and domestic investors in the service and industrial sectors. Sadly, this investment will meet its "waterloo" if the Nigerian power situation does not improve. Nigeria's poor energy situation results from the national grid network with problems ranging from inefficient power plants which are few in numbers to lack of renewables to support peak load, physical deterioration of the long transmission lines to distribution facilities which are inadequately maintained, lack of communication facilities, illegal electricity connections and outdated meters used by the consumers [1]. However, this paper tends to address the research question: what is the way forward for the Nigerian poor electricity situation? In order to solve the above research question, the paper reviews the current electricity sector in Nigeria and the concept of smart grid technology. Furthermore, we suggest a future direction on the integration of smart grid into the Nigerian electricity grid network in conjunction with renewable energy.

This paper is divided into seven main sections: the first section is the introduction and the second is an overview of the power sector where we reviewed the history of the power sector and its various sectors which comprise of the generation, transmission and distribution after which we go into the regulatory agencies, key institutions and the current electricity situation in Nigeria. The third section focused on the renewable energy resource potential and the areas where they are located within the country and their possible application. The forth section introduces the smart grid technology while the fifth section discussed about the integration of smart gird technology into the Nigerian grid network and the various technological upgrades required to make the smart grid system in Nigeria possible. In the sixth section, we give our discussion and present a proposed smart grid model after which we outline some factors that may affect smart grid integration in Nigeria followed by a review on the Nigerian Energy Policy and finally the benefits of smart grid technology are presented. Our policy recommendation to facilitate the proper integration of renewables and smart grid technology into the Nigerian electricity grid system are given in section seven which also includes the conclusion to this paper.

We believe that this paper will contribute immensely to the possibility of smart grid and renewable integration into the Nigerian electricity grid.

2. Overview of the Nigerian Power Sector

2.1. History

In Nigeria, electricity generation rose from few kilowatts that were used in Lagos by the colonial masters when the first generating plant was installed in 1898 [3]. By the Act of Parliament in 1951, the Electricity Corporation of Nigeria (ECN) was established. Niger Dams Authority was set up in 1962 to develop hydroelectricity and was merged with ECN to form the National Electric Power Authority (NEPA) in 1972. Despite various effort by NEPA (which operated a monopolized market) to manage the power sector by providing electricity to the increasing population, it became clear that NEPA was losing the battle to meet up with the electricity demand in the 1990s. Hence, in 2001, the National Electric Power Policy (NEPP) was introduced to kick-off the power sector reform and this lead to several other reforms in the past years [4]. The NEPP in 2001 created the roadmap for Nigeria's Power Sector Privatization, but due to government bureaucracy; the policy was not signed into law until 2005. This signed document was the Electric Power Sector Reform (EPSR) Act in 2005 which was expected to level the playing ground for potential investors and improve the wellbeing of its citizens. The EPSR Act led to the incorporation of the Power Holding Company of Nigeria from NEPA, which was later defunct and divided into sub-sectors [3] [4].

2.2. Sub-Sectors

The power sector in Nigeria is divided into three major subsectors which are; Generation (NESI), Transmission (TCN) and Distribution (DISCOS).

2.2.1. Generation

The splitting of the power sector led to the formation of Nigerian Electricity Supply Industry (NESI) which currently has 23 grid-connected generating plants in operation with a total installed capacity of 10,396 MW and available capacity of 6056 MW. Thermal based installed capacity is 8457.6 MW with available capacity of 4996

MW. Hydropower from three major plants accounts for 1938 MW of the total installed capacity with 1060 MW as available capacity. The generation segments of the Nigerian power sector are divided into [4]:

- Successor Generation Companies (Gencos): with six successor companies.
- Independent Power Producers (IPPs): they are owned and managed by the private sector and have three generating facilities.
- National Integrated Power Projects (NIPP): they are owned by the government and have 10 generating facilities.

2.2.2. Transmission

The splitting also formed the Transmission Company of Nigeria (TCN) which was a successor of the PHCN. It is made up of two departments namely System Operator and Market Operator. The transmission capacity is made up of about 5523.8 KM of 330 KV lines and 6801.49 KM of 132 KV lines [4].

2.2.3. Distribution

The third sub-sector which is the distribution comprises of eleven electricity distribution companies (DISCOS). The DISCOS are listed below in alphabetical order [4]:

- Abuja Distribution Company
- Benin Distribution Company
- Eko Distribution Company
- Enugu Distribution Company
- Ibadan Distribution Company
- Ikeja Distribution Company
- Jos Distribution Company
- Kaduna Distribution Company
- Kano Distribution Company
- Port Harcourt Distribution Company
- Yola Distribution Company

2.3. Regulatory Agencies and Key Institutions

We will now look at the various regulatory agencies and key institution in the power sector.

2.3.1. Regulatory Agencies

The regulatory agencies are [4]:

- The Federal Ministry of Power (FMP): an administrative arm of the federal government that deals with policy formulation and provides general direction to other agencies involved in the power sector.
- Nigerian Electricity Regulatory Commission (NERC): an independent regulatory agency mandated to regulate and monitor the Nigerian Power Sector and was established in 2005 by the EPSR Act.
- Energy Commission of Nigeria (ECN): was established in 1988 with a stator mandate for strategic planning and coordination of national policies in the field of energy.
- Rural Electrification Agency (REA): was established by the EPSR Act with the statutory functions of promoting, supporting and providing electricity access to rural and semi-urban areas of the country.
- Presidential Task Force on Power (PTFP): was set up in 2010 to drive the implementation of the reform of Nigeria's Power sector.

2.3.2. Key Institutions

The key institutions are [4]:

- Niger Delta Power Holding Company Limited (NDPHC): is owned by the three arms of the government (i.e. federal, state and local) with the responsibility for the implementation of the National Integrated Power Project (NIPP).
- Nigerian Bulk Electricity Trading PLC (NBET): also government owned and was established by the EPSR Act as an electric trading licensee that engages the purchase of electric power and ancillary services (from the independent power producers and the successor generating companies) and subsequent resale to distribution companies and eligible customers.

- Operator of the Nigerian Electricity Market (ONEM): was licensed to function as a market operator for the wholesale electricity market of the Nigerian Electricity Supply Industry, with the responsibility of operation in the electricity market and settlement arrangements. A key function of the ONEM is the administration of metering system among generation, transmission and distribution companies.
- Nigerian System Operator (NSO): was licensed to provide system operation services to the Nigerian Electricity Supply Industry; its primary responsibility is the planning, dispatch and operation of the transmission system. Also, it's responsible for the security and reliability of the nation's electricity grid network.
- Gas Aggregation Company Nigeria Limited (GACN): was incorporated in 2010 for the purpose of stimulating growth of natural gas utilization in the Nigerian domestic market.
- National Power Training Institute of Nigeria (NAPTIN): was established in 2009 to serve as a focal point for human resource development and work-force capacity building, and also act as a research center on matters relating to power in Nigeria.
- Nigeria Electricity Liability Management Company Limited (NELMCO): was established in 2006 as a company limited by guarantee, to assume and manage the non-core assets, all liabilities and other obligations that would not be taken over by the successor companies.

2.4. Current Electricity Situation

Nigeria is faced with many electricity problems which range from generation, transmission to distribution and marketing. We will review this problems one after the other;

Generation: The total installed capacity of the currently generating plants in Nigeria as discussed earlier is 10,396.0 MW, but the available Capacity is less than 6056 MW as at December 2013. Seven of the twenty-three generation stations are over 20 years old and the average daily power generation is lower than the peak forecast for the current existing infrastructure. Through the planned generation capacity projects for a brighter future, the current status of power generation in Nigeria presents challenges, such as inadequate generation availability, delayed maintenance of facilities, insufficient funding of power stations, obsolete equipment, tools, safety facilities and operational vehicles, obsolete communication equipment, lack of exploration to tap all sources of energy from the available resources and low staff morale [5] [6].

Transmission: The current transmission system in Nigeria comprises 5523.8 km of 330 kV, 6801.49 km of 132 kV, 32No. 330/132 kV Substations with total installed transformation capacity of 7688 MVA. 105No. 132/33/11 kV Substations with total installed transformation capacity of 9130 MVA. The average available capacity on 330/132 kV is 7364 MVA and 8448 MVA on 132/33 kV. The Nigeria 330 KV transmission grid is characterized by high power losses due to the very long transmission lines. Some of these lines include Benin—Ikeja West (280 km), Oshogbo—Benin (251 km), Oshogbo—Jebba (249 km), Jebba—Shiroro (244 km), BirninKebbi—Kainji (310 km), Jos—Gombe (265 km) and Kaduna— Kano (230 km) [6].

Power losses result in lower power availability to the consumers, leading to inadequate power to operate the appliances. Thus, the high efficiency of the power system is determined by its low power losses. Increased power demand pushes the power transmission and distribution networks to their upper limits and beyond, resulting to shortening of the life span of the network or total collapse [6] [7].

The Nigerian transmission system does not cover every part of the country. It currently has the capacity to transmit a maximum of about 6056 MW and it is technically weak, thus very sensitive to major disturbances. Major problems associated with transmission systems include poor funding by the Federal Government, it is yet to cover many parts of the country, it's current maximum electricity wheeling capacity is 6,056 MW which is awfully below the required national needs, some sections of the grid are outdated with inadequate redundancies as opposed to the required mesh arrangement, regular vandalization of the lines, associated with low level of surveillance and security on all electrical infrastructure, technologies used generally deliver very poor voltage stability and profiles, there is a high prevalence of inadequate working tools and vehicles for operating and maintaining the network, there is a serious lack of required modern technologies for communication and monitoring, transformers deployed are overloaded in most service areas, inadequate of spare parts for urgent maintenance, poor technical staff recruitment, capacity building and training programme [5] [6].

Distribution and Marketing: In most regions in Nigeria, the distribution network is poor, the voltage profile is poor and the billing is inaccurate. As the department, which inter-faces with the public, the need to ensure adequate network coverage and provision of quality power supply in addition to efficient marketing and customer service delivery cannot be over emphasized. Some challenges identified are, weak and inadequate network coverage, overloaded transformers and bad feeder pillars, substandard distribution lines, poor billing system, unwholesome practices by staff and very poor customer relations, inadequate logistic facilities such as tools working vehicles, poor and obsolete communication equipment, low staff morale and lack of regular training, insufficient funds for maintenance activities [5] [6].

Power Deficit: The total installed capacity of generating plants in Nigeria is 10,390 MW with available capacity less than 6056 MW, but power generation has been below 4500 MW. Using the rule of Thumb, where 1000 MW is for 1,000,000 people and the Nigerian population is 174,567,539 [8], we should have about 174,508 MW for the Nigerian people but with the power generation that has not exceeded 4500 MW, we can say that Nigeria has a power deficit of 170,008 MW. Surprisingly, Nigeria is still growing at an amazing 6% - 8% [8] and the nation is doing so without almost no power (from the national grid) compared to other rapidly growing nations like South Africa with a population of about 50 million people (in real sense, less than a third of the Nigerian population) and still generates over 45,000 MW of electricity.

The map of the current Nigerian national grid system is shown in **Figure 1** below [9].

3. Renewable Energy Resources

The renewable energy resources in Nigeria are as enormous as they are diverse. The summary of the renewable energy potentials are given in **Table 1** [9].

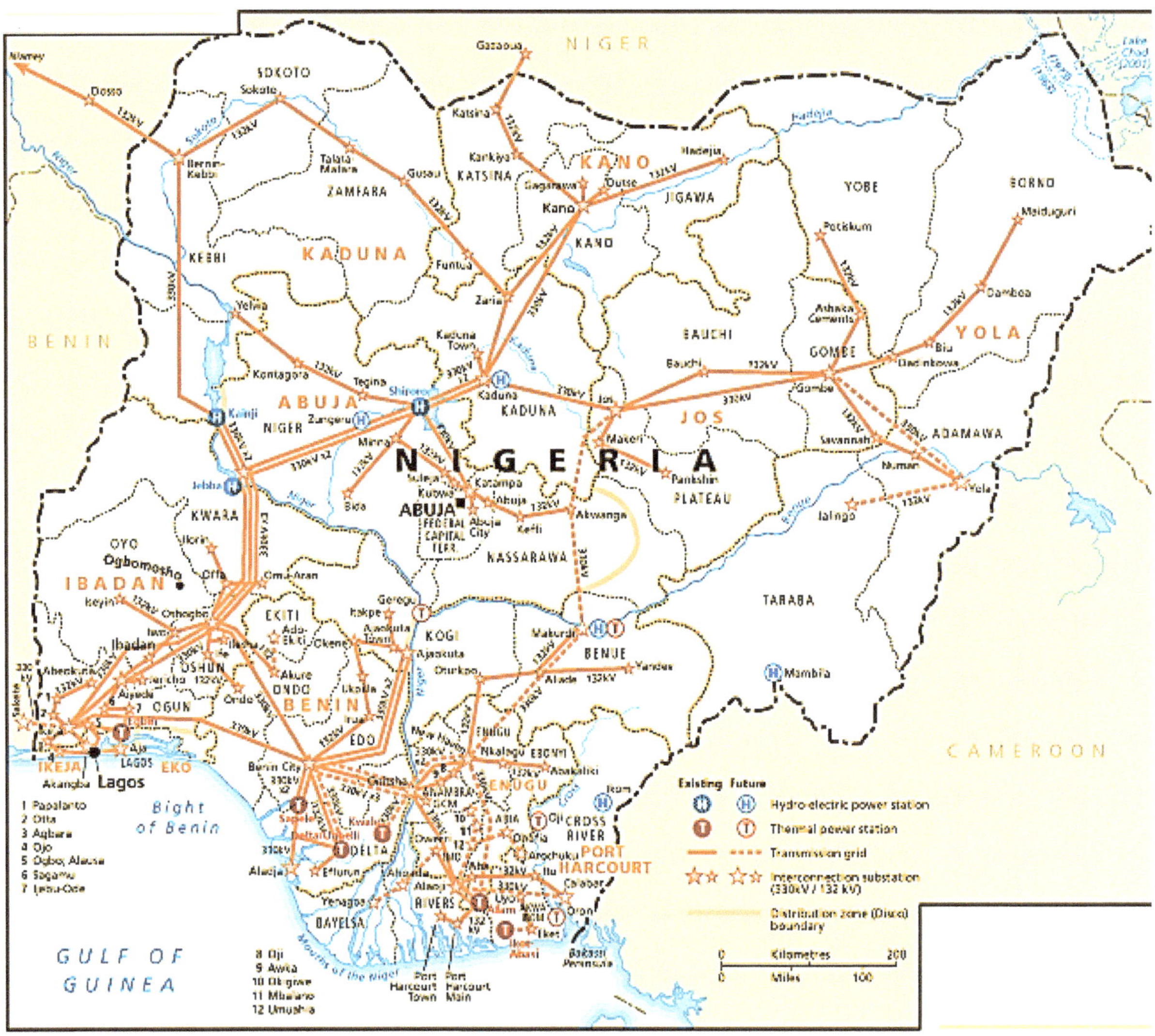

Figure 1. The Nigerian national grid system.

Table 1. Renewable energy resources.

Renewable Energy Source		Capacity
Small Hydropower		3500 MW
Large Hydropower		11,250 MW
Wind		2 - 4 m^2 annually at 10 m height
Solar Radiation		3.5 - 7.0 KMh/m^2/day
Biomass	Fuel Wood	13,071,464 hectares of forest and woodland
	Animal Waste	61 million tons/year
	Crop Residue	83 million tons/year

Source: Energy Commission of Nigeria, 2005.

3.1. Solar Energy

Nigeria is located within the region of high sunshine belt between longitude 3° and 14° east of Greenwich and latitude 4° and 14° north of the equator with a land area of 923,768 km^2 [10]. The mean annual average of total solar radiation varies from about 3.5 kWh/m^2/day in the coastal regions to about 7 kWh/m^2/day along the semi-arid areas in the Northern border region. On the average, the country receives solar radiation at the level of about 19.8 MJ/m^2/day. Average sunshine hours are estimated at 6hrs per day. Solar radiation is fairly well distributed. The minimum average is about 3.55 kWh/m^2/day in Katsina in the far North during the month of January and 3.4 kWh/m^2/day for Calabar in the Southeastern region around August and the maximum average is 8.0 kWh/m^2/day for Nguru in the Northeast during the month of May [1].

At a medium radiation intensity of 5 KWh/m^2/day and convention efficiency and if solar collectors modules were employed to cover 1 % of land areas in Nigeria will generate approximately the daily energy equivalent of 192,000 MW of energy from a gas power plant working at full capacity for 24 hours a day [10] [11].

Solar electricity may be used for power supply to remote villages and locations not connected to the national grid. It may also be used to generate power for feeding into the national grid. Other areas of application of solar electricity include low and medium power application such as: water pumping, village electrification, rural clinic and schools power supply, vaccine refrigeration, traffic lighting and lighting of road signs [1]. The yearly average of daily sun in Nigeria is shown in **Figure 2** [12].

3.2. Biomass Energy

Nigeria's biomass energy resources can be identified as fuel wood with capacity of about 13 million hectares of forest and wood land, animal waste is about 61 million tons/year and crop residue is currently limited to thermal application as fuel for cooking and crop drying. However, these biomass resources could be utilized as fuel for small scale industries to support electricity in the national grid [9].

3.3. Hydropower

Nigeria is reasonable endowed with large rivers and some few natural falls. Small rivers and streams also exist within the present split of the country into eleven river basin authorities. Studies have shown that the country possesses potential hydro-energy resources along her numerous river systems consisting of a total of 70 micro dams, 126 mini dams and 86 small sites [13]. The total exploitable potential of the country's large hydropower is estimated to about 11,250 MW while the small hydropower stands at 3500 MW [9]. These rivers, waterfalls and streams with high hydropower potential, if properly harnessed will lead to decentralized use and provide the most affordable and accessible option to off-grid electricity service especially to the rural communities [8]. **Figure 3** shows the various water ways in Nigeria [14].

3.4. Wind Energy

Wind is produced by the uneven heating of the earth's surface by the sun and its resultant pressure inequalities. Wind energy potential varies with wind speed and is available in Nigeria at annual average speeds of about 2.0 m^2 at the coastal region and 4.0 m^2 at the northern region of the country. Assuming an air density of 1.1 kg/m^3,

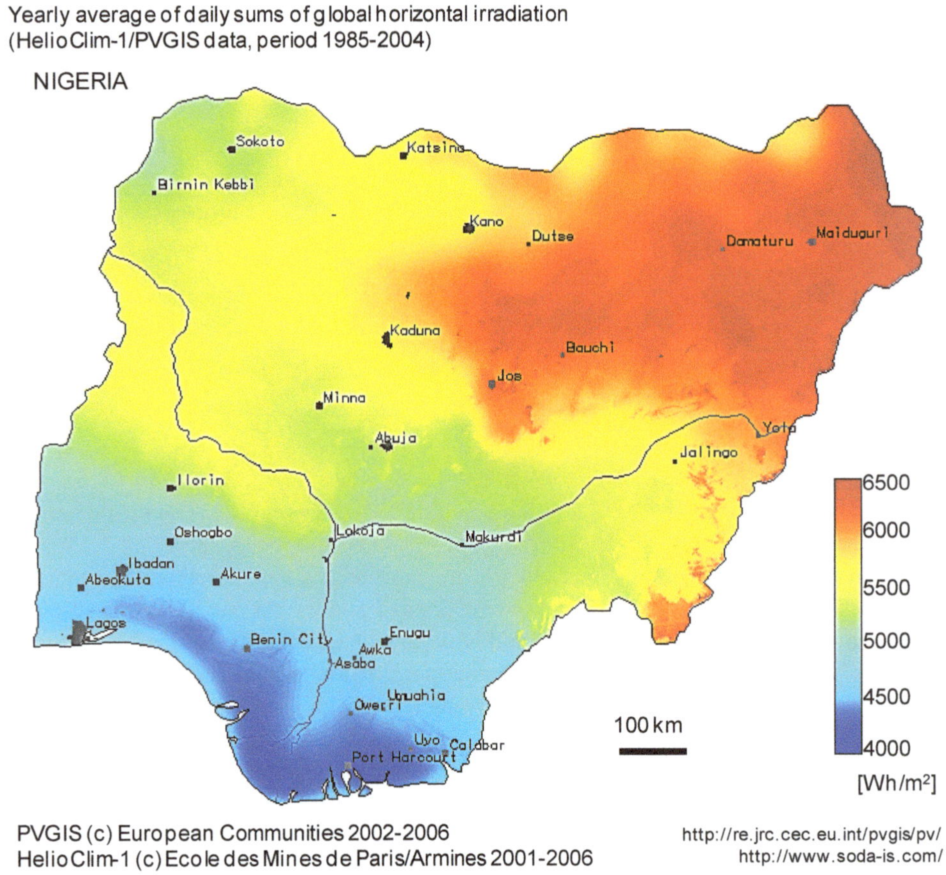

Figure 2. The Yearly average of daily sun in Nigeria.

wind energy intensity, perpendicular to the wind direction, ranges between 4.4 W/m^2 at the coastal areas and 35.2 W/m^2 at the far northern region [1].

Various wind energy conversion systems like wind turbines, wind generators, wind plants, wind machines, and wind dynamos are devices which convert the kinetic energy of the moving air to rotary motion of a shaft, that is, mechanical energy. The technologies for harnessing this energy have, over the years been tried in the northern parts of the country, mainly for water pumping from open wells in many secondary schools of old Sokoto and Kano States as well as in Katsina, Bauchi and Plateau States. It has been reported that an average annual wind speed of not less than 5 m/s at a height of 10 m above ground level is the feasible speed for the exploitation of wind energy at today's cost. In **Figure 4** we can see the various wind energy location across the country [15].

4. Smart Grid Technology

Various authors, government organization and bodies have given numerous definitions of smart grid. A smart grid can be defined as an upgraded electricity grid network enabling two-way information and power exchange between suppliers and consumers, due to the pervasive incorporation of intelligent communication monitoring and management systems [16]. The US Department of Energy (DOE) defined smart grid as a system that will incorporate digital technology to improve reliability, efficiency and security of the electric system [17].

According to the International Energy Agency Technology Roadmap Smart Grid; a smart grid is an electricity network that uses digital and other advance technologies to monitor and manage the transport of electricity from all generation sources to meet the varying electricity demand of end-users. The smart grid will organize various capabilities and needs of the grid operators, generations, end-users and electricity market stakeholders to operate all the parts of the system as efficiently as possible, reducing environmental impacts and cost while improving system reliability, resilience and stability [18]. Besides the smart grid being able to match supply with demand,

Figure 3. Waterways in Nigeria.

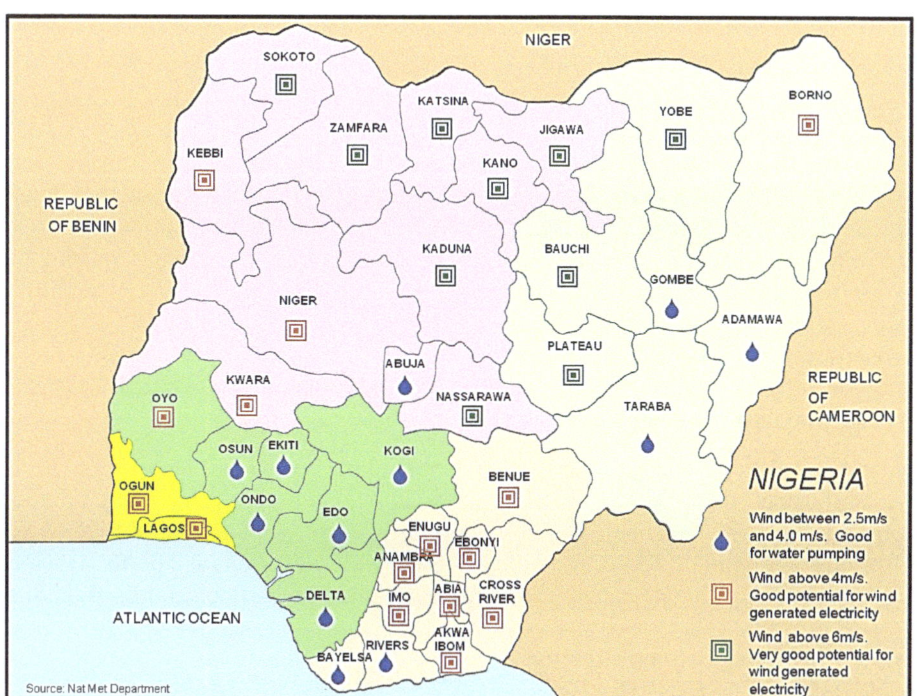

Figure 4. Wind energy location in Nigeria.

it will also reduce the emission of greenhouse gases (GHG) by reducing the heavy dependence on fossil fuel plants through the integration of renewables which will make the system flexible and also electricity use will be more efficiently utilized.

4.1. Wide Area Monitoring and Control

Real-time monitoring and display of power system components and performance, across interconnections and over large geographic areas, help system operators to understand and optimize power system components, behavior and performance. Advanced system operation tools avoid blackouts and facilitate the integration of variable renewable energy resources. Monitoring and control technologies along with advanced system analytics—including wide area situational awareness (WASA), wide-area monitoring systems (WAMS), and wide-area adaptive protection, control and automation (WAAPCA)—generate data to inform decision making, mitigate wide-area disturbances, and improve transmission capacity and reliability [18].

4.2. Information Communication Technology Integration

Underlying communications infrastructure, whether using private utility communication networks (radio networks, meter mesh networks) or public carriers and networks (Internet, cellular, cable or telephone), support data transmission for deferred and real-time operation, and during outages. Along with communication devices, significant computing, system control software and enterprise resource planning software support the two-way exchange of information between stakeholders, and enable more efficient use and management of the grid [18].

4.3. Renewable and Distributed Generation

Integration of renewable and distributed energy resources—encompassing large scale at the transmission level, medium scale at the distribution level and small scale on commercial or residential building can present challenges for the dispatchability and controllability of these resources and for operation of the electricity system. Energy storage systems, both electrically and for thermally based, can alleviate such problems by decoupling the production and delivery of energy. Smart grids can help through automation of control of generation and demand (in addition to other forms of demand response) to ensure balancing of supply and demand [18].

4.4. Transmission and Enhancement Application

There are a number of technologies and applications for the transmission system. Flexible AC transmission systems (FACTS) are used to enhance the controllability of transmission networks and maximize power transfer capability. The deployment of this technology on existing lines can improve efficiency and defer the need of additional investment. High voltage DC (HVDC) technologies are used to connect offshore wind and solar farms to large power areas, with decreased system losses and enhanced system controllability, allowing efficient use of energy sources remote from load centers. Dynamic line rating (DLR), which uses sensors to identify the current carrying capability of a section of network in real time, can optimize utilization of existing transmission assets, without the risk of causing overloads. High-temperature superconductors (HTS) can significantly reduce transmission losses and enable economical fault-current limiting with higher performance, though there is a debate over the market readiness of the technology [18].

4.5. Distribution Grid Management

Distribution and sub-station sensing and automation can reduce outage and repair time, maintain voltage level and improve asset management. Advanced distribution automation processes real-time information from sensors and meters for fault location, automatic reconfiguration of feeders, voltage and reactive power optimization, or to control distributed generation. Sensor technologies can enable condition- and performance-based maintenance of network components, optimizing equipment performance and hence effective utilization of assets [18].

4.6. Advance Metering Infrastructure

Advanced metering infrastructure (AMI) involves the deployment of a number of technologies—in addition to advanced or smart meters that enable two-way flow of information, providing customers and utilities with data on electricity price and consumption, including the time and amount of electricity consumed. AMI will provide a wide range of functionalities which will include; Remote consumer price signals, ability to collect, store and report customer energy consumption data for any required time intervals or near real time, improved energy diagnostics from more detailed load profiles, ability to identify location and extent of outages remotely via a metering function that sends a signal when the meter goes out and when power is restored, remote connection and disconnection, losses and theft detection, ability for a retail energy service provider to manage its revenues through more effective cash collection and debt management [18].

4.7. Customer-Side System

Customer-side systems, which are used to help manage electricity consumption at the industrial, service and residential levels, include energy management systems, energy storage devices, smart appliances and distributed generation. Energy efficiency gains and peak demand reduction can be accelerated with in-home displays/ energy dashboards, smart appliances and local storage. Demand response includes both manual customer response and automated, price-responsive appliances and thermostats that are connected to an energy management system or controlled with a signal from the utility or system operator [18].

Figure 5 shows the smart grid technology areas but this excludes the Electric Vehicle (EV) charging infrastructure but this could also be integrated in the future.

5. Smart Grid Integration into the Nigerian Grid Network

It is important to note that some technologies have been employed in the Nigerian grid system that can be upgraded to improve the "smartness" of the grid.

5.1. Contradiction between the Two Grid Systems

Some contradictions between the Nigerian grid system and smart grid system are given in **Table 2** [6].

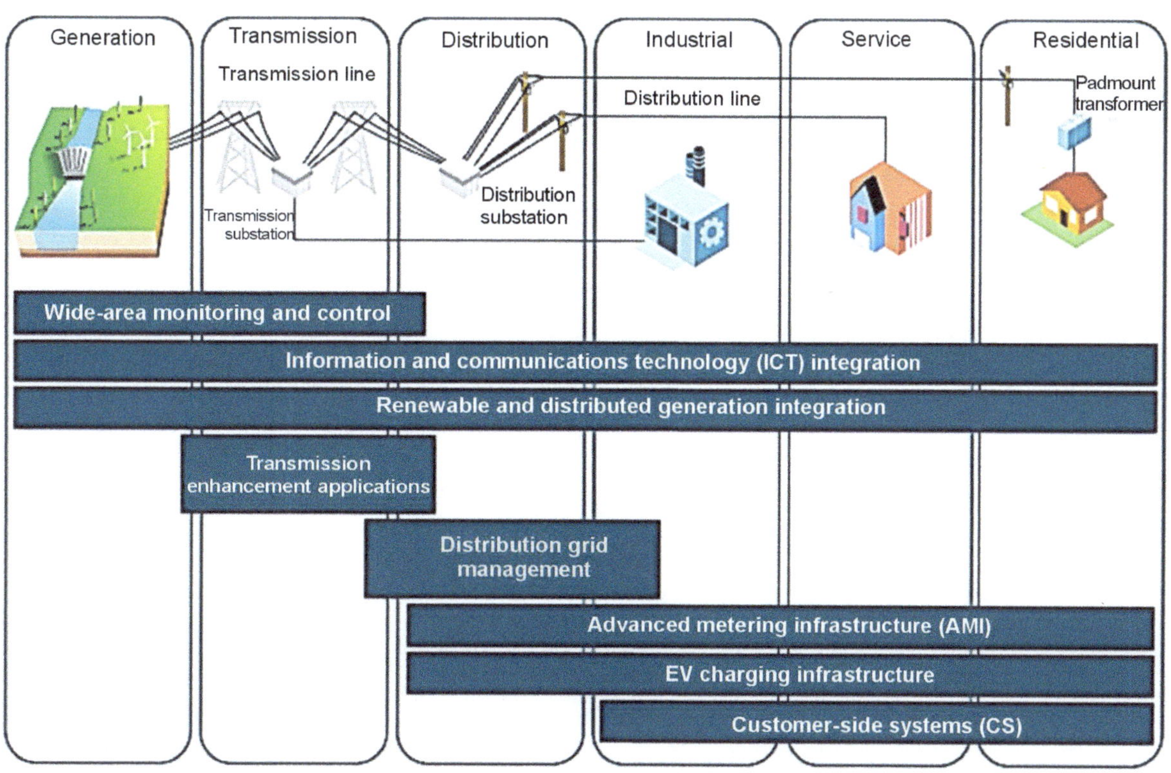

Source: Technology categories and descriptions adapted from NETL, 2010 and NIST, 2010.

Figure 5. Smart grid technology areas [18].

Table 2. Areas of Contradictions between the two grids.

Areas	Nigeria Grid System	Smart Grid
Generation	Majorly centralized stations	Centralized and distributed
Transmission/Distribution line losses	Above ten percent (10%) loss of the total power in the transmission/distribution lines.	About two percent (2%) loss of the total power in the transmission/distribution lines.
System topology	Radial; generally one-way power flow	Closed network; multiple power flow pathways
Operation and maintenance	Manual equipment checks.	Remote monitoring, predictive, time-based maintenance
Reliability	Prone to failures and cascading outages; essentially reactive	Automated, pro-active protection; prevents outages before they start
Power flow control	Limited	Comprehensive, automated
Restoration following disturbance	Manual	Self-healing
Communications	None or one-way; typically not real-time	Two-way, real-time
Metering	Mainly electromechanical	Digital (enabling real-time pricing and net metering)
Customer choice	Few customer choices	Many customer choices

5.2. Integration

The following explains the various enhancements needed in order to fully integrate the smart grid network in the Nigerian electricity grid system;

5.2.1. Wide Area Monitoring and Control

If sensory equipment like Phasor Measurement Units (PMU), strain gauge, infrared sensors, magnetic sensors and accelerometers are integrated into the Nigerian grid system will ensure proper monitoring and enable the system to automatically adapt and respond to changing conditions. Also, distribution devices will become intelligent remote on communication networks providing data collected through sensors back to operations control centers. Current technology on the national grid does not monitor power flows throughout the distribution grid because measurements are usually only available at the distribution substations. The sensors will also be able to collect information. Other types of wide area monitoring and control equipment that can also be mounted as mentioned in Section 4 include: Advance system operator tools, wide area situation awareness (WASA), Wide Area Monitoring System (WAMS) and Wide Area Adaptive Protection, Control and Automation (WAAPCA). The defunct Power Holding Company of Nigeria (PHCN) introduced Supervisory Control and Data Acquisition Systems (SCADA) to its power system to enable data monitoring and data collection of the generating stations. Other useful software includes GridLAB-D, Oracle Utilities Meter Data Management, Distribution Management System (DMS), and Visualizing Energy Resources Dynamically on Earth (VERDE) [19] [20].

5.2.2. Storage Devices

Energy storage devices like Hybrid Air Conditioning systems which come with lithium battery and inverter can convert electrical energy to thermal energy and store it. This is more economical than storing electricity in batteries. Such storage capabilities will be needed to grow the alternative energy sources, where unstable power flow from renewable energy plants such as wind and solar farms can be stored and better controlled. It can also supply power to the grid when it goes into island mode [18].

5.2.3. Transmission Enhancement

Equipment typically found in Transmission Company of Nigeria (TCN) substations includes the following: switch gears, high rupturing capacity fuses (HRC), air-cored ring shielded reactors, lightening arresters, circuit breakers, power transformers, isolators, and bus bars, among other items, while the network components include power lines, cables, circuit breakers, switches, transformers, large size conductors, steel lattice towers, steel tubular poles, wooden poles, earthing equipment. In Nigerian transmission power system, it will become important to fully utilize the existing transmission facilities instead of building new power plants and transmission lines that are costly to implement and involve long construction times. Flexible Alternating Current Transmission Systems (FACTS) controllers can be introduced in power systems to solve the above problems. FACTS make it possible to control the voltage magnitude of a bus, active and reactive power flows through transmission line of a power system. Line sensors can also be introduced to monitor real time situations of the Network [18].

5.2.4. Distributed Generation (Renewables and Micro-Grid)

As previously discussed in Chapter 2 of this article, Nigeria's electricity generating facilities are mainly thermal power plants and large centralized hydro plants scattered across the country. However, electricity must have to get to the users irrespective of where they reside. We also examined the renewable energy resources in Nigeria (Chapter 3), which is abundant in nature and yet not fully tapped. The integration of renewable energy into the national grid is one of the most important areas for infrastructure upgrades.

Considering the high variability and unpredictability of generation from renewable energy sources, the power generated can be safely absorbed in the grid if the frequency is maintained in an appropriate range. Frequency Support Ancillary Service (FSAS) can be used to complement the daily changes in the renewable generation. In the future, based on the renewable forecast for the day, a dispatch schedule for FSAS can be prepared such that the variation in the renewable energy generation can be absorbed easily. In other words, FSAS can be used as a mechanism to facilitate renewable energy integration by reducing the impact of variable generation [21].

Micro-grids can be utilized as a framework of system that can reduce the negative effect of power fluctuation on existing power systems, simultaneously pursuing the coexistence of environmental and supply, and existing power system and distributed power generation. They can also be considered to be flexible load. It keeps the power demand and supply balance by connecting to the external power system for maintenance of frequency and voltage. Micro-grid operators buy power through tie-line from the utility company when power in the micro-grids is deficient, and sell power when the power generated is in excess [6] [22].

5.2.5. Advanced Metering Infrastructure

In the past, electromechanical meters are used by consumers of the defunct Power Holding Company of Nigeria, they had mechanical parts that spin as electricity is consumed in the premises, and they show usage readout on small dials that a utility meter reader reads while on the property of your home. Presently in Nigeria, prepaid meters have been introduced in an effort to provide more efficient service for electricity consumers whereby the buy prepaid cards according to what they can afford [6] [22].

The prepaid meter disconnects from the national grid network once the available credit on the card runs out due to the customer home use. Smart meters, otherwise referred to as Advanced Metering Infrastructure (AMI) or two-way meters are electric meters which utilize two-way communications between the meters and the utility company and enable two way power flows to consumers. This technology enables users to monitor their consumption patterns in real-time and enables features which include time stamping of meter data, outage reporting, communication into the customer premise and on-request reads. It also enables the consumer to supply power back to the grid from solar panels or other local renewable sources and enables load management by responding to fluctuations in demand [6] [22].

5.2.6. Customer-Side System

With the introduction of smart grid technologies in Nigeria, customer-side system will also be introduced to aid in electricity management by residential, service and industrial customers, the include; smart appliances, energy storage devices, energy management devices and energy applications for smart phones and tablets.

6. Discussion, Factors, Energy Policy and Benefits

6.1. Discussion

So far we have briefly reviewed the Nigerian electricity sector, identify the current electricity situation and renewable energy resources available within its boundary. We also reviewed the concept of smart grid technology and then applied the various technological enhancements that could be integrated into the Nigerian electricity grid system. Smart grid as we know it is more of an enhancement to the electricity grid whereby there is an interaction between the producers and consumer due to the introduction of information and communication technology (ICT) as shown in **Figure 5**.

In the current Nigeria national grid system as shown in **Figure 6**, we can see that electricity flows in one direction, *i.e.* from the producer which is the Nigeria Electricity Supply Industry (NESI) to the Transmission Company of Nigeria (TCN) to the Distribution Company of Nigeria (DISCOS) then to the final consumers. In **Figure 6**, we can observe that the communication in the grid is between the TCN and its substation, the DISCOS and its substation, and finally the Nigerian Electricity Regulatory Commission (NERC) which connects the consumers and the DISCOS. As discussed earlier in Section 2, the current national electricity grid lacks properly maintained generating facilities coupled with lack of renewable energy resource to support peak demand and diversify power generation, this we can see in **Figure 6**, other problems include long poorly maintained transmission lines and lack of communication equipment in the both the transmission and distribution centers. The last problem is the metering system used by the electricity consumers.

In our proposed smart grid system (**Figure 7**), we could see that a lot of ICT enhancement has been added to make the grid "smarter" than it was in **Figure 6**. We started by introducing the nuclear power plants which will be introduced in the near future (see Section 6.2), renewables and energy storage system (ESS) were also introduced into the national grid system to enhance distributed generation which will be controlled by Frequency Support Ancillary Service (FSAS). We also observed in **Figure 7** that different communication device have been placed around the grid network to facilitate real time data flow within the grid through the communication lines.

Another important feature in **Figure 7** is the inclusion of mini grid system which will be located in areas that are off-grid. The final observation in **Figure 7** is the include the addition of solar power, energy storage system and smart meter for residential, commercial and industrial customers, these will enable them to be more energy efficient and they will be able to sell back power to the grid, this will make them to be fully involved in the power sector as both consumers and producers.

6.2. Factors That May Affect Smart Grid Integration in Nigeria

Smart grid is possible in Nigeria as we have previously discussed in the previously, the next step is to identify the factors that may affects its integration in Nigeria;

Vandalism of transmission lines and distribution equipment:

The vandalisation of the transmission infrastructures and the distribution equipment has been a common occurrence, even before the on-going power sector reforms. Most times, the sabotage is blamed on the power brokers who imports power generators. Vandalism is one of the current challenges that will likely translate hinder or cripple the integration of Smart grid in Nigeria [2].

Poor maintenance of existing power system:

Over the years of electricity infrastructures in Nigeria, poor maintenance has been one of the major causes of epileptic nature of power supply, even when generation capacity rises and Smart grid is introduced; its utilization is limited by the malfunctioning of the transmission and distribution infrastructure. The only known maintenance carried out by the power sector workers is fault clearing while scheduled maintenance for healthy function and life extension of such infrastructure is non-existent [2].

Information, research and development:

There has been a lack of publicly available data and information on the clean energy potential in Nigeria and potential new players in the sector lack the basic information upon which to make business decisions. The generation of reliable information and effective management of such information are paramount for good planning and implementation. On the issue of R&D, much has not been done to improve research into possible ways of integrating smart grid technology and also proper educational training to Nigerian engineers on smart grid.

Low generating capacity:

The country's present overdependence on gas fired generation plants has resulted in supply disruptions in times of gas shortages which are very common occurrences. This, coupled with the high vulnerability of Nigerians' electricity network system has so far contributed to the unreliability of the overall system. Commercial electricity generation in Nigeria currently comes from power stations and various independent power projects around the country generating available electricity capacity of 6056 MW which translates to 28 watts per capita power capacity. This is grossly inadequate even for domestic consumption. For Nigeria to meet up its energy needs, it requires per capita power capacity of 1000 watts or power generating/handling capacity of 170,000 MW [10].

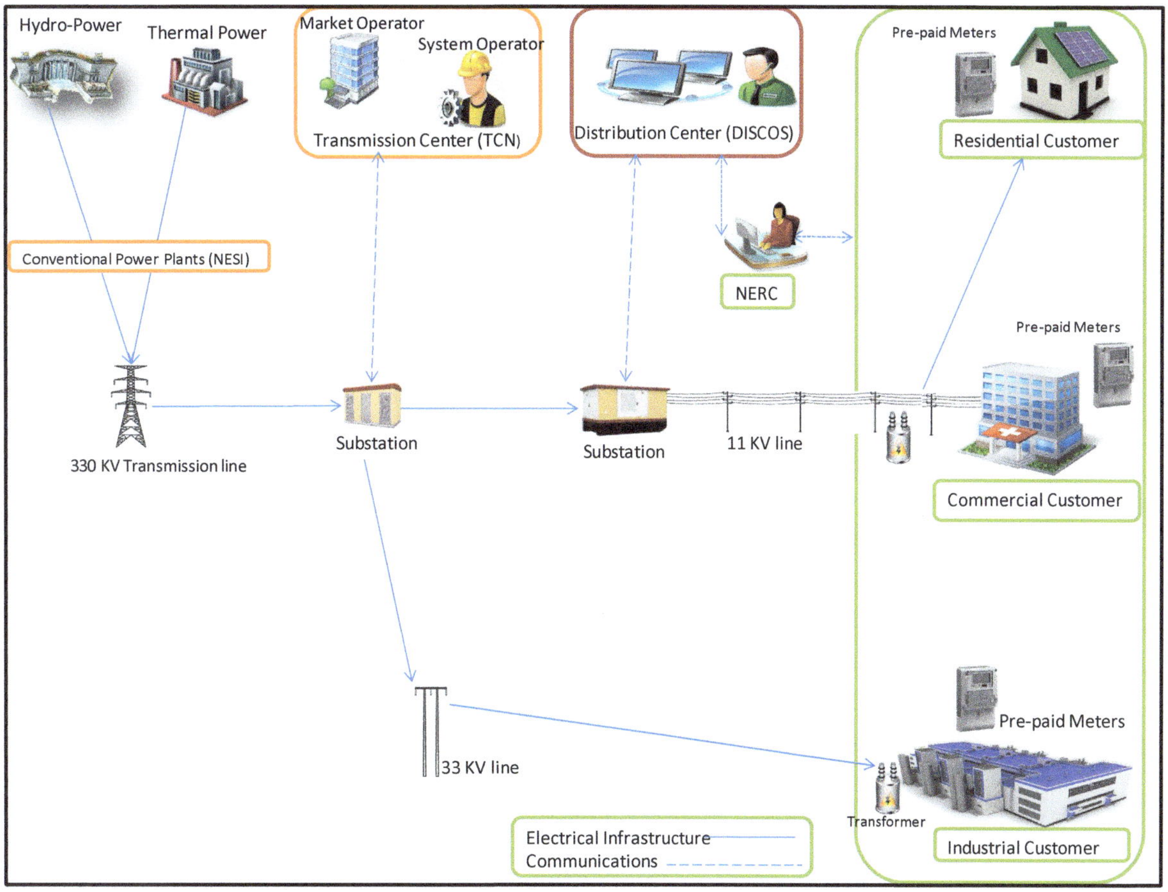

Figure 6. The present Nigeria national grid system.

Lack of institutional arrangement:

The lack of implementation of clean energy policies and the lack of coordination among relevant government agencies has created distortions and makes clean energy development more difficult. Inter-ministerial rivalries, overlap of mandates and outright protection of space often send wrong signals to potential investors. The absence of a clear political vision and leadership, weak regulatory institutions and inadequate human capacity, coupled with limited funding by government, all add up to make Nigeria's renewable energy market look dwarfed compared to that of much smaller countries like Kenya. Currently, there is little hope for a quick transition to a green economy. No program to equip rural women with small-scale renewable energy education and knowledge. The proposed Climate Change Commission under the Presidency would have the potential to increase synergies across ministries, and government agencies at the federal, state and local levels. However, the Bill to establish the Commission remains unsigned by the President. All these will have negative effect on smart grid if the technology is to be introduced in Nigeria [11].

Corruption:

The menace of corruption has set the nation backward among her contemporaries. Corruption is a ubiquitous practice that has ravaged every sector of the economy in Nigeria, including the power sector. Reform is not spared from its venom. The instances of corruptions in the power sector are numerous and provoking to even discuss about.

Consumer behavior:

Due to the unreliable nature of power supply, consumers developed a poor power usage methodology. This method tends towards wasting energy instead of making proper use of it. For instance most consumers don't use energy saving bulbs and rarely switch off portions of energy that are not in use. The inclusion of smart metering will curtail these wastages because energy usage will become more interactive and cost effective, making people purchase what they really need.

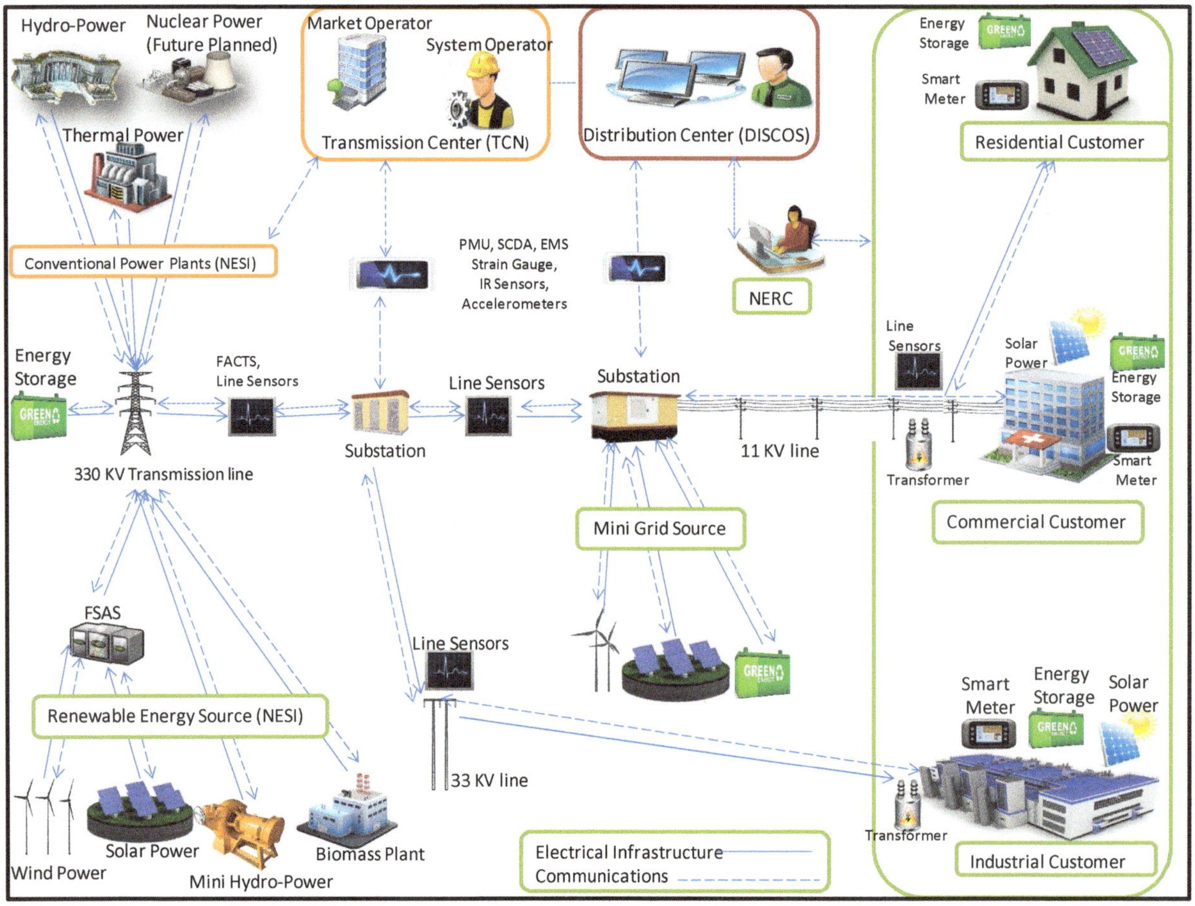

Figure 7. The proposed smart grid system.

6.3. Energy Policy Overview

We will now have a brief overview of the existing policies and required plans on energy to enable us identify the reforms of policy objectives, gaps and options that are unavoidable if Nigeria is to achieve a secure access to renewable energy which will be integrated with the proposed smart grid into the Nigerian grid system.

The National Energy Policy (NEP) of 2006 sets out government policy on the production, supply and consumption of energy reflecting the perspective of its overall needs and options. The main goal of the policy is to create energy security through a robust energy supply mix by diversifying the energy supply and energy carriers based on the principle of "an energy economy in which modern renewable energy increases its share of energy consumed and provides affordable access to energy throughout Nigeria, thus contributing to sustainable development and environmental conservation" [23]. Importantly, the national policy already outlines the key elements for development and application of renewable energy as:

- To promote decentralized energy supply, especially in rural areas, based on RE resources
- To develop, promote and harness Renewable Energy (RE) resources of the country and incorporate all viable ones into the national energy mix
- To promote efficient methods in the use of biomass energy resources
- To de-emphasize and discourage the use of wood as fuel
- To keep abreast of international developments in RE technologies and applications.

In an effort to translate the RE component of the NEP into an actionable plan, the Energy Commission of Nigeria (ECN) in 2005 developed the Renewable Energy Master Plan (REMP) to complement the NEP, which reiterated the government's pledge to support the development, demonstration and implementation of RE sources for both small and large applications. To create the appropriate enabling environment for the promotion of RE, the REMP identified the need for appropriate financial and legal instruments, technology development, awareness raising, capacity building and education as the strategic areas to be paid attention and further sets specific goals for each of those areas [9] [11].

The REMP aims at a 10% RE contribution to the national energy mix by 2020 through the adoption of a renewable portfolio standard (RPS). A RPS is a requirement for electric utilities to supply a specific amount of electricity to customers. This can be achieved through the purchase of renewable energy certificates from suppliers with a larger share of renewables in their energy mix. Other measures considered are the creation of innovative fiscal and market incentives to grow renewable energy industries, as well as preferential customs duty exemptions for imported renewable energy technology components [9] [11].

However, the lack of implementation of the master plan has meant that the 10% target of renewable energy mix in the energy supply cannot be achieved. The REMP is presently being subjected to a review, likely resulting in the setting of new targets. It will be essential that any future targets set for the attainment of a RE energy mix should be backed by legislation to ensure compliance, which is presently lacking. Similarly, the **National Policy and Guidelines on Renewable Electricity** was produced in 2006 with the main aim to expand the market for renewable electricity by 5% of the total electricity generation by 2016. The strategy for achieving this target included: encouraging local manufacture and assembly of renewable energy components, provision of subsidies, and establishment of technical standards for RE components and introduction of feed-in-tariffs. The strategy is yet to be fully adopted as the reforms in the energy sector are still ongoing and decisions on tariffs and subsidies for renewable energy and other incentives have not yet been taken [11].

The Energy Commission also developed a draft **National Energy Master Plan** (NEMP) in 2006. The NEMP acknowledged the imminent dangers of the fossil fuel driven economy, its global environmental concerns and the expediency of adopting a less carbon intensive and environmentally friendly development pathway, and recommended urgent action to diversify the economy and create a sustainable energy supply mix. A **National Gas Master Plan** of 2008 envisages a wholesale transition to decentralized privately held electricity generation gas plants from the erstwhile public power utility. The plan also aims to stem the huge waste associated with gas flaring and to put to more productive use the nation's large gas reserves. What the Gas Master Plan does not clearly mention is that the utilization of Compressed Natural Gas (CNG) can play a major role in transforming the nation's transportation sector if adopted on a large scale. Clear benefits are evident, for example in the Lagos State Bus Rapid System (BRT), which recorded a 13% reduction in CO_2 and 20% in other greenhouse gas emissions [24].

The **Nigerian Atomic Energy Commission** was established to explore Nigeria's nuclear energy potentials for peaceful purposes. Given the expected expansion of electricity demands by the growing population, the ECN has called for an activation of a Nigerian nuclear programme and the Nigerian Atomic Energy Commission is already undertaking preliminary steps [9]. This is however curious when considered against the background that the Vision 2020 does not include atomic energy in its proposed energy mix. After the nuclear accident in Fukushima, Japan, and given the long-term huge costs of producing atomic power, pursuing the nuclear path seems like an outdated strategy. Even though nuclear is said to contribute 16% of global electricity, high cost and liabilities, major safety and waste management issues, exacerbated by the weak regulatory environment, make nuclear energy less attractive for the timely provision of energy access to the poor. Unsurprisingly, countries like Germany are gradually abandoning nuclear energy [9] [24].

6.4. Benefits of Smart Grid Technology

There are various ways Nigeria and her citizens can benefit from the proposed integration of smart grid technology [18]:

- The smart grid will accommodate all generation and storage options. It will seamlessly integrate many types of electrical generation and storage systems with a simplified interconnection process analogous to "plug and play" technology of the retail computer industry
- Optimizes asset utilization and operating efficiency. A smart grid applies the latest technologies to optimize the use of its assets. For example, optimized capacity can be attainable with dynamic ratings, which allow assets to be used at greater loads by continuously sensing and rating their capacities.
- The smart grid will ensure the level of power quality desired by the users. New power quality standards will balance load sensitivity with delivered power quality at a reasonable price. The modernized grid will supply varying grades of power quality at different pricing levels. Not all commercial enterprises, and certainly not all residential customers, need the same quality of power. A smart grid supplies varying grades (and prices)

of power.
- The grid will heal itself. The modernized grid will perform continuous self-assessments to detect, analyze, respond to and as needed, restore grid components or network sections. It will handle problems too large or too fast moving for human intervention. Acting as the grid's "immune system", self-healing will help maintain grid reliability, security, affordability, power quality and efficiency.
- The smart grid will resist attack. Security requires a system-wide solution that will reduce physical and cyber vulnerabilities and recovers rapidly from disruptions. Both its design and its operation will discourage attacks, minimize their consequences and speed service restoration. Resiliency refers to the ability of a system to react to unexpected events by isolating problematic elements while the rest of the system is restored to normal operation. These self-healing actions result in reduced interruption of service to consumers and help service providers better manage the delivery infrastructure.
- The modern grid will enable markets to flourish. Open-access markets expose and shed inefficiencies. The modern grid will enable more market participation through increased transmission paths, aggregated demand response initiatives and the placement of energy resources including storage within a more reliable distribution system that is closer to the consumer.
- It will motivate consumers to be an active grid participant and will include them in grid operations. In the modernized grid, well-informed consumers will modify consumption based on the balancing of their demands and the electric system's capability to meet those demands. Demand for new cost-saving and energy-saving products will benefit both the consumer and the power system. Consumers help balance supply and demand, and ensure reliability by modifying the way they use and purchase electricity. These modifications come as a result of consumers having choices that motivate different purchasing patterns and behavior. These choices involve new technologies, new information about their electricity use, and new forms of electricity pricing and incentives [18].

7. Policy Recommendation and Conclusion

7.1. Policy Recommendation

Some policy recommendation and are given on the short, mid and long term in order to facilitate proper integration of renewables and smart grid into the Nigerian electricity grid;

1) Energy Data

<u>Short term</u>

- Off-grid power production study of small and large generators, rural electrification projects not connected to the national grid, private homes with solar PV installations.
  Result will show areas that needs connection to the national grid and can also support the national grid during peak hours.
- Nationwide power consumption study on residential, service and industries.
  Result will show consumption pattern in regions that are off-grid and on-grid.
- Intensive study on renewable energy resources available in the country.
  Result will show regions of the country with their most promising renewable energy.
- Establishment of a National Energy Databank which should be up-to-date.
  This will aid the government agency, local and international organization to properly analyze energy situation in Nigeria.

<u>Mid term</u>

- Acurate availability of a National Energy Databank.

2) Energy Efficiency

For a proper power management society.

<u>Short term</u>

Introduce and promote:

- Energy education programs in schools, service and industrial sectors.
- Efficiency standards on home, office and industrial appliances and equipment.
- Phase out old energy consuming appliances and electric bulbs with new low energy saving bulbs.
- Replace the old metering still in use in some part of the country to pre-paid Smart meters.
- Create incentives for electricity retailers and utility companies.

Mid term

- Mandatory energy education programs in all the educational levels in Nigeria.
- Complete and mandatory use of low energy consuming appliances, electric bulbs and equipment in homes, offices and industries.
- Incentives for final electricity consumers.

3) Grid-Connected Power

Short term

- Employ movable gas turbines like Single Cycle Gas Turbines (SCGTs) and Combine Cycle Gas Turbines (CCGTs).
- Conduct feasibility studies for 100 MW pilot projects from wind, solar and small hydropower.
- Introduce and encourage micro-grid operators.
- Proper maintenance of power generating facilities including hydropower.
- Proper security of transmission and distribution equipment.

Mid term

- Expand CCGTs electricity generation capacity.
- Build 100 MW demonstration projects for grid connected wind, solar PV and small hydropower.

4) Off-Grid Power

Short term

Introduce, subsides and encourage:

- Solar PV for irrigation, water pumping and lighting.
- Solar PV with battery storage for homes.
- Use natural gas to replace diesel by the use of small gas powered turbines of up to 5 MW.

Mid term

- Promote solar PV and small gas turbines for off-grid and micro-grid.
- Introduce and promote small hydropower plants.

5) Integrated Planning Process

Short term

- Develop a comprehensive electricity system analysis of all generating plants (both current and future planned), transmission networks and distribution centers in order to develop long term plans.
- Consider setting of renewables during grid expansion e.g. during power plant site selection and transmission line expansion, potential renewable energy should be put in mind for integration on the mid and long term.

Mid term

- Integrate planning of CCGTs and more power plants into the national grid.
- Encourage integration of distributed generation into the national grid.
- Introduce wide area monitoring and control device into the national grid.
- Total upgrade of transmission lines and equipment, introduction of transmission enhancement.

6) Policies

Short term

- Allow domestic fossil fuels prices revert to global market prices.
- Allow tariffs to fully reflect in electricity costs.
- Promote CCGT and convert existing SCGTs to CCGTs.
- Introduce policy framework that promotes off-grid renewables; Feed-in-tariffs (FITs), renewable portfolio standards (RPS), and renewable energy certificate (RECs).
- Introduce degree courses on renewable energy and smart grid technology in Nigerian Universities.
- Develop R&D centers on renewable energy and smart grid technology.
- Build demonstration and training projects on renewable energy.
- Develop proper financial mechanisms on renewables like low interest loans and Clean Development Mechanism (CDM).

Short term

- Build small scale pilot projects on smart mini grid technology.
- Government gradually withdraws and private sectors take over.

Long term

Develop a nationwide smart grid connected network.

7.2. Conclusion

In this paper, we have observed that the Nigerian electricity grid is faced with a lot of problems which include the insufficient and inefficient power generation facilities, inadequately maintained long transmission lines and distribution facilities, and outdated metering system used by its customers. The integration of smart grid technology and renewables into the Nigerian electricity grid system remains the only solution to the electricity crisis in the country. We have shown that the smart grid technology is possible in Nigeria through the various technological integrations, enhancement and policy recommendation given in this paper. The benefits of the smart grid technology will not only improve electricity production and efficiency in Nigeria, but will also enable electricity consumers to become producers of electricity and enhance Nigeria's international competiveness. Limitation in this paper includes analysis like cost benefit analysis, Feed-in-Tariffs (FITs), Renewable Portfolio Standards (RPS), Renewable Energy Certificate (REC) which were not discussed in order to determine the best policy mechanism for renewable energy and smart grid integration in Nigeria, therefore this limitation should be addressed in subsequent research.

References

[1] Sambo, A.S. (2009) Strategic Developments in Renewable Energy in Nigeria. International Association for Energy Economics, 16.

[2] Worldbank Databank. http://www.tradingeconomics.com/nigeria/access-to-electricity-percent-of-population-wb-data

[3] Koledoye, T.O., Abdul-Ganiyu, J.A. and Phillips, D.A. (2013) The Current and Future Challenges of Electricity Market in Nigeria in the Face of Deregulation Process. *African Journal of Engineering Research*, **1**, 33-39.

[4] KPMG Nigeria: A Guide to the Nigerian Power Sector, December 2013.

[5] Sambo, A.S., Garba, B., Zarma, I.H. and Gaji, M.M. (2003) Electricity Generation and the Present Challenges in the Nigerian Power Sector. *Energy Resources Review*, **4**, 7-10.

[6] Patrick, O., Tolulolope, O. and Sunny, O. (2013) Smart Grid Technology and Its Possible Applications to the Nigeria 330 kV Power System. *Smart Grid & Renewable Energy*, **4**, 391.

[7] Sunday, O.O. and Friday, O.O. (2010) Empirical Modelling of Power Losses as a Function of Line Loadings and Lengths in the Nigerian 330 kV Transmission Lines. *International Journal of Academic Research*, **2**, 47-53.

[8] Nigeria Population. http://www.cia.gov/library/publications/the-world-factbook/geos/ni

[9] (2005) Energy Commission of Nigeria: 1st Energy Lecture Series.

[10] Aniefiok, E., Igweonu, I.E., Eguzo, C.V. and Robert, B.J. (2013) Integrating Smart Grid Model in Nigeria Power Network. *International Journal of Advances in Engineering & Technology*, **6**, 1760-1768.

[11] Mshelia, I.H. (2012) Energy Access for All: The Role of Clean Energy in Alleviating Energy Poverty. http://us.boell.org/2012/05/18/green-deal-nigeria-greening-economy

[12] The Nigerian Yearly Average of Daily Sun. www.geni.org

[13] Akpu, I.V. Renewable Energy Potentials in Nigeria. Internation Association for Impact Assessment.

[14] Nigeria Wind Energy. http://www.neenigeria.com/Nigeria_wind_NEW.png

[15] Nigeria Water Ways. http://www.safety4sea.com/

[16] Giordano, V. and Bossart, S. (2012) Accessing Smart Grid Benefits and Impacts: EU and U.S Initiatives.

[17] U.S. Department of Energy (DOE) (2009) The Modern Grid System, A Vision for the Smart Grid. National Energy Technology Laboratory.

[18] IEA (2011) Technology Road Map on Smart Grids. International Energy Agency.

[19] (2009) Oracle White Paper: Smart Grids-Strategic Planning and Development.

[20] Smart Grid by PDHengineer.com, Course No. E-5006. http://www.pdhengineer.com/page-not-found-404/

[21] (2013) Integrating Wind and Solar Energy in India for a Smart Grid Platform, Global Energy Network Institute (GENI).

[22] (2009) Integration of Demand-Side Management, Distributed Generation, "Renewable Energy Sources and Energy Storages", International Energy Agency Demand-Side Management Programme (Ieadsm), Finland.

[23] (2006) National Energy Policy (NEP). www.energy.gov.ng

[24] Eleri, E., Ugwu, O. and Onuvae, P. (2011) Low-Carbon Africa: Leapfrogging to Green Future: Low Carbon Africa: Nigeria. www.christainaid.org/uk/resources/policy/climate/low-carbon-africa

5

Increasing the Efficiency of Grid Tied Micro Wind Turbines in Low Wind Speed Regimes

Kshitij Tiwari[1,2], Tirumalachetty Harinarayana[2]

[1]Department of Electrical Engineering, Indian Institute of Technology, Kanpur, India
[2]Gujarat Energy Research and Management Institute, Gandhinagar, India
Email: kshitij537@gmail.com, harinarayana@germi.res.in

Abstract

Major problem with grid tied micro wind turbine is synchronization and wind variability. Due to this problem the stability of available grid gets reduced. The stability can be achieved by output power control of the turbine. Major part of many countries like India, the annual mean wind speed is not high. The rated wind speed of turbine remain around 11 m/s and cut in is around 3.5 m/s. Due to this problem we aimed to develop a sustainable wind energy system that can provide stable power supply even at the locations of low wind speed of 2 - 4 m/s. To address this issue, a momentary impulse or external torque to the rotor by external motor is one of the good options to maintain the momentum of blades and thus provide stability for sufficient time. Various theoretical calculations and experiments are conducted on the above method. This would increase the output power and also the efficiency of wind turbine. We show that Return-On-Investment will be high as compared with other grid connected turbines. Our proposed concept in the present study, if implemented properly, can help the installation of number of wind turbines even at domestic level. It also makes the consumers energy independent and promotes the use of wind as a source of energy and may enter as a rooftop energy supply system similar to solar.

Keywords

Grid Tied Rooftop Micro Wind Turbine, Small External DC Motor, Wind Variability, Tip-Speed Ratio (TSR), Stabilized Grid Supply

1. Introduction

Wind energy development is picked up as a part of electricity generation in different parts of India. Although it was initiated a decade back, there is still enough scope for its improvement in this technology. The reality is that our life-style in the early twenty-first century requires huge energy and wind turbines are one of the least environmentally harmful ways of generating electricity. Of the installed capacity, wind turbines have a large share among the other grid connected renewable sources, close to 66.66%[1]. Obviously wind turbines need to be built where the wind blows reliably and strongly. The society is certainly not willing to give up private cars, air conditioning, computers, television, refrigeration, and all the other energy consuming equipment of the modern life-style. Accordingly, we must build sustainable energy supply systems. Earlier works promoted wind energy in many ways [1] [2]. In our study we have worked on grid tied rooftop micro wind turbine. The major problem with such a system is synchronization largely due to wind variability. Due to this problem, the stability of available grid gets reduced. This can somehow be achieved by the output power control of the turbine. India is the fifth largest wind power producer in the world. As of March 2014, the installed capacity of wind power in India is 21136.3 MW [3] [4], mainly spread across Tamil Nadu (7154 MW), Gujarat (3093 MW), Maharashtra (2976 MW), Karnataka (2113 MW) and Rajasthan (2355 MW). Wind power accounts for 8.5% of India's total installed renewable power capacity, and it generates 1.6% of the country's power. Still it is not very much popular at domestic level due to the high initial investment and other problems as discussed above. Our goal is to achieve stable power supply to the grid from wind source. To this issue, we suggest a new method of small external motor that will provide momentary impulse or external torque to the rotor through relay switch or voltage controller that helps to maintain the momentum of blades for sufficient time and provide stable power supply. In this method, the light weight external motor has been attached at the hub of the wind turbine and it works on the method of spring balance system. The motor aligns itself with the axis of rotation of blades when a mass of around 10 pounds is put to the groove attached to motor. This is done so as to not disturb the aerodynamics of the blades and the yaw system. The motor provides external torque to the rotor only during moderately low wind speeds, *i.e.*, around 2.5 - 4 m/s. At other times, when the wind speed is self sufficient to produce stable power supply, the motor remains off-axis from the rotor and doesn't interfere in its motion. This mechanism is currently mechanical but could be automated for better performance. The concept presented in this paper is implemented and tested.

2. Materials and Methods

In our method, we attached an external motor to the turbine rotor that provides sufficient external torque to the rotor to rotate at its rated rpm value when the wind speed is less than the rated speed so that the wind generator can generate a stabilized power supply.

2.1. Theory

It is well known that the amount of power extracted from the wind is given by the equation:

$$P_{\text{Wind}} = \frac{1}{2}\rho A V^3 C_p \quad [5]\text{-}[7] \tag{1}$$

$$P_{\text{Generated}} = \frac{1}{2}\rho \pi R^2 V^3 C_p C_t C_g \left(\frac{p}{2}\right) \quad [5]\text{-}[7] \tag{2}$$

$$\text{Tip Speed Ratio}(\lambda) = \frac{\omega R}{V} \quad [5]\text{-}[7] \tag{3}$$

where, P_{Wind} = Power Delivered by Wind to the rotor below the rated value, $P_{\text{Generated}}$ = Output Power of the generator, C_p = Power Coefficient, C_t = Transmission Coefficient, C_g = Generator Efficiency, p = Number of poles in Generator.

Motor supply sufficient torque to the rotor so that generator operates at rated power. The following is the mathematical condition:

$$\tau_{\text{Wind}} + \tau_{\text{Motor}} = \tau_{\text{Rated}} \tag{4}$$

$$\frac{P_{\text{Wind}}}{\omega_{\text{Rotor}}} + \frac{P_{\text{Motor}}}{\omega_{\text{Motor}}} = \frac{P_{\text{Rated}}}{\omega_{\text{Rated}}}$$

[1]http://www.re-solve.in/wp-content/uploads/2014/05/Installed-capacity-31March-2014.png

$$\frac{\frac{1}{2}\rho\pi R^2 V_{\text{Wind}}^3 C_P}{\omega_{\text{Rotor}}} + \frac{P_{\text{Motor}}}{\omega_{\text{Motor}}} = \frac{\frac{1}{2}\rho\pi R^2 V_{\text{Rated}}^3 C_P}{\omega_{\text{Rated}}}$$

Using Equation (3)

$$P_{\text{Motor}} = \frac{1}{2}\frac{\rho\pi R^3 C_P\left(V_{\text{Rated}}^2 - V_{\text{Wind}}^2\right)\omega_{\text{Motor}}}{\lambda}$$

For the above suggested mechanism to be useful, the generated power should be greater than the power supplied by the motor:

$$P_{\text{Generated}} \geq P_{\text{Motor}}$$

$$V_{\text{Rated}}^3 C_t C_g\left(\frac{p}{2}\right) \geq \frac{R\omega_{\text{Motor}}}{\lambda}\left(V_{\text{Rated}}^2 - V_{\text{Wind}}^2\right) \quad (5)$$

In our study, SNT-1[2] 700 W micro grid tied turbine is used. Its parameters are, $V_{\text{Rated}} = 11$ m/s (according to industry standards), $C_t = 1$ (since no gearbox system is used), $C_g = 0.25$ (assumed to be constant for easier calculations), $C_P = 0.33$ (assumed to be constant), $p = 8$, $R = 920$ mm, $V_{\text{Cut-in}} = 2$m/s, $(\omega_{\text{Motor}})_{\text{Max}} = 70$ rpm, $(P_{\text{Motor}})_{\text{Max}} = 18$ Watts.

SNT-1[2] turbine is although designed for maximum rotational speed of 300 rpm at a wind speed of 6 m/s. So, TSR = 4.81 (assumed as constant for easier calculations) [8]-[10]. Above the wind speed of 6 m/s, it dampens extra energy to the dummy load.

Accordingly, the calculations have been carried out considering the wind speed of 6 m/s as rated. Another reason for this is that the generator is not designed for higher wind speeds.

2.2. Theoretical Results

Following are the results observed by calculating using theoretical considerations:

1) Following are the results to power available to the rotor shaft at different wind speeds (**Figure 1**);

2) Following is the plot of external torque to be provided by motor to achieve the rated rpm (**Figure 2**). Maximum torque that can be provided by our installed DC Motor is 2.45 N-m. In the **Figure 2** green region corresponds to desirable region of operation and red region corresponds to undesirable operational region.

2.3. Experimental Results

1) An experiment with power output is conducted on 700 W SNT-1 micro wind turbine at the premises of

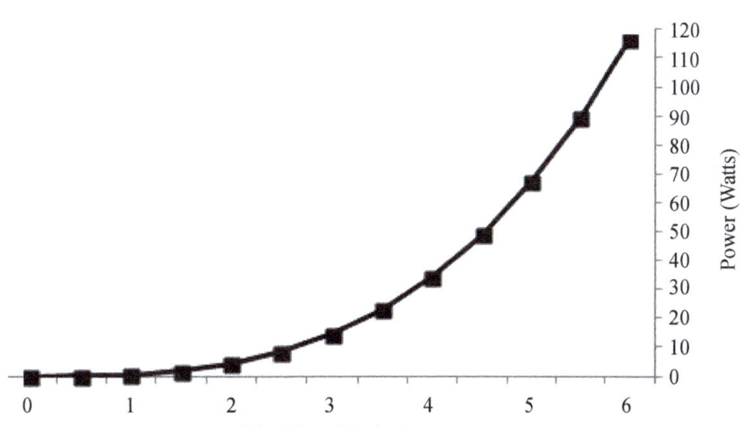

Figure 1. Available rotor power at various wind speeds.

[2]SNT-1 Wind Turbine: Supernova Technologies Pvt. Ltd. Turbine Model No.1 (**Figure 10**).

SNT Unit, Udyognagar, Anand, Gujarat and a comparative study of theoretical [11]-[13] and practical results are shown in **Figure 3**.

2) However in a controlled environment on a test bench of SNT, a 3HP motor is used to rotate the shaft of 400 W generator. The results obtained are contradictory to our theoretical results. We obtained power consumed is greater than power generated for every value of generator rpm. However, this is not our intention to do, because, the entire energy to rotate the generator shaft is provided by motor in this case. In our method, partial energy would be provided by the wind. Experimental results using external motor without wind are as shown below (**Figure 4** & **Figure 5**).

3) When the above curve of power generated was studied at various equivalent wind speeds (assuming TSR as constant according to the rated value) following results were obtained (**Figure 6**).

4) Experiment is conducted on Rooftop Wind SNT-1 Wind Turbinein normal mode with low wind speed. Grid Tied Rooftop Windmill installed on the rooftop of GERMI is SNT-1 Wind Turbine with generator Rated Power of 700 W and external Motor Rated Power of 18 W. Results obtained are as follows (**Figure** 7).

5) In another experiment on SNT-1 Rooftop wind turbine at GERMI, it is observed that power consumed by motor is much higher compared to power generated by the turbine. This is due to the fact that the rotor is purely running on motor and there is no effect of wind on rotor when the motor is in switched on mode. This is due to the current mechanism of spring balance which is being used to align motor with generator axis which prohibits any affect of wind on rotor. Experimental results when the external motor is used are as follows (**Figure 8**).

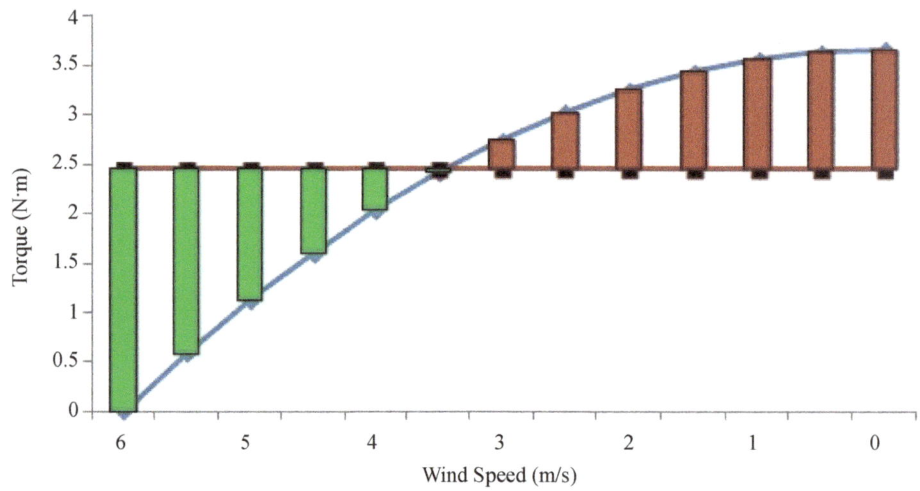

Figure 2. Torque to be provided by external motor at various wind speeds.

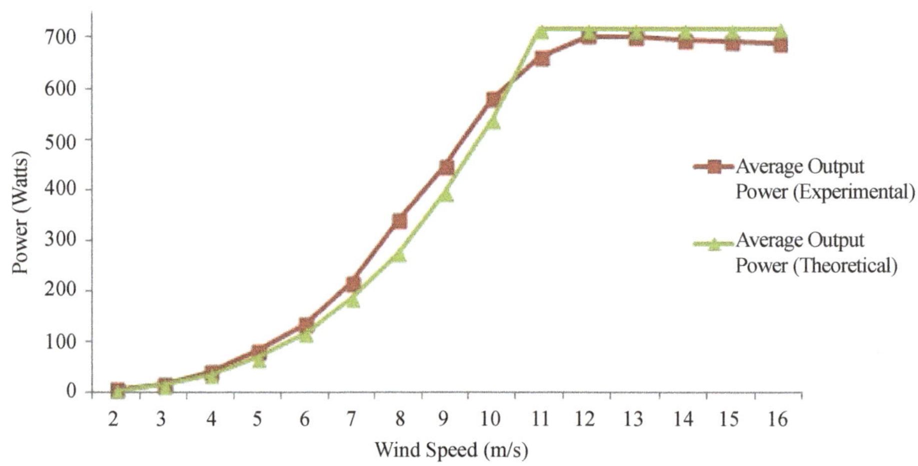

Figure 3. Comparative study of average output power (theoretical & experimental).

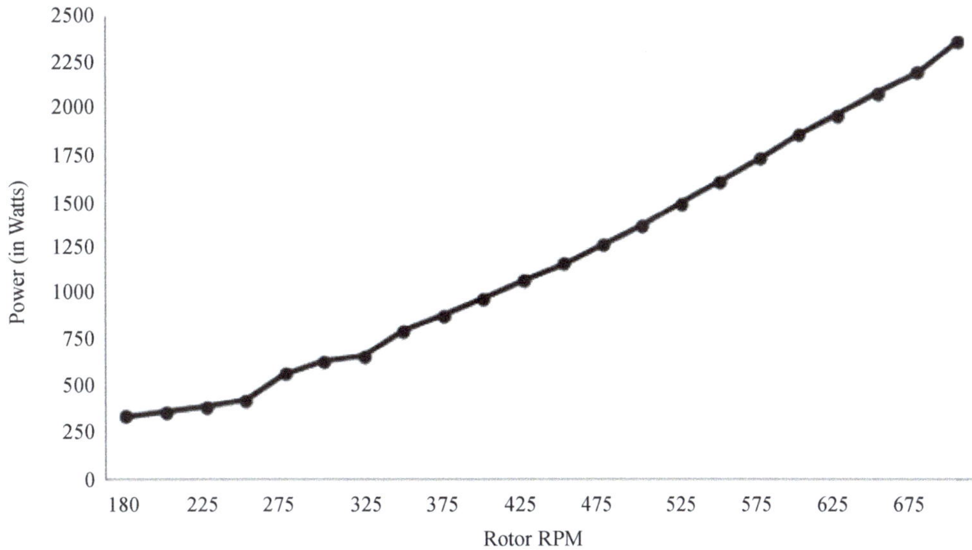

Figure 4. Power consumed by motor at various rotor RPM.

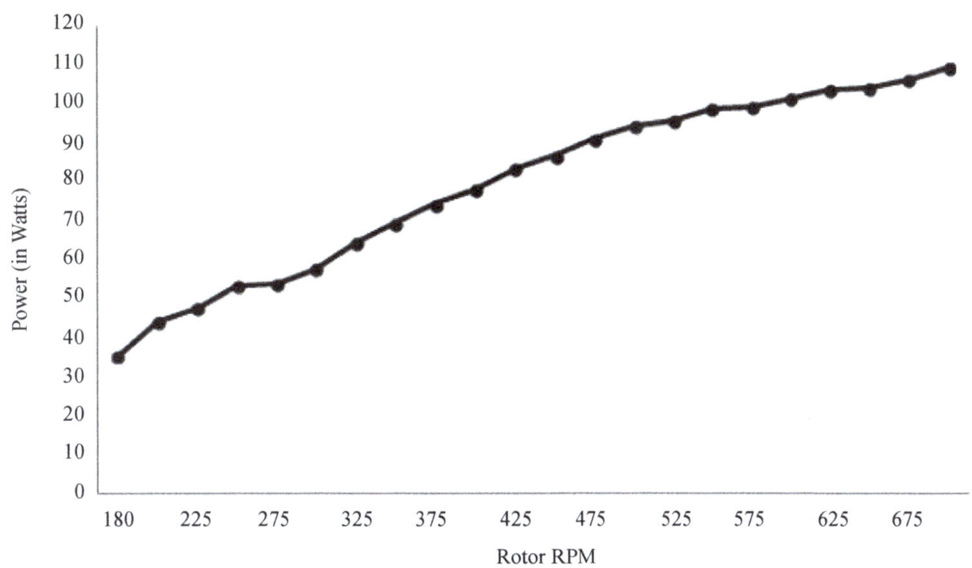

Figure 5. Power generated by generator at various rotor RPM.

2.4. Theoretical Power Gain

The results in **Figure 9** are obtained on comparing the results of power generated and power consumed for live wind speed data.

The wind data is recorded for every 2 minutes and the difference between the the two curves actually demonstrate the net power produced by the wind turbine system. The above results clearly show that we gain power in our discussed method and hence our method proves to be fruitful theoretically whereas the experimental feasibility has been partially established but it is yet to be worked upon more extensively.

3. Conclusions

We have addressed a few issues related to wind turbines and suggested solutions for the same. Hopefully our suggested method will yield into a product that would help changing the wind power scenario at domestic level. SNT-1 wind turbine installed on rooftop of our energy building—GERMI at PDPU campus (**Figures 10-12**)—is

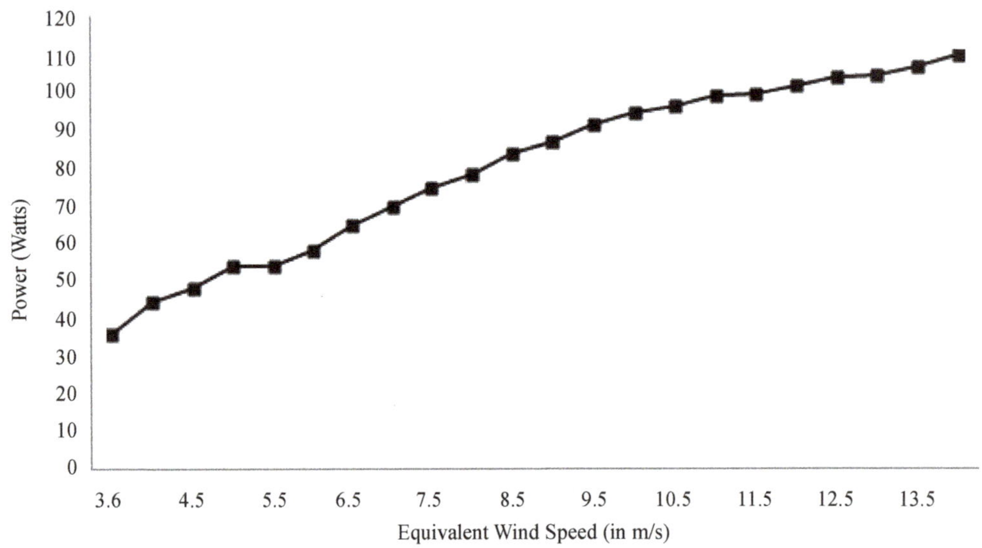

Figure 6. Power generated by generator at various equivalent wind speeds.

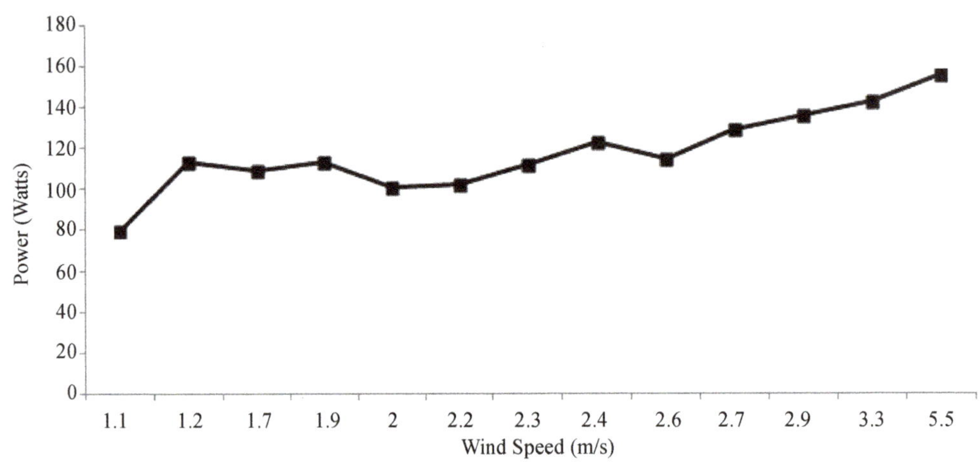

Figure 7. Power generated at various wind speeds (site results).

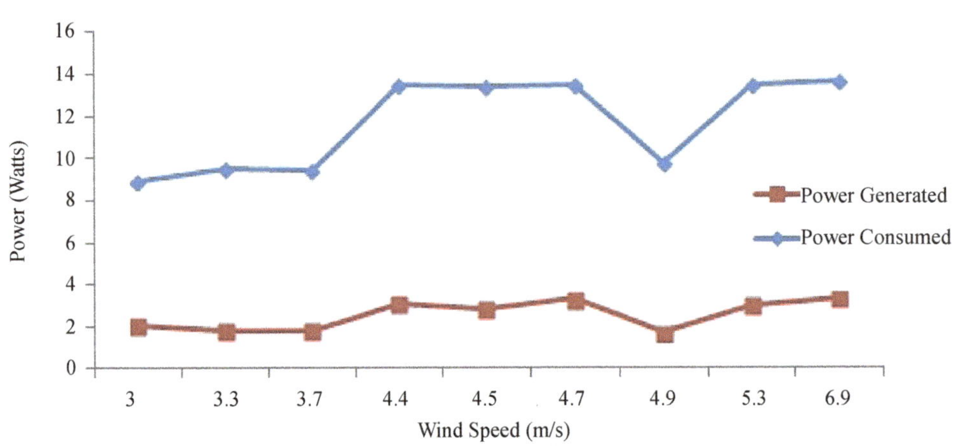

Figure 8. Power generated and consumed at various wind speeds.

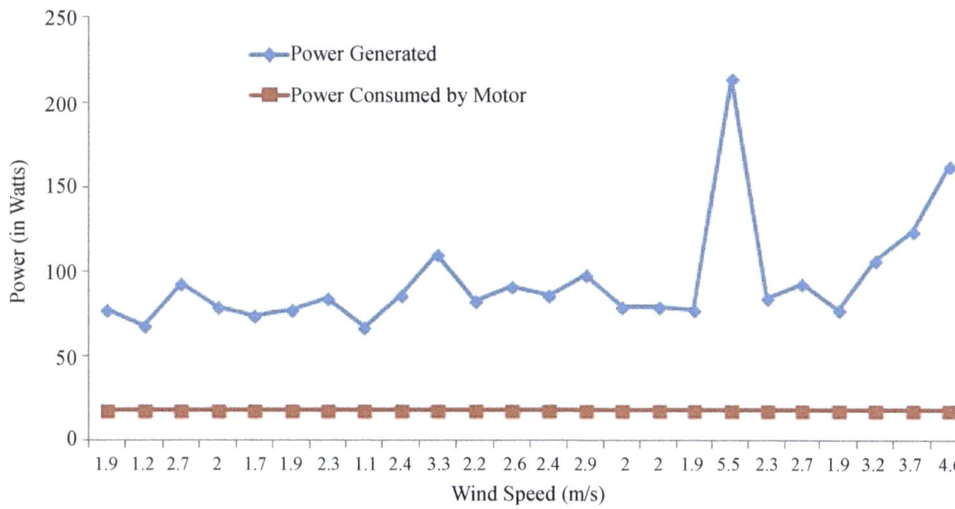

Figure 9. Power generated and consumed at live wind speeds.

Figure 10. 0.7 kW SNT-1 rooftop wind turbine installed at GERMI rooftop.

specially designed for low wind speed regime and for the *first time* attempted and installed a **grid tied** rooftop micro wind turbine. Further improvement is needed to reach to a level to work on concrete proof. We need to test our suggested method's usefulness to the society as well as to the industry. In this respect, more exhaustive experiments could be devised with data logging facilities. The wind braking and motor system will be upgraded and automated in near future for better results. A control system could also be attached which disconnects the generator to grid when the grid supply is off. During the monsoon season, when the wind velocity is fair enough, our method can be a real asset at least for rooftop wind turbines [14]. The only limitation being it will be commercially beneficial only when the wind speed ranges from 3 to 6 m/s.

Figure 11. Grid tied inverter for SNT-1 wind turbine.

Figure 12. Variac for external 12V DC Motor.

Acknowledgements

We would like to thank all the GERMI staff for their full cooperation to execute this task. Full support provided by Dr. Sagar Agarvat, Scientist C, Solar Research Wing, GERMI is greatly acknowledged. His coordination and help is very much appreciated without which we could have not completed this work. Thanks are due to Mr. Manyam Telukuntla from UPES Dehradun, a summer intern student at GERMI for assisting in the experimental work. Mr. Ankur Bhatnagar, Mr. Dharmik Lay, Mr. Prashant Gopiyani, Mr. Nilesh Patel and Mrs. Sandhya Radhakrishnan have helped in many ways for successful completion of this project.

References

[1] Christiner, M., Dobbins, R., Ndegwa, A. and Sivak, J. (2010) Rooftop Wind Turbine Feasibility in Boston, Massachusetts. Worcester Polytechnic Institute, Worcester.

[2] Jha, S.N. and Gandhi, J.R. (2006) Low RPM, High Torque Small Wind Generator for Hybrid System. http://www.ibrarian.net/navon/paper/Low_RPM__High_Torque_Small_Wind_Generator_for_Hy.pdf?paperid=7651162

[3] Indian Wind Energy and Economy. www.Indianwindpower.com

[4] Centre for Wind Energy Technology (C-WET), Chennai. http://www.cwet.tn.nic.in

[5] Burton, T., *et al.* (2001) Wind Energy Handbook. John Wiley & Sons, Ltd., Hoboken.

[6] Ahmed, S. (2010) Wind Energy: Theory and Practice, Department of Mechanical Engineering, MNNIT Bhopal, Pren-

tice Hall of India, New Delhi.

[7] Johnson, G.L. (2001) Wind Energy Systems. Kansas State University

[8] Ragheb, M. (2014) Optimal Rotor Tip Speed Ratio. http://mragheb.com/NPRE%20475%20Wind%20Power%20Systems/Optimal%20Rotor%20Tip%20Speed%20Ratio.pdf

[9] McCosker, J. (2012) Design and Optimization of a Small Wind Turbine. Rensselaer Polytechnic Institute, Troy, NY.

[10] Ragheb, M. and Ragheb, A.M. (2011) Wind Turbines Theory—The Betz Equation and Optimal Rotor Tip Speed Ratio. Fundamental and Advanced Topics in Wind Power, Dr. Rupp Carriveau (Ed.), InTech. http://www.intechopen.com/books/fundamental-and-advanced-topicsin-wind-power/wind-turbines-theory-the-betz-equation-and-optimal-rotor-tip-speed-ratio

[11] Predescu, M., Bejinariu, A., Mitroi, O. and Nedelcu, A. (2009) Influence of the Number of Blades on the Mechanical Power Curve of Wind. *International Conference on Renewable Energies and Power Quality* (*ICREPQ*'09), Valencia, 15-17 April 2009, 6p.

[12] Sarkar, A. and Behera, D.K. (2012) Wind Turbine Blade Efficiency and Power Calculation with Electrical Analogy. *International Journal of Scientific and Research Publications*, **2**, 1-5.

[13] Martinez, J. (2007) Modelling and Control of Wind Turbines. https://workspace.imperial.ac.uk/centreforprocesssystemsengineering/Public/MSc%20Thesis/2007%20MSc%20Thesis/2007.2%20-%20Martinez%20Jasmin.pdf

[14] Rivkin, D.A. and Silk, L. (2013) Wind Energy (Art and Science of Wind Power). Jones & Bartlett Learning, Burlington, Mass.

6

Economic Evaluation Method of Photovoltaic Power Generation Installed in Ordinary Homes

Yasushi Iwasaki, Kayoko Yamamoto

Graduate School of Information Systems, University of Electro-Communications, Tokyo, Japan
Email: iwasaki@si.is.uec.ac.jp, k-yamamoto@is.uec.ac.jp

Abstract

This study aims to develop an economic evaluation method for installing photovoltaic power generation in ordinary homes using GIS (Geographic Information Systems). The conclusions can be summarized in the following three points: 1) This method determines the profit and loss and payback period in order to evaluate the installation of photovoltaic power generation, taking into account the price of equipment, solar battery module conversion efficiency, subsidy, electricity purchase price, service life and rate for selling electricity. 2) The proposed evaluation method was applied to Kanagawa Prefecture in Japan, providing plural scenarios. Using a solar battery module conversion efficiency of more than 15%, it is possible to make the payback period shorter than the 20-year service life and anticipate a profit in almost the whole area. 3) The areas suitable for photovoltaic power generation are Kawasaki City and Ninomiya-machi. It is necessary to adopt measures to increase the subsidy and install photovoltaic power generating systems in specific places in areas where subsidies are not provided in enough amounts.

Keywords

Photovoltaic Power Generation, Economic Evaluation, Subsidy, Profit and Loss and Payback Period, GIS (Geographic Information Systems)

1. Introduction

The nuclear power plant accident that arose from the Great East Japan Earthquake (2013) has become an opportunity to consider a new power supply method in Japan. Therefore, photovoltaic power generation, which can also be performed at ordinary households and has low environmental risks, is gaining attention. In Japan, economic support is provided for the installation of photovoltaic power generation. The "Residential photovoltaic power generation system installation support reconstruction project", in which national government subsidies are provided when photovoltaic power generation systems are installed in ordinary households, and the "Feed-in Tariff Scheme for Renewable Energy" in which power companies purchase surplus generated power, have been in place since 2012. However, photovoltaic power generation is greatly influenced by the amount of solar radiation, and there is a problem in that the amount of electricity generated differs depending on the place of installation. Further, the subsidy amount differs between local governments, and there are districts where the advantages of installing a photovoltaic power generation system cannot be gained. Therefore, this study aims to propose a method of evaluating the suitability of installing photovoltaic power generation systems, focusing on the profit and loss and the payback period.

To achieve these, this study focuses on ordinary household residences and uses Geographic Information Systems (GIS) to estimate the amount of electricity generation that can be obtained by installing a photovoltaic power generation system. The estimate is based on the amount of solar radiation in the region subject to evaluation, and the area available for installation of a photovoltaic power generation system. The estimate of the amount of electricity generation, the cost of installing and maintaining a photovoltaic power generation system, subsidies and the electricity purchase price are then compared, and considered in order to evaluate the suitability of installing a photovoltaic power generation system. This evaluation method is applied to the entire area of Kanagawa Prefecture in Japan. As mentioned below, Kanagawa Prefecture and cities, towns and villages in the prefecture have implemented unique subsidy schemes, and are tackling the installation of photovoltaic power generation systems in a positive manner.

In this study, first, Section 2 introduces the originality of this study comparing the related works. Section 3 describes the evaluation method of this study, and Section 4 shows the results of application of the evaluation method. Based on these, Section 5 proposes the policies for the promotion of the spread of photovoltaic power generation. Finally, Section 6 shows the conclusions and future works of this study.

2. Related Work

Research which uses GIS to estimate the amount of electricity generated by photovoltaic power generation systems has been conducted from various viewpoints. Luca *et al.* (2011) [1] proposed a method of estimating the amount of electricity generation, and compared different power generation amounts that resulted from differences in solar cell modules. In addition to solar energy, Furuichi *et al.* (2011) [2] obtained information regarding the amount of natural resources and the amount of energy reserves for biomass energy, temperature difference energy, and small and medium-scale hydraulic energy, and analyzed their availability. Further, Shinohara *et al.* (2010) [3] considered the installation area ratio of solar energy collectors and power generators from the perspective of energy saving and CO_2-reducing effects.

An example of a study in which an evaluation of economic efficiency relating to the installation of photovoltaic power generation systems was conducted using GIS is that of Sugihara *et al.* (2011) [4]. This study clarified differences in the amount of electricity generated by photovoltaic power generation systems which are caused by differences in the amount of subsidy money invested. Further, Tamura *et al.* (2010) [5] focused on surplus power and analyzed economic efficiency for both the case of consumers who do install and the case of consumers who do not install photovoltaic power generation systems. The former study focused on the whole of Japan, and estimated power generation amount in broad regions. This study indicated that a detailed evaluation of economic efficiency which focuses on smaller units of spatial scale was a topic for further research. Further, neither of the studies grasped the extent of the profit that could be anticipated by users who introduced photovoltaic power generation systems. Nor did they consider the differences in profit and loss, payback period and so on that occur depending on where the photovoltaic power generation system is installed.

Therefore, this study is unique in that the perspective of users who install photovoltaic power generation systems is taken, and the extent of the profit and loss that installing photovoltaic power generation systems will bring to users in ordinary households is estimated. Further, differences in profit and loss and payback period caused by differences in installation location are shown. Moreover, this study is useful in that it evaluates the entire region of the prefecture, and uses GIS to estimate the extent of differences in profit and loss and in payback period, based on differences in the amount of solar radiation caused by differences in installation location, and differences in the subsidy amount provided by local governments. The estimation focuses on detailed regional areas. Estimating the differences enables policies to be formulated for promoting the spread of photovoltaic power generation at the basic local government level.

3. Evaluation Method

3.1. Evaluation Outline and Method

Figure 1 shows the flow of the evaluation method of this study. In this study, according to **Figure 1**, Section 3 describes the evaluation method, and Section 4 shows the results of application of the evaluation method. The details of these sections are as follows. Firstly, in Section 3, economic evaluations of each scenario in the region for evaluation are conducted. In derivations of profit and loss and payback period for photovoltaic power generation system installation, multiple scenarios are prepared, and profit and loss and payback period are estimated for each scenario. The conditions of scenarios where profit can be anticipated when a photovoltaic power generation system is installed and the conditions of scenarios in which installing a photovoltaic power generation system results in loss occurring are clarified. Doing this enables the suitability of installing a photovoltaic power generation system to be evaluated.

In Section 4, based on the evaluation results in Section 3, in order to propose policies for promoting the spread of photovoltaic power generation, photovoltaic power generation system installation in different local government areas is compared and considered. Firstly, all local governments are classified, based on comparisons of the profit and loss which result from photovoltaic power generation system installation and the payback period. A problem associated with installing a photovoltaic power generation system is the high initial cost. Taking this into account, the installation of small-capacity photovoltaic power generation systems (2 kW systems which cost 1,000,000 yen or less are considered to be small-capacity systems) is considered. Based on the results of the comparisons and the consideration, policies for promoting the spread of photovoltaic power generation that are suited to each group are proposed.

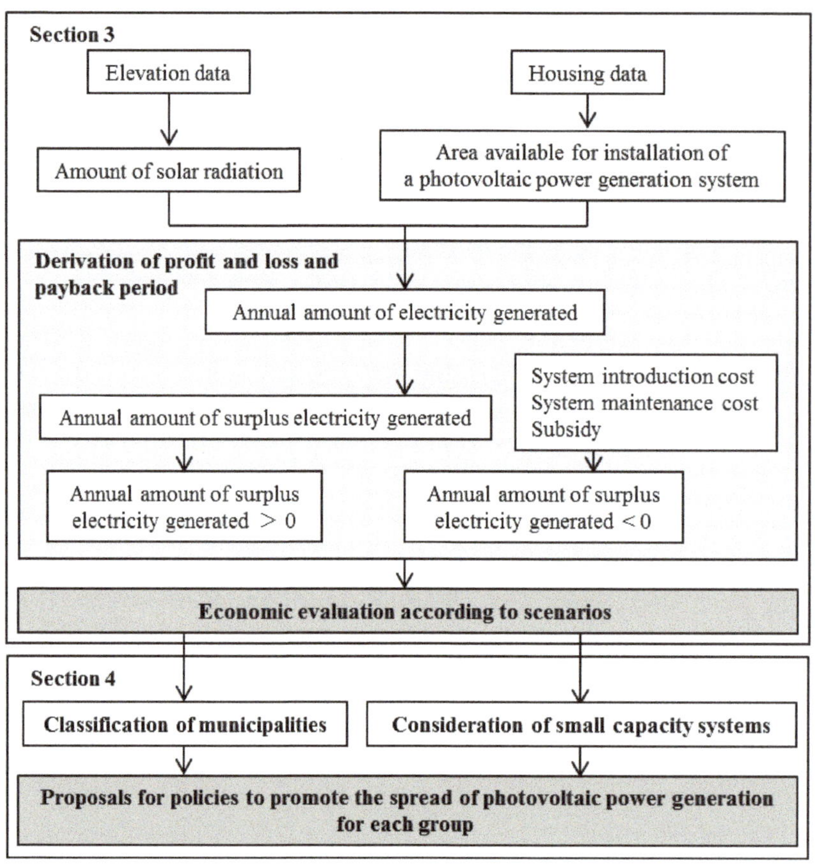

Figure 1. Evaluation method flow.

3.2. Method of Analysis

In this study, analysis was conducted according to the outline shown in **Figure 1**. Esri, Inc.'s Arc GIS ver. 10.1 was used as the GIS. The GIS was used in converting and processing all data into 500 m mesh units with the same format, organizing all analysis and evaluation results using similar units as well, and displaying the data on a digital map. Also, the GIS was used in further collecting and adding these results and grouping them by local government, and in estimating the annual amount of solar radiation.

3.2.1. Annual Amount of Solar Radiation

The amount of solar radiation is divided into three types: the amount of direct solar radiation, the amount of diffuse solar radiation, and the amount of reflected solar radiation. The Arc GIS tool for analyzing the amount of solar radiation was used to estimate the amount of direct solar radiation, the amount of diffuse solar radiation, and the amount of global solar radiation, which was the total amount of direct solar radiation and diffuse solar radiation. Estimation was performed taking into account the influence of the atmosphere, the latitude and elevation of the location, the angle of inclination and the direction of inclination, the angle of the sun, and the influence of shade resulting from the geographical features of the surroundings. A solar radiation amount map was then generated. However, concerning the amount of reflected solar radiation, with the exception of the surface of snow and the like, the amount of reflected solar radiation is minute; therefore, reflected solar radiation is not included in the calculation of the amount of global solar radiation performed by the tool for analyzing the amount of solar radiation. Further, elevation data was required in this analysis, so base map information (a digital elevation model) from the Geospatial Information Authority of Japan [6] was used. This digital map data was converted into raster data, the tool for analysis of the amount of solar radiation was used to output a map of solar radiation amounts for all regions evaluated, and the data was divided into 500 m mesh units.

3.2.2. Area Available for Installation of a Photovoltaic Power Generation System

In the derivation of the area available for installation of a photovoltaic power generation system on a single house (S), values from the "Housing and Land Survey (2003)" on e-stat (the Portal Site of Official Statistics of Japan) [7] for the average building area of a single house (S_h) in different local government areas were used. The value for each local government area was used for all 500 m mesh units in the corresponding local government area in this study. The value was multiplied by roof shape efficiency (C_R) and system installation efficiency (C_S) in order to derive the area available for installation of a photovoltaic power generation system (S). In this study, it was assumed that the roof would be a gabled roof, which is the form of roof most suited to the installation of a photovoltaic power generation system, and is found on many houses in Japan. It was considered that it would only be possible to use the southern half of the roof, so the roof shape efficiency (C_R) was set as approximately 50% (Photovoltaic Power Generation/Photovoltaic Panel Comparison/Installation Guide) [8]. It is recommended that when installing a photovoltaic power generation system, the solar cell module be installed so as to avoid the edge of the eaves, the edge of the gabled roof, and the ridges of the roof. Therefore, these parts were taken into consideration in the system installation efficiency (C_S), and it was assumed that installation was possible on about 70% of the roof (The Panel for Improving the Quality of Construction of Residential Solar Power Generation Systems of the Ministry of Economy, Trade and Industry's Agency of Natural Resources and Energy, 2007) [9]. Further, the angle of the roof (θ) was set as 30°, the angle at which the most amount of solar radiation can be expected in Japan (The Panel for Improving the Quality of Construction of Residential Solar Power Generation Systems of the Ministry of Economy, Trade and Industry's Agency of Natural Resources and Energy, 2007) [9].

$$S = C_S \times C_R \times S_h / \cos\theta \quad (1)$$

S: Area available for installation of a photovoltaic power generation system on a single house (m^2)
C_R: Roof shape efficiency (%)
C_S: System installation efficiency (%)
S_h: Average building area of a single house (m^2)
θ: Roof angle (°)

3.2.3. Derivation of Annual Amount of Electricity Generated

The annual amount of electricity generated (G) was derived for each 500 m mesh unit based on the annual amount of solar radiation (H) and the area available for the installation of a photovoltaic power generation system (S). In performing this derivation, the solar cell module conversion efficiency (C_M) and loss coefficient (C_T) were taken into account. For the former, there are national government subsidy provision conditions and electricity purchasing system conditions; however, in this study, investigation was made using three patterns of module conversion efficiency (silicon single crystal type—16%; silicon polycrystalline type—15%; and chemical compound type—12%) which meet the national government subsidy provision conditions, which are stricter than the electricity purchasing system conditions. According to the Japan Photovoltaic Energy Association (2013) [10], the silicon polycrystalline-type solar cell module with a module conversion efficiency of 15% is the most widely used in Japan. Next, based on information from the New Energy and Industrial Technology Development Organization (2008) [11], the loss coefficient (C_T) for loss caused by temperature rise and loss that occurs when a power conditioner is used to convert direct-current power to alternating-current power which can be used in household products was considered. Taking into account loss caused by temperature rise (15%), loss caused by the power conditioner (8%), and loss caused by soiling of wiring, solar panel surfaces and so on (7%), the total for the loss coefficient (C_T) was 73%.

$$G = C_M \times C_T \times H \times S \tag{2}$$

G: Annual amount of electricity generated (Wh)
C_M: Solar cell module conversion efficiency (%)
C_T: Loss coefficient
H: Annual amount of solar radiation (Wh/m^2)
S: Area available for installation of a photovoltaic power generation system on a single house (m^2)

3.2.4. Derivation of Profit and Loss and Payback Period

1) Annual amount of surplus electricity generated

In order to derive the profit and loss (P) that results from installation of a photovoltaic power generation system, firstly, the annual amount of surplus electricity generated (G_S) was found. In the case where the amount of electricity generated annually (G) exceeds the amount of electricity consumed annually (G_C), during the time outside the surplus electricity purchase period, the amount of electricity generated that exceeds the amount of electricity consumed annually is not utilized and does not lead to a profit. In order to take account of that amount of electricity, in Section 4, which concerns the derivation of profit, the calculation method is changed depending on the difference in the annual amount of surplus electricity generated (G_S). The annual amount of surplus electricity generated (G_S) is the difference between the annual amount of electricity generated (G) obtained by installing a photovoltaic power generation system and the annual amount of electricity consumed (G_C) by an ordinary household. Data from the "2010 National Census" on e-stat [7] concerning the number of households in each 500 m mesh unit, and data from "Electric Power Consumption by Prefecture (Electricity for Light)" in the "Japan Statistical Yearbook 2010" of the Statistics Bureau of the Ministry of Internal Affairs and Communications of Japan [12] were used to calculate the annual amount of electricity consumed (G_C). For each mesh unit, electricity consumption was distributed proportionally based on the number of households to find the annual amount of electricity consumed (G_C).

$$G_S = G - G_C \tag{3}$$

G: Annual amount of electricity generated (Wh)
G_S: Annual amount of surplus electricity generated (Wh)
G_C: Annual amount of electricity consumed (Wh)

2) System price and cost of maintenance of system

The system price (PV) is derived by dividing the area available for installation of a photovoltaic power generation system (S) by the system area per 1 kW (S_{PV}), and then multiplying the result of that by the system price per 1 kW (PV_i). According to the Ministry of Economy, Trade and Industry's Agency of Natural Resources and Energy (2013) [13], the cost of a photovoltaic power generation system includes the price of solar cell modules, a power conditioner, a support, construction costs, and so on. In the period October to December, 2012, the cost was approximately 466,000 yen/kW; therefore, that average price was used as a standard in this study. Further, the expected life of a solar cell module is 20 to 30 years, and that of a power conditioner is 10 to 15 years; therefore, if the solar cell module is used for the whole period of its life, it is necessary to replace the power conditioner with a new one once during that time. Therefore, in this study, it is assumed that the life of a photovoltaic power generation system is 20 years long. To derive the cost of maintaining the system (PV_M), based on the results of an interview survey held with the Japan Photovoltaic Energy Association that were presented to the Ministry of Economy, Trade and Industry's Agency of Natural Resources and Energy (2012) [14], as the cost of replacing a power conditioner with a new one, 1% of the system price (PV) was included in the calculation and multiplied by the life of the system (T).

$$PV = \frac{S}{S_{PV}} \times PV_i \tag{4}$$

$$PV_M = PV \times 0.01 \times T \tag{5}$$

PV: System price (yen)

S: Area available for installation of a photovoltaic power generation system on a single house (m^2)

S_{PV}: System area per 1 kW (m^2/kW)

PV_i: System price per 1 kW (yen/kW)

PV_M: System maintenance cost (yen)

T: System life (years)

3) Subsidies

As subsidies for the installation of a photovoltaic power generation system, subsidies provided by the national government, prefectures and municipalities were taken into account. The sum of each of the subsidies was multiplied by the capacity of the photovoltaic power generation system to be installed in order to derive the total subsidy amount. The subsidy amount per 1 kW provided by the national government is 20,000 yen when the cost per 1 kW to be subsidized is more than 20,000 yen and equal to or less than 410,000 yen, and 15,000 yen when the cost per 1 kW to be subsidized is more than 410,000 yen and equal to or less than 500,000 yen.

$$M = \left(M_x + M_y + M_z\right) \times \frac{S}{S_{PV}} \tag{6}$$

M: Total subsidy amount (yen)

M_x: Subsidy from the national government per 1 kW (yen/kW)

M_y: Subsidy from the prefecture per 1 kW (yen/kW)

M_z: Subsidy from the municipality per 1 kW (yen/kW)

S: Area available for installation of a photovoltaic power generation system on a single house (m^2)

S_{PV}: System area per 1 kW (m^2/kW)

4) Profit and loss

In the derivation of the profit and loss (P) resulting from installation of a photovoltaic power generation system, the system price (PV), the system maintenance cost (PV_M), and the total subsidy amount (M) are taken into account in the sum of the profit during the electricity purchasing period (T_M) and the profit in the period outside the electricity purchasing period ($T - T_M$). In the case where the annual amount of surplus electricity generated (G_S) is positive (Equation (7))—that is, in the case of a 500 m mesh unit where the annual amount of electricity generated (G) is more than the annual amount of electricity consumed (G_C)—a calculation in which profit from the annual amount of surplus energy generated ($G_S \times M_S$) is included in the profit during the electricity purchasing period is conducted. The electricity rate was set as 24 yen per 1 kWh, which is the same as in the electricity rate system of TEPCO (Tokyo Electric Power Company) [15]. Further, according to the "Feed-in Tariff Scheme for Renewable Energy", the electricity purchasing price for residential photovoltaic power generation (less than 10 kW) for the fiscal year 2013 is fixed at 38 yen, and the electricity purchasing period is fixed at 10 years. This study also uses that electricity purchasing price and electricity purchasing period. Concerning the rate for selling electricity (B), an average rate for selling electricity of 60% which is used in a provisional calculation by the Ministry of Economy, Trade and Industry's Agency of Natural Resources and Energy (2009) [16], is used in this study.

$$\begin{aligned} P = &\left\{G_S \times M_S + G_C \times B \times M_S + G_C \times (1-B) \times M_C\right\} \times T_M \\ &+ \left\{G_C \times (1-B) \times M_C\right\} \times (T - T_M) - (PV + PV_M - M) \,\middle|\, G_S \geq 0 \end{aligned} \tag{7}$$

$$P=\{G\times B\times M_S+G\times(1-B)\times M_C\}\times T_M+\{G\times(1-B)\times M_C\}\times(T-T_M)-(PV+PV_M-M)|G_S<0 \quad (8)$$

G: Annual amount of electricity generated (Wh)
G_S: Annual amount of surplus electricity generated (Wh)
G_C: Annual amount of electricity consumed (Wh)
P: Profit and loss from installation (yen)
M_C: Electricity rate (yen)
M_S: Electricity purchasing price (yen)
PV: System installation cost (yen)
PV_M: System maintenance cost (yen)
M: Total subsidy amount (yen)
B: Rate for selling electricity (%)
T_M: Electricity purchasing period (years)
T: System life (years)
5) Payback period

The payback period (Y), which is the period of time it takes until the initial cost of installing a photovoltaic power generation system is recovered by the annual profits, is derived. In this derivation, the profit which occurs during the electricity purchasing period is deducted from the necessary costs, which take into account the system price (PV), the system maintenance cost (PV_M), and the total subsidy amount (M). The result of this is divided by the profit gained in one year by electricity generated in the period outside the electricity purchasing period. The electricity purchasing period (T_M) is then added to the result of that to obtain the payback period. It can be determined that if the payback period (Y) is less than or equal to the device life (T), the region is suited to the installation of photovoltaic power generation systems, and if the payback period (Y) exceeds the length of the device life (T), the region is not suited to the installation of photovoltaic power generation systems.

$$Y=\frac{\left((PV+PV_M-M)-(G_S\times M_S+G_C\times B\times M_S+G_C\times(1-B)\times M_C)\times T_M\right)}{G_C\times(1-B)\times M_C}+T_M|G_S<0 \quad (9)$$

$$Y=\frac{\left((PV+PV_M-M)-(G\times B\times M_C+G\times(1-B)\times M_S)\times T_M\right)}{G\times(1-B)\times M_C}+T_M|G_S<0 \quad (10)$$

G: Annual amount of electricity generated (Wh)
G_S: Annual amount of surplus electricity generated (Wh)
G_C: Annual amount of electricity consumed (Wh)
Y: Payback period (years)
M_C: Electricity rate (yen)
M_S: Electricity purchasing price (yen)
PV: System cost (yen)
PV_M: System maintenance cost (yen)
M: Total subsidy amount (yen)
B: Rate for selling electricity (%)
T_M: Electricity purchasing period (years)
T: System life (years)

3.3. Subsidy Systems in the Regions Evaluated

3.3.1. Prefectural Subsidy Systems

Kanagawa Prefecture offers the "Kanagawa Prefecture residential smart energy facility installation subsidy" (started in fiscal 2013) for the introduction of photovoltaic power generation systems. In this system, people are eligible for a subsidy if they install a photovoltaic power generation facility together with a HEMS (home energy management system) device. A subsidy of 15,000 yen per 1 kW (maximum 50,000 yen) is provided. It is necessary to install a HEMS device in order to receive the Kanagawa prefectural subsidy. In this study, both the case where the cost of a HEMS device is included in the price of the photovoltaic power generation system and the case where it is not included are imagined. Calculations are performed taking into account both the case where the prefectural subsidy can be used and the case where it cannot be used.

3.3.2. Municipal Subsidy Systems

Differences can be observed in the provision of subsidies for the installation of photovoltaic power generation systems in the municipalities in Kanagawa Prefecture. For example, in Kawasaki City, a subsidy of 25,000 yen per 1 kW is provided, whereas in Miura City there is no municipal subsidy system and a subsidy is not provided. Similarly, the maximum amount of subsidy money provided also differs. In Minamiashigara City the maximum is 28,000 yen, whereas in Ebina City it is 200,000 yen. Meanwhile, in Ninomiya Town, Isehara City, Yokohama City, Yokosuka City and Hiratsuka City, a fixed sum subsidy provision system is used. The amount of money provided differs among the fixed sum system regions too. In Ninomiya Town it is 70,000 yen, whereas in Yokosuka City it is 10,000 yen and in Hiratsuka City it is 30,000 yen.

4. Application of the Evaluation Method to the Regions for Evaluation

4.1. Outline of Evaluation Scenarios

Multiple scenarios were imagined, and evaluation was performed in order to find the differences in profit and loss and in payback period caused by differences in the price of photovoltaic power generation systems, solar cell module conversion efficiency, and subsidies provided by the national government, prefectures and municipalities. For the photovoltaic power generation system price, two patterns were used: a system with a price of 466,000 yen per 1 kW, which is the standard in this study, as mentioned in Section 3.2.4; and a system with a price of 410,000 yen per 1 kW, which is the maximum price allowed in order to be eligible for the national subsidy of 20,000 yen per 1 kW. For the solar cell module conversion efficiency, three patterns of conditions for which the national subsidy is provided were used, as mentioned in Section 3.2.3. Concerning whether subsidies are provided or not, three patterns were used. These are the case where none of the national, prefectural and municipal subsidies is provided; the case where only the prefectural subsidy is not provided; and the case where all the subsidies are provided. Details of each scenario are summarized in **Table 1**.

Table 1. Outlines of scenarios.

Scenario	System price per 1 kW (yen/kW)	Solar cell module conversion efficiency (%)	Subside per 1 kW		
			National government	Prefecture	Municipality
1	466,000	12	×	×	×
2	466,000	12	○	×	○
3	466,000	12	○	○	○
4	466,000	15	×	×	×
5	466,000	15	○	×	○
6	466,000	15	○	○	○
7	466,000	16	×	×	×
8	466,000	16	○	×	○
9	466,000	16	○	○	○
10	410,000	12	×	×	×
11	410,000	12	○	×	○
12	410,000	12	○	○	○
13	410,000	15	×	×	×
14	410,000	15	○	×	○
15	410,000	15	○	○	○
16	410,000	16	×	×	×
17	410,000	16	○	×	○
18	410,000	16	○	○	○

4.2. Results of Application of the Evaluation Method

4.2.1. Evaluation of Profit and Loss for Each Scenario

As mentioned in Section 3.2.4, 60% was used as the rate for selling electricity. Average profit and loss for each scenario for regions evaluated is shown in **Figure 2**. From the results for each scenario shown in **Figure 2**, it can be seen that there was a difference of over 900,000 yen between the largest loss, which was approximately 130,000 yen in Scenario 1 (system price 466,000 yen; module conversion efficiency 12%; no national, prefectural or municipal subsidy provided) and the largest profit, which was approximately 800,000 yen in Scenario 18 (system price 410,000 yen; module conversion efficiency 16%; national, prefectural and municipal subsidies all provided). In Scenarios 1, 2 and 3, which had a module conversion efficiency of 12%, only loss occurred. Scenarios 10, 11 and 12, which also had a module conversion efficiency of 12%, also made less profit than other scenarios that had a module conversion efficiency of 15% or 16% but otherwise shared the same conditions as Scenarios 10, 11 and 12. Comparing scenarios which differed only in regard to module conversion efficiency (for example, Scenarios 1, 4 and 7), there was a difference in profit and loss of approximately 430,000 yen between Scenario 1 (module conversion efficiency 12%) and Scenario 4 (module conversion efficiency 15%), and a difference in profit and loss of approximately 140,000 yen between Scenario 4 and Scenario 7 (module conversion efficiency 16%).

Meanwhile, concerning system price, pairing and comparing scenarios which had the same conditions except for system price, it can be seen that there was a difference of approximately 200,000 yen in profit and loss between scenarios which have a system price of 466,000 yen (Scenarios 1 to 9), and scenarios which have a system price of 410,000 yen (Scenarios 10 to 18). Comparing scenarios which differed only in regard to whether subsidies were provided or not (e.g., Scenarios 1, 2 and 3), there was a difference of approximately 50,000 yen between Scenario 1 (in which no subsidies were provided) and Scenario 2 (in which only the prefectural subsidy was not provided), and a difference of approximately 50,000 yen between Scenario 2 and Scenario 3 (in which all subsidies were provided) as well. Therefore, because there are regions where it is possible that a loss will become a profit depending on whether or not subsidies are provided, when purchasing a photovoltaic power generation system, it is necessary to thoroughly check whether conditions for receiving the national, prefectural and municipal subsidies are fulfilled.

4.2.2. Evaluation of Payback Period for Each Scenario

The average payback period for each scenario for regions evaluated when the rate for selling electricity is 60% (as in the previous section) is shown in **Figure 3**. From the results for each scenario in **Figure 3**, it can be seen that the shortest payback period is the approximately 12 year payback period of Scenario 18, and the longest payback period is the approximately 22-year payback period of Scenario 1. There is a difference of about 10 years between these. The payback period for Scenarios 1, 2 and 3 (which have a module conversion efficiency of 12%) exceeds the 20-year life of the system assumed in Section 3.2.4, and it can be said that these scenarios are not suited to the installation of photovoltaic power generation systems. With a payback period of over 15 years, Scenarios 10, 11 and 12 (which also have a module conversion efficiency of 12%) also have a longer payback period than other scenarios which have a module conversion efficiency of 15% or 16% but otherwise share the same conditions as Scenarios 10, 11 and 12. When comparing scenarios which differ only in regard to module conversion efficiency, as in the previous section, it can be seen that there is a difference of approximately 5 years between the payback periods of Scenarios 1 and 4, and a difference of approximately one year between the payback periods of Scenarios 4 and 7.

Meanwhile, concerning system price, pairing and comparing scenarios which have the same conditions except for system price, it can be seen that there is a difference in payback period of two to three years between scenarios which have a system price of 466,000 yen (Scenarios 1 to 9), and scenarios which have a system price of 410,000 yen (Scenarios 10 to 18).Comparing scenarios which differed only in regard to whether subsidies were provided or not (e.g., Scenarios 1 and 2), although there is variance between regions due to subsidy amounts differing between municipalities, a difference of approximately one year in payback period can be observed depending on whether or not subsidies are provided. However, if conditions for receiving all subsidies are satisfied, as in Scenario 6, even if the system has a silicon polycrystalline-type solar cell module with a conversion efficiency of 15%, the payback period is about the same as that of Scenario 7, in which no subsidies are received and the system has a silicon single crystal-type solar cell module with a conversion efficiency of 16%.

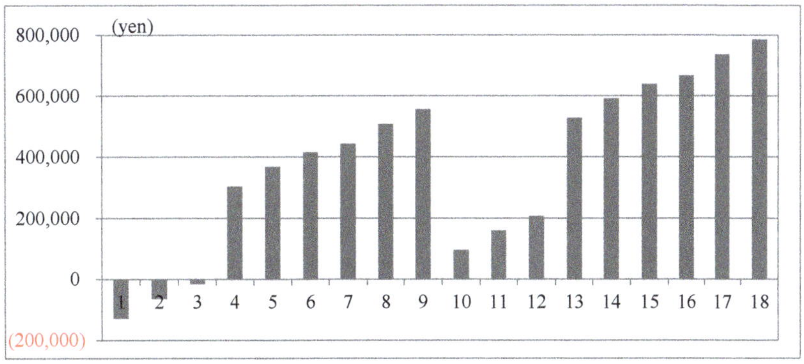

Figure 2. Average profit and loss for each scenario for regions evaluated.

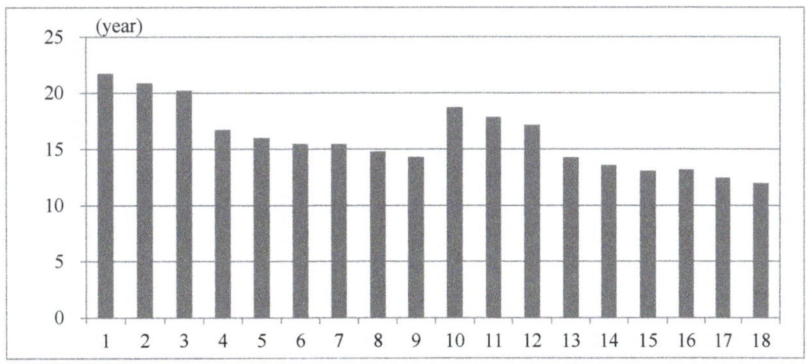

Figure 3. Average payback period for each scenario for regions evaluated.

4.3. Summary of Evaluation Results

As an example of evaluation results in all regions evaluated, **Figure 4** and **Figure 5** show results for profit and loss and payback period for Scenario 6 (system price 466,000 yen; module conversion efficiency 15%; national, prefectural and municipal subsidies provided) when the rate for selling electricity is 60%. Taking account of what is outlined in Sections 3.2.3. and 3.2.4, it can be considered that as things stand now, Scenario 6 has the conditions that are most likely to apply when a photovoltaic power generation system is installed. Concerning the installation of photovoltaic power generation systems in Kanagawa Prefecture, **Figure 4** and **Figure 5** clearly show that in the case of a system with a silicon polycrystalline-type solar cell module with a conversion efficiency of 15% and an estimated system life of 20 years, a profit can be expected in almost all the regions in the prefecture. Therefore, it can be said that Kanagawa Prefecture is a region suited to the installation of photovoltaic power generation systems. In **Figure 4**, a trend for profit to be large in the southwest of the prefecture but small in the eastern part of the prefecture can be observed. Further, in **Figure 5** it can be seen that depending on the location of installation, the payback period differs by several years. It is short in the northern part of the prefecture, but the further south the location of installation, the longer it is. However, the results of this study are based on the case of an estimated photovoltaic power generation system life of 20 years. Systems may actually operate for longer than the estimated life, and in such cases an increase in profit proportional to the extra operating time can be obtained.

5. Proposals for Policies for the Promotion of the Spread of Photovoltaic Power Generation

5.1. Classification of Municipalities

Based on the results described in the previous section concerning evaluation of profit and loss and payback period for Scenario 6 when the rate for selling electricity is 60%, using the average of the largest and smallest values of these results as standards, the municipalities were classified into four groups (**Figure 6**). The Most Suitable type is a region in which a large profit can be anticipated and the payback period is relatively short. Regions

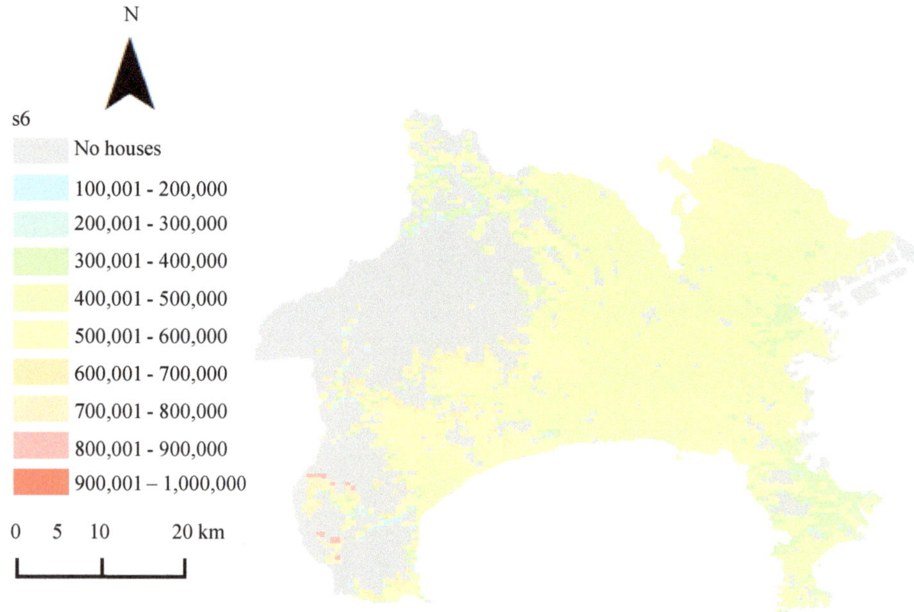

Figure 4. Results for profit and loss (yen) for scenario 6 in all regions evaluated.

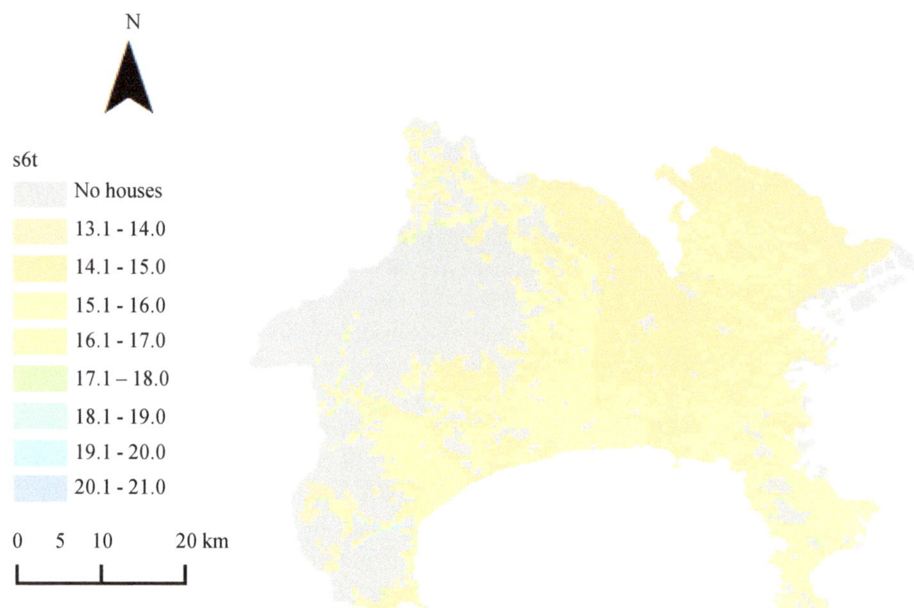

Figure 5. Results for payback period (year) for scenario 6 in all regions evaluated.

with small populations such as Ninomiya Town and Aikawa Town belong to this category. Meanwhile, some regions which are ordinance-designated cities, such as Asao Ward and Miyamae Ward in Kawasaki City and Aoba Ward in Yokohama City also belong to this group. In these regions, there is a large area available for the installation of a photovoltaic power generation system, and a large amount of subsidy money is provided as well. Therefore, the profit is large and the payback period is short, and these regions are the most suited to the installation of photovoltaic power generation systems in Kanagawa Prefecture.

Next are the High-profit/Long-term Payback-type regions. In these regions, a large profit can be anticipated, but on the other hand, the payback period is long. In these regions there is a large area available for the installation of a photovoltaic power generation system, so it is possible to install a high-capacity photovoltaic power generation system; however, doing this requires a large initial payment, and subsidies provided by local gov-

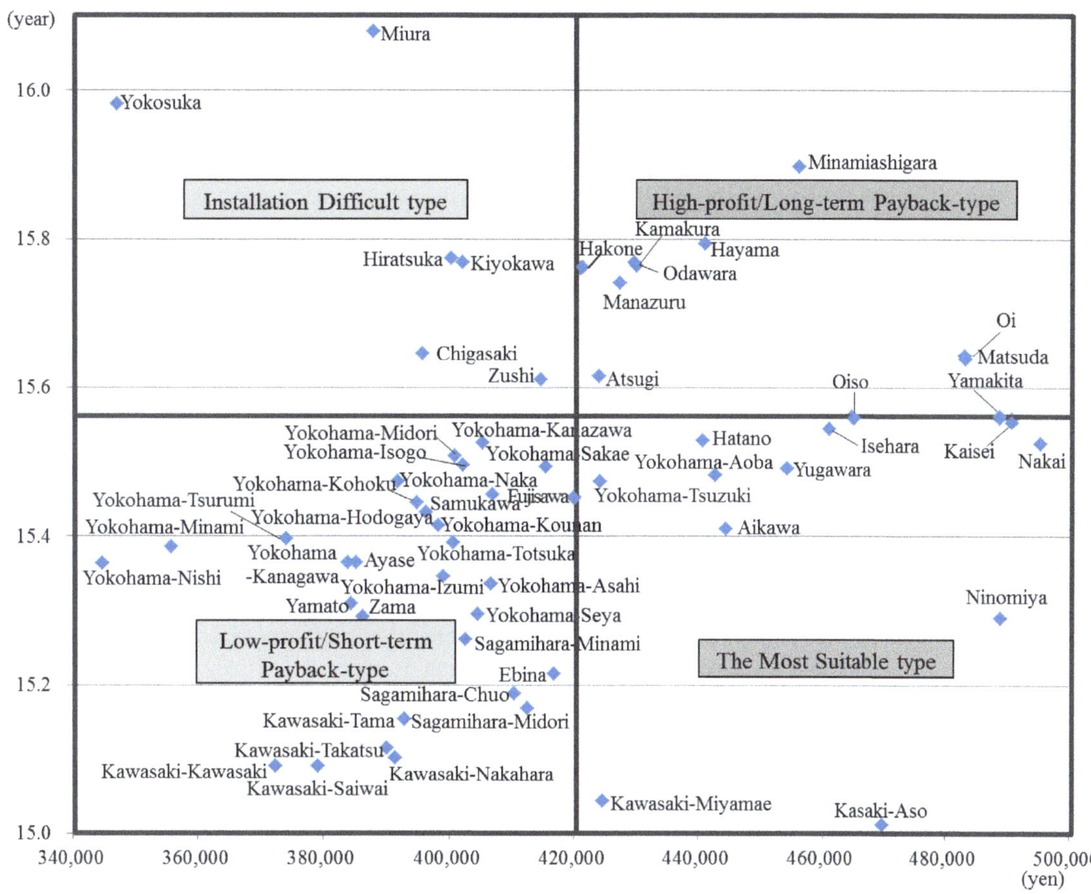

Figure 6. Comparison of the profit and loss and payback period in each municipal.

ernments in these regions are less than those in other regions; therefore, the payback period is long. Minamiashigara City has a particularly low subsidy of 8000 yen/kW (maximum subsidy: 20,000 yen), so it has an even longer payback period than the other regions in this group. Further, of the regions which are mostly mountainous, such as Hakone Town, because of the high elevation, in regions which receive sunlight there is a large amount of solar radiation and a large ultimate profit can be anticipated. On the other hand, in locations between mountains there are also places where there is not much solar radiation and not much profit can be anticipated. Therefore, in regions where the profit differs greatly depending on the installation location, it is necessary to pay attention to this point when installing a photovoltaic power generation system. However, in this group the payback period is long but the annual profit is large. Therefore, if the photovoltaic power generation system operates for longer than the estimated 20-year life, more profit can be anticipated than for other groups.

In the Low-profit/Short-term Payback-type regions, the area available for the installation of a photovoltaic power generation system is small, so not much profit can be anticipated; however, the payback period is shorter than that of other groups. Many districts which are ordinance-designated cities and have large populations belong to this group, such as Yokohama City, Kawasaki City and Sagamihara City. Because the area available for the installation of a photovoltaic power generation system is small, the initial cost can be kept down. Further, subsidies provided by the municipalities are larger than in other regions. Therefore, the payback period is short.

Finally, in the Installation Difficult type of region, as things stand now, while it cannot be said that these regions are unsuited to the installation of photovoltaic power generation systems, compared to other regions the profit is low and the payback period is long. Miura City, which does not provide a municipal subsidy, and Yokosuka City, which provides a fixed sum of only 10,000 yen, are regions which belong to this group. They have particularly long payback periods even compared to other regions in this group. Further, regions such as Chigasaki City, where the subsidy is notvery large (10,000 yen/kW (maximum subsidy: 40,000 yen)) and there is also not much area available for the installation of a photovoltaic power generation system belong to this group.

5.2. Consideration of Small Capacity Systems

The profit and loss and payback period in the case where the small capacity photovoltaic power generation system introduced in Section 2.1 is installed are estimated, and the suitability of installing this system is also considered. Concerning whether or not a region is suited to the installation of small capacity systems, in addition to considering the issue of whether a profit can be anticipated within the life of a regular photovoltaic power generation system, it is necessary for differences in profit and loss and payback period to be small compared to the case where a regular system is installed on all the area available for installation of a photovoltaic power generation system. If a small-capacity system is installed, it is possible that almost no surplus electricity generation can be expected, profit will decrease, and the payback period will be long. Therefore, even if a profit can be expected within the life of the system, for regions where there is a large difference in the profit and loss and the payback period compared to the case of installing a regular system, it is difficult to say that they are suited to the installation of a small-capacity system.

Based on the above, the results of consideration of the installation of small-capacity systems are shown in **Figure 7**. The vertical axis shows the difference in payback period between the case of installing a small-capacity system and the case of installing a regular photovoltaic power generation system on all the area available for installation, and the horizontal axis shows the difference in profit and loss between these cases. Therefore, the closer a region is to the origin, the smaller the differences are in both profit and loss and payback period between the case of installing a small-capacity system and the case of installing a regular system on all the area available for installation of a photovoltaic power generation system, and the more suitable the area is for the installation of a small capacity system. Regions close to the origin are Yokohama City, Ninomiya Town, Isehara City, Hiratsuka City and so on. They are all regions with a fixed-sum subsidy system. Further, most districts in Kawasaki City and Yokohama City belong to the Low-profit/Short-term Payback-type group, but Yokohama City is more suited to the installation of small capacity systems.

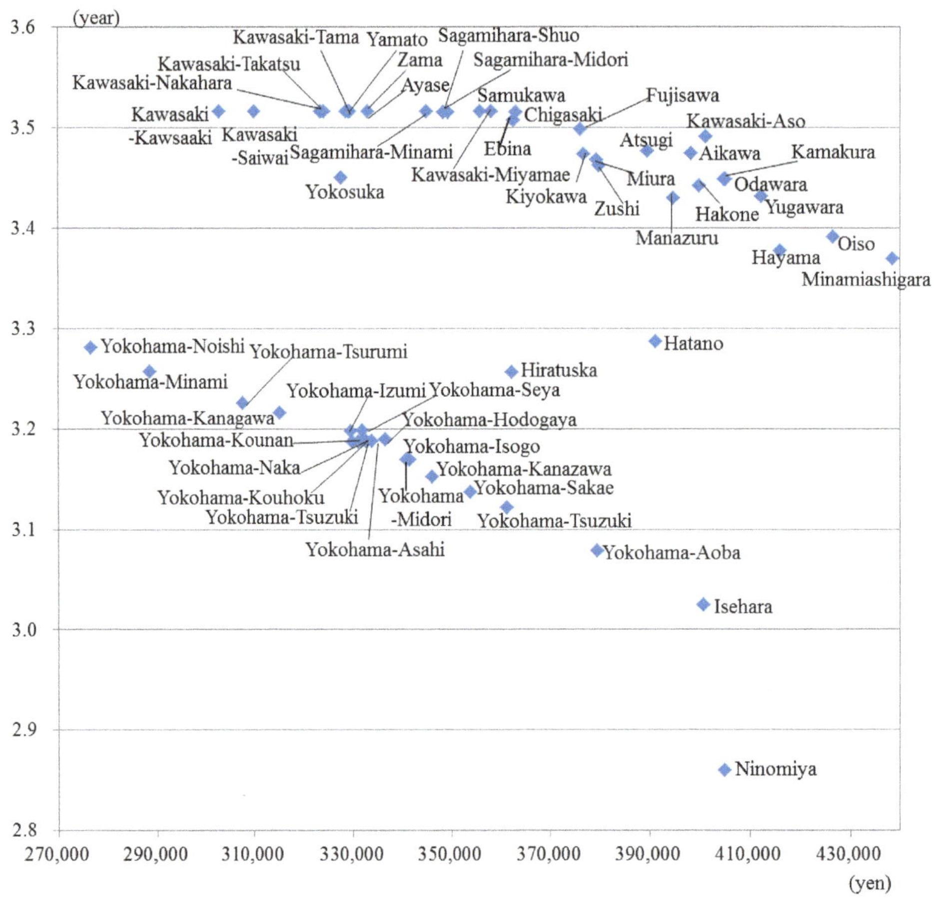

Figure 7. The results of consideration of the installation of small-capacity systems.

5.3. Proposals for Policies for Each Group

Based on the classification of the local governments and the results of consideration of the installation of small capacity systems in this section, policies to promote the spread of photovoltaic power generation are proposed for each of the four groups. As a policy for the Most Suitable type of region, rather than by trying new subsidy systems or the like, it is more likely that the installation of photovoltaic power generation can be promoted by widely publicizing the fact that this type of region is suited to the installation of photovoltaic power generation. In particular, because the installation of small-capacity systems can also be recommended in Aoba Ward and Tsuzuki Ward in Yokohama city and in Ninomiya Town, if there is a desire to strongly promote this, a method where firstly, installation is tested in the above-mentioned regions and model cases are created can be considered. In the High profit/Long-term Payback type of region, there are many regions where the subsidy provided by local governments is less than that of other regions. Therefore, for these regions, raising the amount of the subsidies provided by the local governments and implementing other similar measures can lead to promoting the spread of photovoltaic power generation.

In the Low-profit/Short-term Payback type of region, the subsidy provided by each local government is large, but the area available for the installation of photovoltaic power generation systems is small. Therefore, the subsidy provided by the prefecture is small in comparison to other regions. However, as indicated in the previous section, in Yokohama city, the subsidy is based on a fixed sum system. Therefore, if the subsidy provided by the prefecture is also based on a fixed sum system, an increase in profit can be expected. Further, Yokohama City is suited to the installation of small-capacity systems; therefore, by strongly recommending this, the number of households which install photovoltaic power generation systems can be increased. Among the Installation Difficult-type regions, there are many regions where a subsidy is not provided or the subsidy amount is small, such as Miura City and Yokosuka City. Therefore, for this group, the provision of some subsidy money from each municipality or an increase in the amount of the subsidy provided by each municipality can be considered. Further, in this group, in regions where the profit is relatively large and the payback period is relatively short, a method where firstly, installation of photovoltaic power generation systems is encouraged in only the places where the profit is larger and the payback period is shorter can be considered.

6. Conclusions and Future Work

The conclusion of this study can be summarized into the three points below.

1) A method of evaluating the suitability of introducing a photovoltaic power generation system was proposed. In the method, profit and loss resulting from the installation of a photovoltaic power generation system and the payback period are derived taking into account the price of the photovoltaic power generation system, sunlight conversion efficiency of solar cell modules, subsidies, electricity purchase price, the life of the photovoltaic power generation system, the rate for selling electricity and so on.

2) Kanagawa Prefecture was selected as the region for evaluation. Multiple scenarios were prepared using different photovoltaic power generation system prices, solar cell module conversion efficiencies, and subsidies, and the evaluation method was applied. It was found that in Kanagawa Prefecture, if a system with a module conversion efficiency of 12% is used, there are few regions in which a profit can be anticipated. However, if a system with a module conversion efficiency of 15% or more is used, in almost all regions in the prefecture, the payback period is shorter than the 20-year life of the photovoltaic power generation system, and a profit can be expected. Further, it was found that in the southwest part of the prefecture, there are many regions where a large profit can be expected, and in the eastern part of the prefecture, there are many regions where the profit is small.

3) Based on the classification of each municipality and the results of consideration of the installation of small-capacity systems, policies for the promotion of the spread of photovoltaic power generation in each municipality were proposed. Regions which particularly suit the installation of photovoltaic power generation systems include Asao Ward and Miyamae Ward in Kawasaki City and Ninomiya Town. Ninomiya Town is also suited to the installation of small-capacity systems. In contrast, it was found that in regions where a subsidy is not provided or the amount of the subsidy provided is small, such as Miura City and Yokosuka City, the profit is small and the recovery period is long compared to other regions. In such regions, policies which increase subsidy amounts and encourage the installation of photovoltaic power generation systems in a limited selection of places are required.

An example of a future research topic is to take account of detailed data on the decrease in performance of solar cell modules, if such data is available. Another example is to apply the evaluation method of this study to other regions.

References

[1] Luca, B. and Pietro, A. (2011) Scalable Methodology for the Photovoltaic Solar Energy Potential Assessment Based on Available Roof Surface Area: Application to Piedmont Region (Italy). *Solar Energy*, **85**, 1041-1055. http://dx.doi.org/10.1016/j.solener.2011.02.022

[2] Furuichi, H., Sadohara, S. and Yoshida, S. (2011) Study on the Possibility of Renewable Energy Use in Hadano. *Technical Paper Summaries from the* 2011 *Convention of the Architectural Institute of Japan*, 969-970.

[3] Shinohara, S., Yoshida, S. and Sadohara, S. (2010) Study on the Use of Solar Energy Based on the Mesh Clustering of Regional Energy Supply and Demand Characteristics. *Technical Paper Summaries from the* 2011 *Convention of the Architectural Institute of Japan*, 703-704.

[4] Sugihara, H., Yamashita, J., Ikoma, Y., Akisawa A. and Kashiwagi, T. (2011) A Study on the Potential of Photovoltaic Power Generation Systems in All Houses in Japan Using Climatic Mesh Data. *Solar Energy*, **37**, 41-48.

[5] Tamura, S., Uchiyama, Y. and Okajima, K. (2010) Analysis of Local Power Load and Economic Efficiency Resulting from the Spread of Photovoltaic Power Generation: A Case Study in Tsukuba City. *Socio-Economic Research Center*, **58**, 15-25.

[6] The Ministry of Land, Infrastructure, Transport and Tourism's Geospatial Information Authority of Japan (2013) Base Map Information Digital Elevation Model elevation Data. Accessed 2013.2.5. in Japanese. http://www.gsi.go.jp/kiban/

[7] The Statistics Bureau of the Ministry of Internal Affairs and Communications of Japan (2013) the Portal Site of Official Statistics of Japan (E-Stat). Accessed 2013.2.5. in Japanese. http://wwwe-stat.go.jp/SG1/estat/eStatTopPortal.do

[8] Photovoltaic Power Generation/Photovoltaic Panel Comparison/Installation Guide (2013) Accessed 2013.2.5. in Japanese. http://xn--3ck6al9gm5gib0355ejqq630cny0ahx0ej4vacwb.com/myroof/

[9] The Panel for Improving the Quality of Construction of Residential Solar Power Generation Systems of the Ministry of Economy, Trade and Industry's Agency of Natural Resources and Energy (2007) Residential Solar Power Generation System Design and Construction Guideline, 33.

[10] Japan Photovoltaic Energy Association (2013) Solar Cell Module Shipment Statistics, 1.

[11] New Energy and Industrial Technology Development Organization (2008) Technical Guidebook for Installation of Regional-Focused New Energy Systems and So Forth, 144.

[12] The Statistics Bureau of the Ministry of Internal Affairs and Communications of Japan (2013) Japan Statistical Yearbook 2010. Accessed 2013.2.5. http://www.stat.go.jp/data/nenkan/10.htm

[13] The Ministry of Economy, Trade and Industry's Agency of Natural Resources and Energy (2013) Supporting Data for Considering Acquisition Cost in FY.

[14] The Ministry of Economy, Trade and Industry's Agency of Natural Resources and Energy (2012) Data 2. Results of Interview, 36.

[15] TEPCO (Tokyo Electric Power Company) (2013) Accessed 2013.2.5. http://www.tepco.co.jp/index-j.html

[16] The Ministry of Economy, Trade and Industry's Agency of Natural Resources and Energy (2009) New Purchasing System for Solar Power Generation, 15.

7

Control System Development and Test for the Operation of a Micro-Grid System—PART I

Razzaqul Ahshan, Mohammad Tariq Iqbal, George K. I. Mann, John E. Quaicoe

Faculty of Engineering and Applied Science, Memorial University of Newfoundland, St. John's, Canada
Email: razzaqul.ahshan@cna.nl.ca

Abstract

This paper presents design, analysis and simulation performance of an active power controller for stable and reliable operation of a micro-grid system. Power balance between generation and consumer is a critical issue for stable and reliable operation of the micro-grid systems. This issue becomes more critical when a micro-grid system contains stochastic nature distributed generations such as wind and solar because their output power changes non-uniformly. In order to achieve accurate and fast power balance in such a micro-grid system, power in the system has to be regulated continuously. Such an objective can be achieved using droop based alternating current control technique. Because the droop characteristic employed into the developed controller initiates determining the power deviation in the system which is continuously regulated by controlling the current flow into dump power resistors. The designed controller is simulated for the operation of a micro-grid system in stand-alone mode under various operating conditions. The simulated results show the ability of the developed controller for stable and reliable operation of the micro-grid that contains renewable sources. The experimental development of the micro-grid system and the testing of the developed active power controller are presented in PART II of this paper.

Keywords

Power Balance, Frequency Control, Renewable Energy Sources, Micro-Grid

1. Introduction

Increasing energy demand, environmental concern, de-regulated market policy and technology advancement are main causes to integrate distributed generation (DG) units into power system network. Such integration process leads the conventional power system to the new power system domain called micro-grid system. A micro-grid system generally composes of DG units with diverse generating capacity, availability, primary energy and operational costs. The application of such DG units in a micro-grid domain requires delivering stable, reliable and sustainable power generation, conversion, storage as well as distribution and exchanging during the entire operational time of the micro-grid system. In order to achieve such operational objectives, coordination and control in a micro-grid system are essential with the consideration of a series of factors such as availability of power generation, operational modes, voltage and frequency limitation, micro-grid load demands, renewable power penetration as well as energy cost [1] [2].

Operation and control of an isolated micro-grid system consisting of renewable power generations become critical because of the stochastic variation in primary energy sources such as wind speed, solar isolation etc. Such variation always leads to the change in active power generated by the DG units in the micro-grid system. As a result the active power generation in the micro-grid system varies randomly that results in an active power mismatches between generation and consumption, which is stochastically varying in nature. Moreover, low inertia in the isolated micro-grid system as compared to a conventional power system along with stochastic behavior in active power mismatches can easily lead the micro-grid system instability. Hence, continuous and smooth active power regulation in an isolated micro-grid system has become an important issue to achieve stable and reliable operation of the system that comprises of renewable power generations such as wind turbines, photovoltaic panels etc.

The growing interest of utilizing renewable power generations in micro-grid domain has turned researchers to address a variety of issues such as architecture, modeling, reliability and cost, operation and control strategies related to micro-grid systems. Several researchers have studied the issues related to frequency control in the micro-grid systems. Power management strategies in a micro-grid system proposed by R. H. Lasseter are carried out by controlling the power angle between the voltage at the voltage source converter and the micro-grid voltage for each micro-source [3], where micro-sources are assumed to be converter interfaced in the micro-grid domain. The successful laboratory test of micro-grid system presented in [3] is performed and reported in [4]. The application of non-inverter interfaced DG units and storage systems in such micro-grid systems is also studied [4]. Power sharing technique among multiple DGs in a micro-grid system is also presented and discussed, where DGs are assumed as inverter interfaced with dispatchable energy sources [5]. In [5], one of the limitations of feeder flow control introduced in [3] is addressed to achieve improved power sharing among DGs which are connected to a common feeder in series during transition between grid connected to island system. Such limitation is solved by employing a method of determining the new droop coefficients during the transition period. Droop based parallel inverter control concept is also adopted to achieve frequency regulation in a wind-solar power source based micro-grid system [6]. A DC/AC pulse width modulated inverter interfaced battery storage system that is used to maintain the micro-grid frequency is presented in [7]. The battery-inverter arrangement is controlled to regulate the magnitude and frequency of its voltage output during the isolated operation of the micro-grid system, according to the power-frequency droop curve. Frequency control in the New Energy and Industrial Technology Development Organization micro-grids is also performed by controlling the active power component of an inverter interfaced battery storage system using power versus frequency droop characteristics [8]. In [9] [10], frequency controller for a micro-grid system is presented, where droop methods and frequency restoration technique are employed to regulate active power component of the inverter interfaced DG units. The DG unit in such micro-grid system is assumed as a DC source. A primary frequency controller is proposed in [11] for variable speed wind energy conversion system in micro-grid operation. The control scheme determines the new operating point of the wind generator unit based on the frequency variation in the micro-grid system. The power at the new operating point provides current reference of the rotor side converter. Parallel inverters control concepts using droop methods are proposed for micro-generations control in a micro-grid system to maintain the active power balance between generation and consumption [12].

A central automatic control system based on estimated load curve is presented for a micro-grid system. Such controller maintains the system frequency by controlling the individual generation units based either on or off depending on the state of the micro-grid frequency [13]. Fuel cell and electrolyzer hybrid system is employed with a dynamic fuzzy proportional-integral controller to ensure real power balance between generation and consumption is studied in [14].

According to the technical literature, parallel inverter control technique based on droop methods has been extensively used to regulate frequency in the micro-grid system. However, it has been reported in [15] because of the droop characteristics; the system frequency may drop to a value that all micro-generation units will be operating in a newer operating frequency which may differ from the nominal frequency of the micro-grid system. Thus, a power sharing strategy using static droop characteristics and an adaptive transient droop function is investigated in [15]. Moreover, one of the limitations of parallel inverter control concept is identified and an idea to overcome such limitation is presented in [5].

In [16], auction based theory is applied in pricing strategy for solar-powered micro-grid system, where supply and demand issue is considered as a Potluck problem. A dynamic model of a hybrid system along with an optimal control issue for real time operation is discussed and presented in [17]. A STATCOM based control technique for power quality improvement of a grid connected wind turbine system is proposed, and the proposed control technique showed the significant improvement in power quality in an integrated wind generator, STATCOM and battery energy storage system [18].

In this paper, a micro-grid system comprising of wind power generation system and a small hydro generation unit is presented. The system is located at Fermeuse, Newfoundland, Canada. An active power regulation scheme for stable and reliable operation of such a micro-grid system in an isolated condition has been developed and presented. The proposed scheme is designed to accommodate excess active power in the micro-grid system generated by variable speed wind power generation system in a continuous fashion. Stochastically varying active power in the isolated micro-grid system can be accommodated either by storing energy as water head potential or by using heat load (or resistive dump load) or a combination of both. The performance of the designed controller is tested through simulation in stand-alone mode under various operating conditions of the micro-grid system. The results indicate that the developed active power controller is capable to maintain smooth power balance during stand-alone mode of operation. Also, the controller shows the ability to maintain stable and reliable operation of the micro-grid system that contains only renewable sources.

2. The Micro-Grid System

A micro-grid system is a pliable combination of loads, DG units, storage systems either centrally or individually, and interfaced power conditioning units. Such combination can be operated as a single controllable system to provide electric power or electric power and heat to loads within the micro-grid service domain. Wind turbines, solar panels, fuel cells, hydro units and/or non-renewable energy sources are the commonly used DG units in the micro-grid system. However, the combination of the DG units in a micro-grid domain depends upon sites and the availability of the primary energy sources in the site. Micro-grid systems provide benefits to the utility operators and investors, DG owners and customers. Such benefits may include reliable power supply to loads, efficient power transmission, reduction in transmission system expansion and increases renewable power penetration [3] [9] [19].

Fermeuse Micro-Grid System

This research considers a micro-grid system located at Fermeuse, Newfoundland, Canada. The Fermeuse system consists of 9 WTs, a hydro generation unit (HGU), transmission lines, transformers, a storage unit and two loads. Also, the Fermeuse system includes a link for utility-connected mode of operation. It is important to note that the combination of DG units in the micro-grid system is from renewable energy sources. **Figure 1** shows the single line diagram of the Fermeuse system. Data for the buses of the study micro-grid are provided in **Table 1**. Also, the data for transmission lines and transformers are provided in **Table 2**.

In order to study and observe the entire operational behavior of the micro-grid system, three operational modes are considered and are conceptually shown in **Figure 2**. Such operational modes are: 1) utility grid connected, 2) isolated micro-grid with wind power generation system, and 3) isolated micro-grid without wind power generation system [20]. During any operating mode, the active and reactive powers balance in a micro-grid system can be stated as:

$$\Delta P_E = P(v_\omega) + P(h) - P_{L1} - P_{L2} \tag{1}$$

$$\Delta Q_E = Q(v_\omega) + Q(h) - Q_{L1} - Q_{L2} \tag{2}$$

where v_ω is the wind speed, $P(h)$ and $Q(h)$ are active and reactive power outputs of the hydro unit, and P_{L1}, Q_{L1}, P_{L2} and Q_{L2} are active and reactive power of loads $L1$ and $L2$ respectively.

The study concludes that when the micro-grid is operated in utility-connected mode, the excess powers, $\Delta P > 0$ and $\Delta Q > 0$ are supplied to the grid, while $\Delta P < 0$ and $\Delta Q < 0$ are delivered from the utility grid.

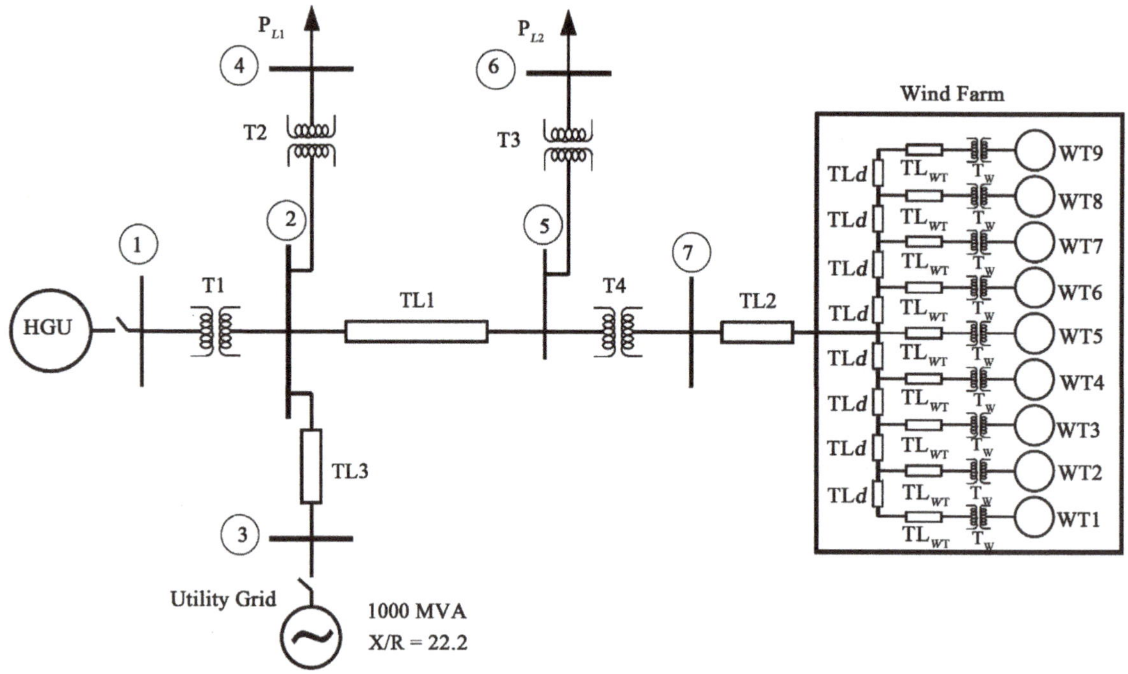

Figure 1. The single-line diagram of the study micro-grid system located in Fermeuse, Newfoundland, Canada.

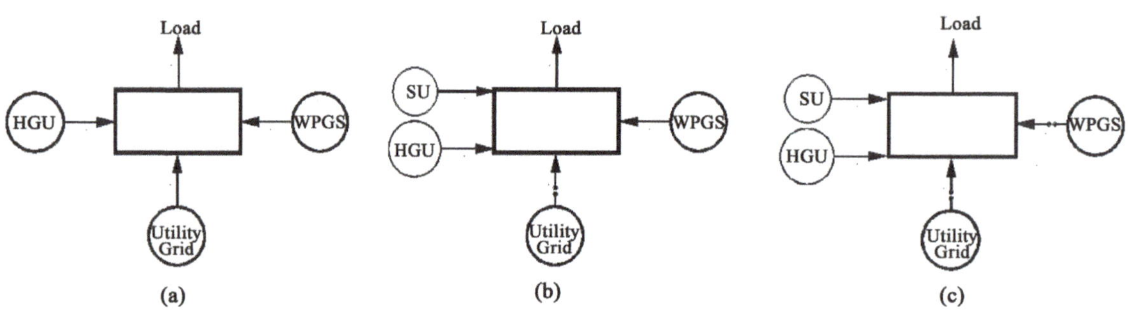

Figure 2. Conceptual presentation of micro-grid system: (a) Utility-grid connected system; (b) Isolated micro-grid with wind power generation system; (c) Isolated micro-grid without wind power generation system.

Table 1. Fermeuse micro-grid system: Buses.

Bus	P [MW]	Q [MVAR]	V [kV]
1	7.3	6.8	6.9
2	ΔP_E	ΔQ_E	66
3	ΔP_E	ΔQ_E	66
4	3.94	0.95	12.5
5	0	0	66
6	2.82	0.6	12.5
7	27	12	12.5

During isolated micro-grid with wind power generation system operating mode, it also shows that when $\Delta P > 0$ and $\Delta Q > 0$, the excess powers are required to accommodate continuously either by storing or by supplying to the dump load, and when $\Delta P < 0$ due to low wind speed availability, more than one wind turbine is required to switch to supply power demand along with dump load. Also, during isolated micro-grid without wind power generation system operating mode, while $\Delta P < 0$ and $\Delta Q < 0$, the additional powers are required

Table 2. Fermeuse micro-grid system: transmission lines and transformers.

Components	V [kV]	Z [Ω]	S [MVA]
TL1	66	$0.0416 + j0.0663$	40
TL2	12.5	$0.0374 + j0.3741$	40
TL3	66	$0.002 + j0.0032$	50
TLd	12.5	$0.00072 + j0.0012$	40
TL_{WT}	12.5	$0.00006 + j0.000096$	5
T1	6.9/66	$0.0854 + j0.8541$	10
T2	66/12.5	$0.0493 + j0.493$	5
T3	66/12.5	$0.061 + j0.62$	4
T4	66/12.5	$0.05 + j0.5$	40
TL_{W}	1/12.5	$0.006 + j0.0625$	5

to supply by the SU [20]. It is to be noted that the proposed active power controller in this paper is investigated for an isolated micro-grid with wind power generation system operating mode (**Figure 2(b)**).

3. Active Power Controller

The main objective of the proposed active power control scheme is to maintain continuous and smooth active power balance between generation and micro-grid load. It is worth mentioning that during isolated micro-grid with wind power generation system operating mode, the active power generation in the system varies due to the stochastic variation in wind speeds. Moreover, the power balance in such a micro-grid system has to be achieved by allowing wind generator to operate in optimal power extraction criteria rather forcing wind generator to produce power within a limited capacity of generation. Forcing wind turbine to generate power less than their available power will reduce the turbine efficiency as well as will require additional time to get pay back, which may not be attractive option for investors. Therefore, power control scheme is required to design in a micro-grid system that contains renewable energy sources such as wind turbines by allowing them to operate in their maximum power extraction criteria so that wind generators can produce power at their optimal efficiency. **Figure 3** shows the active power control strategy for the operating mode of anisolated micro-grid system with wind power generation system.

During isolated micro-grid with wind power generation system operating mode, the control principle lies into two folds. One is to store energy as water head potential and the other is to shape excess energy using dump load such as heating load. Due to the availability and opportunity to use existing reservoir in the selected site (Fermeuse, Newfoundland, Canada), pumped hydro storage system is proposed for the micro-grid operation. A motor equivalent load is used in the system to present motor-pump set during this operational mode. However, as the wind power varies stochastically and motor-pump set consumes a finite amount of power, the isolated micro-grid system will still have excess active power which is required to shape to maintain the system frequency in an acceptable range (59.5 - 60.3 Hz).

The functional block diagram of the developed controller is shown in **Figure 4**. The controller comprises of phase locked loop, droop calculator, proportional-integral regulator, firing angle calculator, firing pulse generator, switching devices, and resistive dump load. Two thyristor switches are connected in anti-parallel and used in one phase to achieve bi-directional operation. The thyristor switches are selected because of their capability to operate in high power applications [21].

The 3-phase system voltages $V_{s,abc}$ are utilized in phase locked loop to measure the micro-grid frequency. The measured frequency is compared with the micro-grid base frequency and then input to the frequency-active power droop characterizer. The power variation in the micro-grid system due to the change in the system frequency can be expressed as

$$\Delta P = -\frac{1}{D_c}\left(f_{\text{base}} - f_{\text{mes}}\right) \tag{3}$$

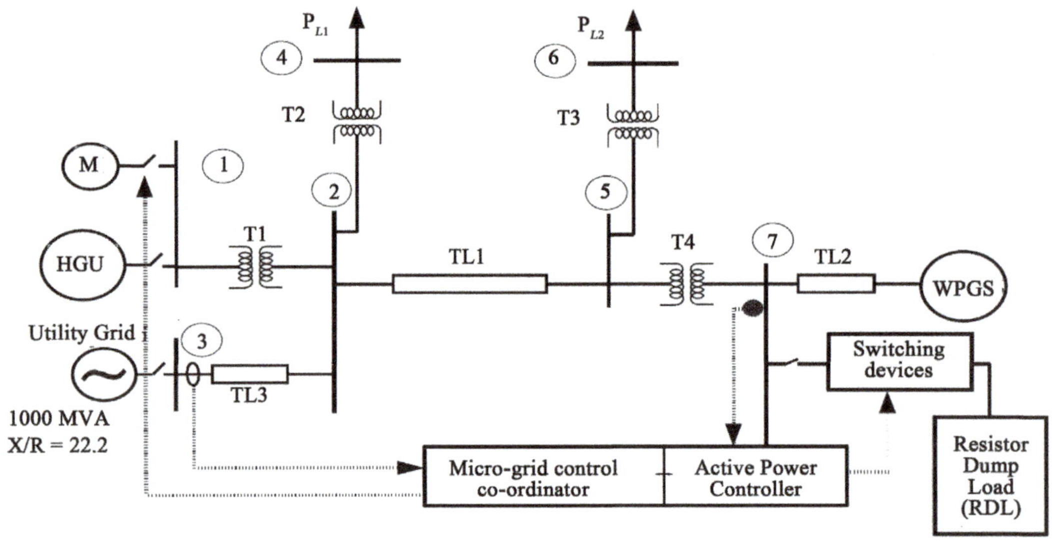

Figure 3. Conceptual diagram of the active power controller along with the micro-grid system.

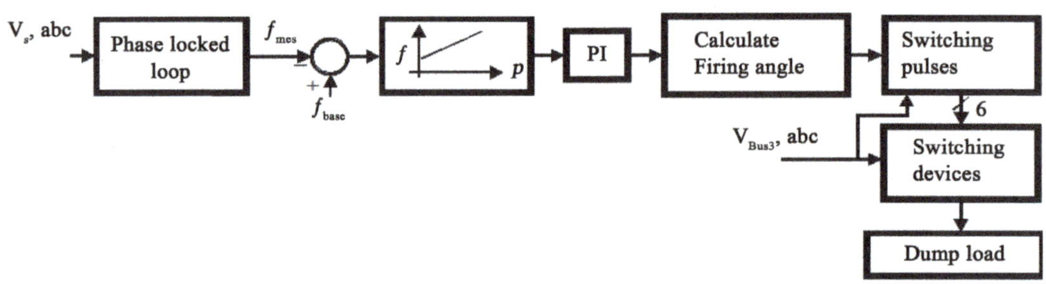

Figure 4. Functional block diagram of the active power controller.

where, ΔP is the power variation in the micro-grid, D_c is the droop coefficient, f_{base} is the nominal micro-grid frequency and f_{mes} is the measured frequency from the system. The power ΔP is passed through the PI regulator to generate the control signal u_{cs} according to (4).

$$u_{cs} = \Delta P\left(k_p + \frac{k_i}{s}\right) \tag{4}$$

where, k_p and k_i are the proportional and integral gain of the PI regulator. The signal u_{cs} obtained from the PI regulator is employed to calculate the firing angle for the switching devices. The change in control signal create variation in firing angles of switches, which results a change in conduction angle of the switches. The change in conduction angle of the SCR switches causes a change in RMS value of the current flowing into the dump load and results in a change in the amount of real power to be dumped into the dump load. The firing angle generator is represented by a linear equation using the intuition that the higher value of control signal requires lower value of firing angle and vice versa, because lower value of firing angle means allowing more power into the dump load and vice versa. Conceptually, higher value control signals represent larger error or larger power deviation in the micro-grid system which requires absorbing more power from the micro-grid system and vice versa. The relation between the firing angle and control signal based on the principle of SCR switched three phase full wave AC voltage controller is expressed as follows.

$$\alpha = \begin{cases} -69^\circ \times u_{cs} + 87^\circ & \text{for } 1 \ge u_{cs} \ge 0 \\ 69^\circ \times u_{cs} + 87^\circ & \text{for } 0 \ge u_{cs} \ge -1 \end{cases} \tag{5}$$

The firing angle α and the 3-phase reference voltages are utilized to generate synchronized switching pulses for the switching devices.

4. Modeling and Simulation

The designed active power control scheme described in the preceding section is modeled and simulated in the Matlab/Simulink environment in order to observe its performance. The available model of a phase locked loop and a firing pulse generation unit in the Simpower system library are used in this simulation. The models of power variation coefficient, control signal and firing angle estimator are developed using the available components in Simulink. The simulation is performed for a 20 seconds interval. The designed control scheme is simulated for this case, while a step load is applied to or released from the micro-grid system. The amount of load for a step change is equivalent to a load of a motor-pump set. Since it is necessary to run motor-pump sets to pump water into the upper reservoir for generating power using the stored water, a load equivalent to a motor-pump set is used for step change in the simulation. The simulation for a step load change is also a 20 seconds interval. The grid disconnection occurs at $t = 5$ seconds due to a fault or regular maintenance.

Simulation Performance

A simulation study is carried out on a system consisting of the active power controller along with the micro-grid system based on renewable power generation shown in **Figure 3**, and the performance results are presented in following figures. The performances of the designed controller are observed for various operating conditions. These conditions are utility grid disconnection and subsequent operation, variable power generation by the wind generators, and step increase and decrease in loading conditions. The micro-grid frequency using the designed active power controller and without the controller are shown in **Figure 5** and **Figure 6**, respectively. Until $t = 5$ seconds, the system frequency remains at its rated value as the micro-grid system is connected to the utility grid. At $t = 5$ seconds, the utility grid is disconnected from the micro-grid system and the designed active power controller is activated for the operating mode of an isolated micro-grid system with wind power generation.

The frequency of the isolated micro-grid system during utility grid disconnection and in subsequent operation while wind power generation system is connected is shown in **Figure 5**. Due to the control action of the active power controller, the micro-grid frequency settles to its nominal value after a smooth dynamic change during the time of transition from grid-connected to isolated micro-grid operation. On the other hand, the frequency of the isolated micro-grid system with wind power generation and without an active power controller, shown in **Figure 6**, starts with a large overshoot at the time of grid disconnection. This overshoot occurs as no control action is

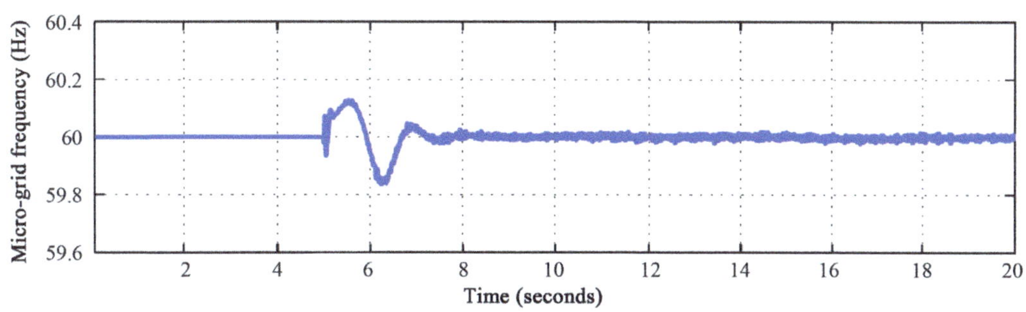

Figure 5. Micro-grid system with active power controller.

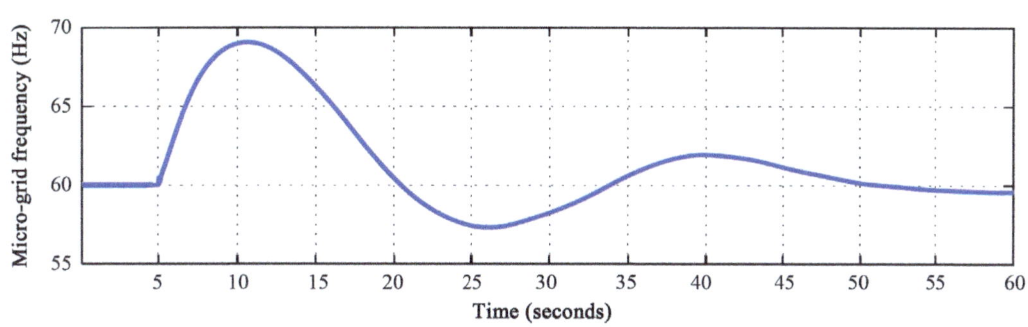

Figure 6. Micro-grid system without active power controller.

applied to the system. Thus the system frequency response indicates a significant ability of the developed controller to maintain an active power balance between the generation and the loads in the isolated micro-grid system with wind power generation.

The firing angle for the thyristor switches that is derived from control signals at the proportional-integral regulator output, and the average value of the current flowing into the dump load, are shown in **Figure 7** and **Figure 8**, respectively. **Figure 7** illustrates that there is a fast, smooth and wide range of change in the firing angles at the beginning of the grid disconnection. These changes allow the controller to be able to accommodate the initial power mismatches during the transition from the utility connection to the isolated mode of operation. The rest of the time, t = 10 - 20 seconds, the firing angle varies dynamically to adjust the power in the micro-grid system that is supplied by the wind power generation system, since the power generation by the wind generator varies according to the change of the wind speed. Furthermore, **Figure 8** demonstrates the change in the current flow into the dump load with the variation in firing angles. The variation in the average value of the current flow dictates the variation in power delivered to the dump load, which results an active power balance in an isolated micro-grid system with wind power generation.

The active power controller performances are also verified using step changes in the loading condition, and the simulation results are presented in following Figures. During the utility grid disconnection at t = 5 seconds, the system frequency varies and then settles down to its rated value smoothly. The frequency of the isolated micro-grid system with wind power generation along with the active power controller is recorded during step changes in load and is shown in **Figure 9**. The figure illustrates that a dip in system frequency occurs due to the

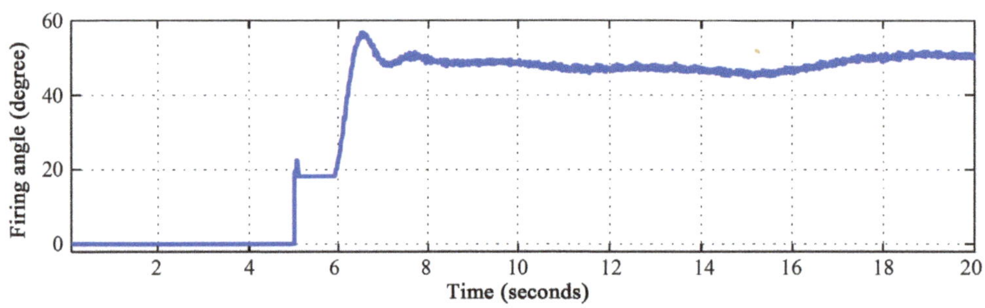

Figure 7. Firing angle for thyristor switches.

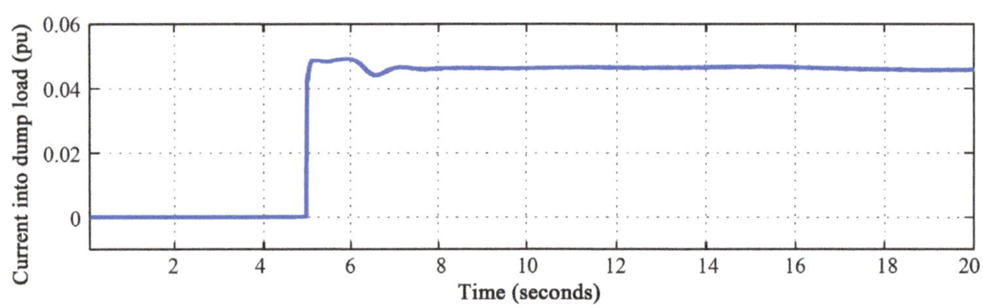

Figure 8. Current flowing into dump load.

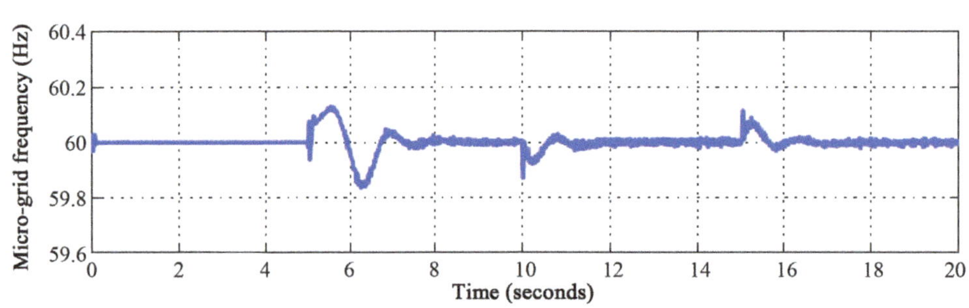

Figure 9. Micro-grid frequency using active power controller with step changes in load.

change in load demand every time; however, after a short while the frequency settles back to its rated value. Frequency regulation in these operating conditions by adjusting the active power in the system also indicates a significant ability of the designed controller in the isolated micro-grid operation.

An increase step change in load is applied to the system at t = 10 seconds in an amount of 7.4 percent of the total load demand of the micro-grid system. Similarly, a decrease step change in load is also created in the system at t = 15 seconds in the same quantity of the load. These step changes in load are applied at the same bus where load-II is connected and are shown in **Figure 10**. Furthermore, the firing angles for thyristor switches derived from control signals and the current flowing into the dump load during step changes in the load are also shown in **Figure 11** and **Figure 12**, respectively.

Figure 11 illustrates that the firing angle increases at t = 10 seconds because the micro-grid load demand increases. This condition indicates that there is a requirement to send less power into the dump load. This requirement is achieved by the action of the designed controller and is shown in **Figure 12** because less power is supplied to the dump load starting at t = 15 seconds. **Figure 11** also illustrates that the firing angle decreases at t = 15 seconds because micro-grid load demand decreases. This case indicates that there is a need to deliver more power into the dump load. This goal is reached by applying the developed controller action and is shown in **Figure 12**, because more power is delivered into the dump load starting at t = 15 seconds.

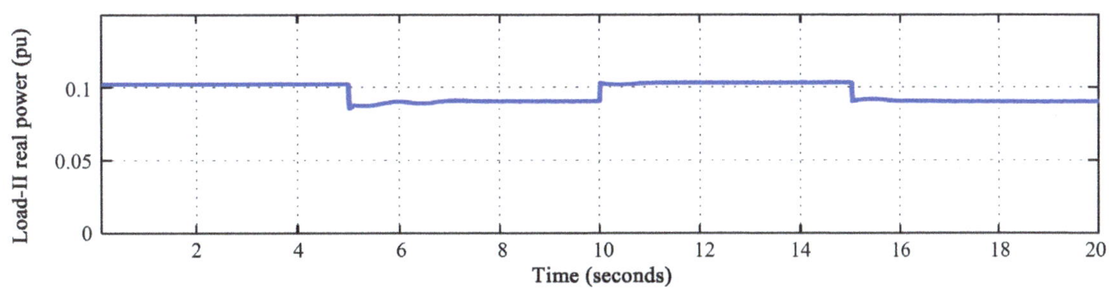

Figure 10. Power consumed by load-II during step change in micro-grid load.

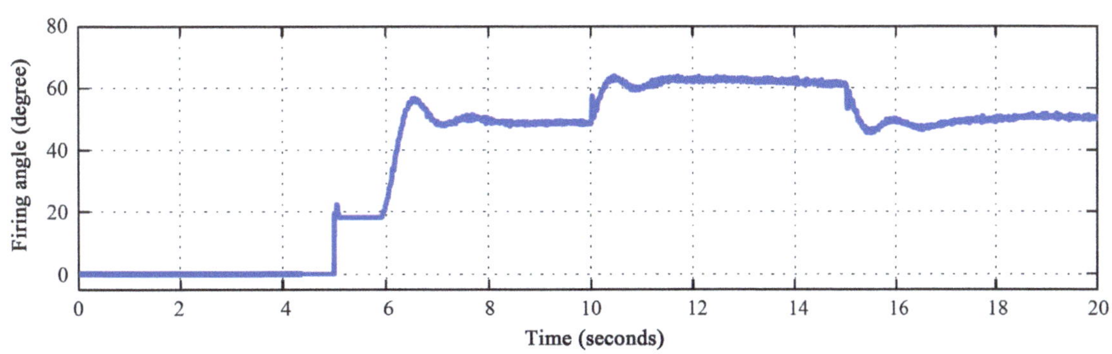

Figure 11. Variation in firing angle for thyristor switches during step changes in load.

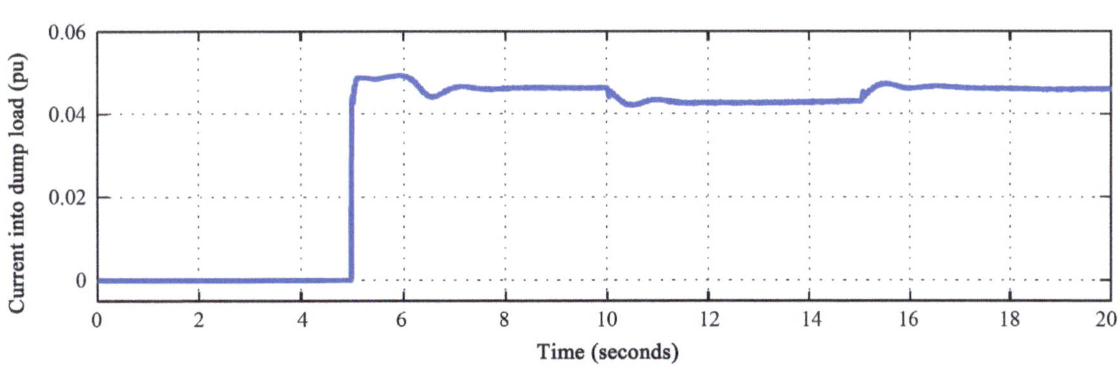

Figure 12. Change in current flow into the dump load during step changes in load.

5. Conclusion

This paper has presented the design, simulation and performance testing of an active power controller for an isolated micro-grid system which contains only renewable energy sources. The controller is developed based on alternating current control technique, where dump load resistor and power-frequency droop characteristics are employed. The controller is modeled and simulated using MATLAB/SIMULINK simulation package. The controller performances are tested through simulation under various operating conditions and satisfactory simulation results are presented in this paper. It is found that the design active controller is capable to maintain the fast and accurate power balance between the generation and load, while a stable and reliable operation was maintained during transition between the grid connected to the isolated micro-grid domain, and in the subsequent operation.

Acknowledgements

This work is supported by a research grant from the National Science and Engineering Research Council (NSERC) of Canada, the Atlantic Innovation Fund (AIF) Canada, and Memorial University of Newfoundland. The authors would like to acknowledge the utility company, Newfoundland Power, Canada for providing the system information and data.

References

[1] Mahat, P., Chen, Z. and Bak-Jensen, B. (2010) Under Frequency Load Shedding for an Islanded Distribution System with Distributed Generators. *IEEE Transactions on Power Delivery*, **25**, 911-918. http://dx.doi.org/10.1109/TPWRD.2009.2032327

[2] Katiraei, F. and Iravani, M. (2006) Power Management Strategies for a Micro-Grid with Multiple Distributed Generation Units. *IEEE Transactions on Power Delivery*, **21**, 1821-1831. http://dx.doi.org/10.1109/TPWRS.2006.879260

[3] Lasseter, R.H. (2002) Micro-Grids. *IEEE Power Engineering Society Winter Meeting*, New York, January 2002, 305-308.

[4] Lasseter, R.H., Eto, J.H., Schenkman, B., Stevens, J., Vollkommer, H., Klapp, D., Linton, E., Hurtado, H. and Roy, J. (2011) CERTS Micro-Grid Laboratory Test Bed. *IEEE Transactions on Power Delivery*, **26**, 325-332. http://dx.doi.org/10.1109/TPWRD.2010.2051819

[5] Ahn, S., Park, J., Chung, I., Moon, S., Kang, S. and Nam, S. (2010) Power-Sharing Method of Multiple Distributed Generators Considering Control Modes and Configurations of a Micro-Grid. *IEEE Transactions on Power Delivery*, **25**, 2007-2016. http://dx.doi.org/10.1109/TPWRD.2010.2047736

[6] Ribeiro, L.A.de.S., Saavedra, O.R., Lima, S.L. and Matos, J.G.de. (2011) Isolated Micro-Grids with Renewable Hybrid Generation: The Case of Lenis Island. *IEEE Transaction on Sustainable Energy*, **2**, 1-11.

[7] Georgakis, D., Papathanasiou, S.A., Hatziargyriou, N., Engler, A. and Hardt, Ch. (2004) Operation of a Prototype Micro-Grid System Based on Micro-Sources Equipped with Fast-Acting Power Electronics Interfaces. *IEEE 35th Annual Power Electronics Specialists Conference*, **4**, 2521-2526.

[8] Kojima, Y., Koshio, M., Nakamura, S., Maejima, H., Fujioka, Y. and Goda, T. (2007) A Demonstration Project in Hachinohe: Micro-Grid with Private Distribution Line. *IEEE International Conference on System of Systems Engineering*, San Antonio, 16-18 April 2007, 1-6.

[9] Katiraei, F. and Iravani, M.R. (2005) Transients of a Micro-Grid System with Multiple Distributed Energy Resources. *International Conference on Power Systems Transients*, Montreal, 19-23 June 2005, Paper No. IPST05-080.

[10] Katiraei, F., Iravani, M.R. and Lehn, P.W. (2005) Micro-Grid Autonomous Operation during and Subsequent to Islanding Process. *IEEE Transaction on Power Delivery*, **20**, 248-257. http://dx.doi.org/10.1109/TPWRD.2004.835051

[11] Shahabi, M., Haghifam, M.R., Mohamadian, M. and Nabavi-Niaki, S.A. (2009) Microgrid Dynamic Performance Improvement Using a Doubly Fed Induction Wind Generator. *IEEE Transactions on Energy Conversion*, **24**, 137-145. http://dx.doi.org/10.1109/TEC.2008.2006556

[12] Majumder, R., Ghosh, A., Ledwich, G. and Zare, F. (2009) Load Sharing and Power Quality Enhanced Operation of a Distributed Micro-Grid. *IET Renewable Power Generation*, **3**, 109-119. http://dx.doi.org/10.1049/iet-rpg:20080001

[13] Kawasaki, K., Matsumura, M., Iwabu, K., Fujimuram, F. and Iima, T. (2009) Autonomous Dispersed Control System for Independent Micro-Grid. *Journal of Electrical Engineering, Japan*, **166**, 1121-1127.

[14] Li, X., Song, Y. and Han, S. (2008) Frequency Control in Micro-Grid Power System Combined with Electrolyzer System and Fuzzy PI Controller. *Journal of Power Sources*, **180**, 468-475.

http://dx.doi.org/10.1016/j.jpowsour.2008.01.092

[15] Mohamed, Y.A.I. and El-Saadany, E.F. (2008) Adaptive Decentralized Droop Controller to Preserve Power Sharing Stability of Paralleled Inverters in Distributed Generation Micro-Grids. *IEEE Transactions on Power Electronics*, **23**, 2806-2816. http://dx.doi.org/10.1109/TPEL.2008.2005100

[16] Maity, I. and Rao, S. (2010) Simulation and Pricing Mechanism Analysis of a Solar-Powered Electrical Micro-Grid. *IEEE Systems Journal*, **4**, 275-284. http://dx.doi.org/10.1109/JSYST.2010.2059110

[17] Dagdougui, H., Minciardi, R., Ouammi, A., Robba, M. and Sacile, R. (2010) A Dynamic Decision Model for the Real-Time Control of Hybrid Renewable Energy Production Systems. *IEEE Systems Journal*, **4**, 323-333. http://dx.doi.org/10.1109/JSYST.2010.2059150

[18] Mohod, S.W. and Aware, M.V. (2010) A STATCOM-Control Scheme for Grid Connected Wind Energy System for Power Quality Improvement. *IEEE Systems Journal*, **4**, 346-352.

[19] Barklund, E., Pogaku, N., Prodanovi´c, M., Aramburo, C.H. and Green, T.C. (2008) Energy Management in Autonomous Micro-Grid Using Stability-Constrained Droop Control of Inverters. *IEEE Transactions on Power Electronics*, **23**, 2346-2352. http://dx.doi.org/10.1109/TPEL.2008.2001910

[20] Ahshan, R., Iqbal, M.T., Mann, G.K.I. and Quaicoe, J.E. (2010) Micro-Grid System Based on Renewable Power Generation Units. *Proceedings of the* 23*rd Canadian Conference on Electrical and Computer Engineering*, Calgary, 2-5 May 2010, 1-4.

[21] Rashid, M.H. (2005) Power Electronics: Circuits, Devices and Applications. 3rd Edition, Prentice-Hall, New Delhi.

8

Control System Development and Test for the Operation of a Micro-Grid System—PART II

Razzaqul Ahshan, Mohammad Tariq Iqbal, George K. I. Mann, John E. Quaicoe

Faculty of Engineering and Applied Science, Memorial University of Newfoundland, St. John's, Canada
Email: razzaqul.ahshan@cna.nl.ca

Abstract

This paper presents experimental development and performance testing of an active power controller for stable and reliable operation of a micro-grid system. In order to achieve accurate and fast power balance in a micro-grid system that contains renewable energy sources, power in the system has to be regulated continuously. Such an objective can be achieved using droop based alternating current control technique. Because the droop characteristic employed into the developed controller initiates to determine the power deviation in the system which is continuously regulated by controlling the current flow into dump power resistors. The designed controller is tested and validated using a micro-grid prototype in the laboratory environment for stand-alone mode of operation under various operating conditions. The key development in the micro-grid prototype is the development of a wind turbine simulator. A dSPACE ds1104 DSP board is used to implement and interface the designed controller with the micro-grid system. The experimental investigation of the developed controller presents the significant capability to achieve continuous power balance in the micro-grid system, while it maintains stable and reliable operation of the system. Finally, the power quality of the isolated micro-grid system is presented and discussed under the operation of the developed controller.

Keywords

Active Power Controller, Micro-Grid Prototype, Renewable Energy Sources, Wind Turbine Simulator

1. Introduction

Increasing energy demand, environmental concern, de-regulated market policy and technology advancement are main causes to integrate distributed generation (DG) units into power system network. Such integration process leads the conventional power system to the new power system domain called micro-grid system. A micro-grid system generally composed of DG units with diverse generating capacity, availability, primary energy and operational costs. The application of such DG units in a micro-grid domain requires delivering stable, reliable and sustainable power generation, conversion, storage as well as distribution and exchanging during the entire operational time of the micro-grid system. In order to achieve such operational objectives, coordination and control in a micro-grid system is essential with the consideration of a series of factors such as availability of power generation, operational modes, voltage and frequency limitation, micro-grid load demands, renewable power penetration as well as energy cost [1] [2].

Operation and control of an isolated micro-grid system consisting of renewable power generations becomes critical because of the stochastic variation in primary energy sources such as wind speed, solar isolation etc. Such variation always leads to the change in active power generated by the DG units in the micro-grid system. As a result the active power generation in the micro-grid system varies randomly that results an active power mismatches between generation and consumption, which is stochastically varying in nature. Moreover, low inertia in the isolated micro-grid system as compared to a conventional power system along with stochastic behavior in active power mismatches can easily lead the micro-grid system instability. Hence, continuous and smooth active power regulation in an isolated micro-grid system has become an important issue to achieve stable and reliable operation of the system that comprises of renewable power generations such as wind turbines, photovoltaic panel etc.

Power management strategies in a micro-grid system proposed by R. H. Lasseter is carried out by controlling the power angle between the voltages at the voltage source converter and the micro-grid voltage for each micro-source [3], where micro-sources are assumed to be converter interfaced in the micro-grid domain. The successful laboratory test of a micro-grid system presented in [3] is performed and reported in [4]. The application of non-inverter interfaced DG units and storage systems in such micro-grid systems is also studied [4]. Power sharing technique among multiple DGs in a micro-grid system is also presented and discussed, where DGs are assumed as inverter interfaced with dispatchable energy sources [5]. In [5], one of the limitation of feeder flow control introduced in [3] is addressed to achieve improved power sharing among DGs are connected to a common feeder in series during transition between grid connected to island system. Such limitation is solved by employing a method of determining the new droop coefficients during the transition period. Droop based parallel inverter control concept is also adopted to achieve frequency regulation in a wind-solar power source based micro-grid system [6]. A DC/AC pulse width modulated inverter interfaced battery storage system that is used to maintain the micro-grid frequency is presented in [7]. The battery-inverter arrangement is controlled to regulate the magnitude and frequency of its voltage output during the isolated operation of the micro-grid system, according to the power-frequency droop curve. Frequency control in the New Energy and Industrial Technology Development Organization micro-grids is also performed by controlling the active power component of an inverter interfaced battery storage system using power versus frequency droop characteristics [8]. In [9] [10], frequency controller for a micro-grid system is presented, where droop methods and frequency restoration technique is employed to regulate active power component of the inverter interfaced DG units. The DG unit in such micro-grid system is assumed as a DC source. A primary frequency controller is proposed in [11] for variable speed wind energy conversion system in micro-grid operation. The control scheme determines the new operating point of the wind generator unit based on the frequency variation in the micro-grid system. The power at the new operating point provides current reference of the rotor side converter. Parallel inverters control concepts using droop methods is proposed for micro-generations control in a micro-grid system to maintain the active power balance between generation and consumption [12].

The design, analysis and simulation performance of an active power controller for a micro-grid system in an isolated condition has been presented in [13]. The presented scheme is designed to accommodate excess active power in the micro-grid system generated by variable speed wind power generation system in a continuous fashion under various operating conditions. The performance of the designed controller is tested through simulation and the simulation results were found satisfactory. However, in order to validate the theoretical investigation, a complete test bench examination is required to support theoretical and simulation analysis of the developed power regulation scheme. Hence, a small micro-grid prototype is constructed in the laboratory to verify the performance of the developed active power controller. The development of the micro-grid test setup and the performance test results of the active power regulation scheme under various operating conditions of the isolated micro-grid system is presented and discussed. The performance test results proof the ability of the developed controller to maintain stable and reliable operation of an isolated micro-grid system. It is to be noted that the scaled version micro-grid test setup is developed within the scope, ability and facilities available in the lab.

2. Active Power Controller

A micro-grid system located at Fermeuse, Newfoundland, Canada is considered in this research. The Fermeuse system consists of 9 WTs, a hydro generation unit (HGU), transmission lines, transformers, a storage unit and two loads. Also, the Fermeuse micro-grid system includes a link for utility-connected mode of operation. It is important to note that the combination of DG units in the micro-grid system is from renewable energy sources. **Figure 1** shows the conceptual diagram of an active power control strategy for the Fermeuse micro-grid with wind power generation system. The main objective of the designed active power control scheme is to maintain continuous and smooth active power balance between generations and micro-grid loads. The modeling and simulation study of the designed active power controller is presented in [13]. The controller performances are tested through simulation under various operating conditions and satisfactory simulation results were found. The major focus of this paper is to develop a micro-grid test bench to validate the performances of the active power controller those obtained through simulation study.

3. Experimental Test Setup

In order to investigate the experimental performances of the developed active power controller, a micro-grid test setup is developed in the laboratory environment. The layout of a micro-grid prototype is shown in **Figure 2** which consists of a wind turbine simulator (WTS), a hydro generation simulator (HGS), utility grid, loads and lines. The laboratory WTS is developed, where a separately-excited DC machine is simulated as wind turbine rotor that is coupled with a wound rotor induction generator. The developed WTS is 3-phase, 1.6 kW rated power at 208 V (L-L) terminal voltages. A 3-phase, 1.8 kVA, 4-pole, 208 V (L-L) synchronous generator is used as a HGS. The synchronous generator is coupled with a separately-excited DC machine that is operated at constant speed and constant field voltage. It is to be noted that the HGU in the micro-grid system is able to generate its rated power constantly due to the advantage of availability of the water potential. Such advantage is the main reason to operate the HGS at its rated power. A 3-phase, 208 V, 60 Hz supply is utilized as the utility grid. There are three small lengths of lines are used in the micro-grid prototype and their specifications are as follows:

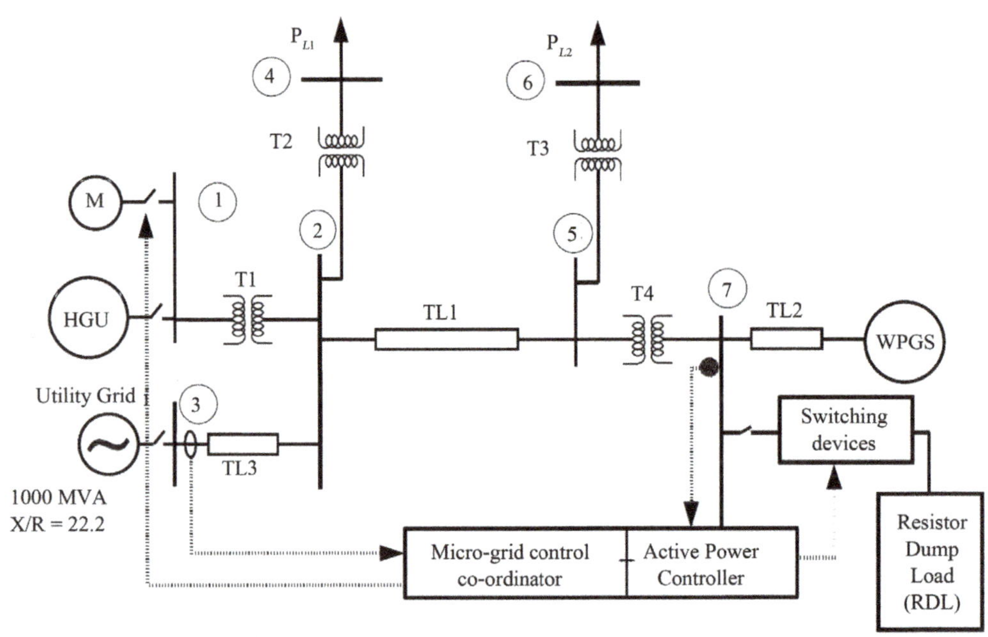

Figure 1. Micro-grid system located in Fermeuse, Newfoundland, Canada along with active power controller.

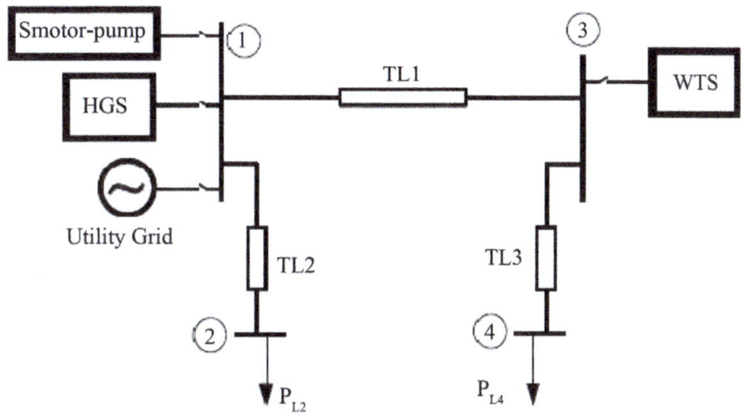

Figure 2. Experimental layout of a micro-grid test system.

- TL1 = 6 m and $Z1 = (0.0341 + j0.0413)\ \Omega$
- TL2 = 2 m and $Z2 = (0.0143 + j0.0212)\ \Omega$
- TL3 = 2.5 m and $Z3 = (0.0161 + j0.0271)\ \Omega$

Three different loads are employed in the micro-grid prototype. Two of such loads ($P_{L2} = 1.24$ kW and $P_{L4} = 1.24$ kW) represent the regular consumer loads in the micro-grid system. The other load $(S_{\text{motor-pump}} = 0.5\ \text{kVA})$ is utilized as load equivalent to a motor-pump set. The photograph of the micro-grid test setup is shown in **Figure 3**.

Wind Turbine Simulator

A separately excited DC motor is simulated to reflect the characteristics of a 1.6 kW wind turbine, which follows the theoretical rotational torque of the wind turbine rotor. Inertia of the DC motor and induction generator is considered as the inertia of the wind turbine rotor. Aerodynamic torque of the wind turbine is calculated from the rotational speed of the DC motor using wind turbine model equations. Such torque is considered as the reference torque of the WTS and can be expressed as:

$$T_{\text{ref}} = 0.5\rho A v_w^2 C_q(\lambda) r_t \tag{1}$$

where ρ is the air density, A is the area swept out by the turbine blades, v_w is the wind velocity, r_t is the radius of the rotating turbine, and C_q is the torque coefficient and can be expressed as:

$$C_q = -0.02812 + 0.038576\lambda - 0.0045912\lambda^2 + 0.0001489\lambda^3 \tag{2}$$

where, tip speed ratio, $\lambda = \frac{\omega_m r_t}{v_w}$, and ω_m is the angular velocity of the turbine rotor. The torque at the shaft of the DC motor is acquired by feedback mechanism. Based on the reference torque and the torque produced by the motor, the DC motor armature voltage is controlled using an analog input power controller (AIPC). The control signal for the AIPC is generated using a digital signal processing (DSP) board, where a recursive discrete PI controller is implemented along with the wind turbine model. Such controller ensures that the actual rotational speed of the motor is same as that of the wind turbine rotor. Moreover, the developed wind turbine simulator is featured with a maximum power controller. The rotor of the induction generator is connected to the external resistor through a 3-phase full bridge rectifier.

The experimental formation of the WTS is shown in **Figure 4**. The field of DC motor is excited by a single phase AC supply using a bridge rectifier and a series power resistor. To calculate the reference torque of the WTS, it is required to acquire the speed information of the DC motor shaft. A tacho-generator is used to collect the speed information. The tacho-generator output is an AC signal which passes through the instrumentation circuitry and finally acquired by the dSPACE ds1104 DSP board. A calibration equation between speed and corresponding voltage at the sensor output in (3) is employed in the DSP memory to obtain the rotational speed of the DC motor shaft. Torque sensor output is fed to a high gain instrumentation amplifier because of the low

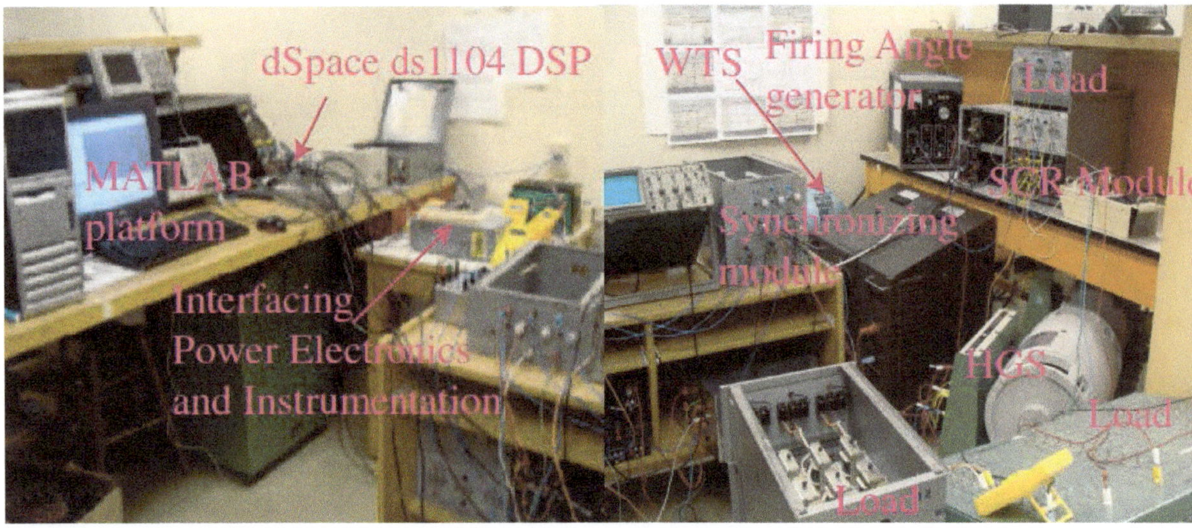

Figure 3. A photograph of the micro-grid test setup for the experimental testing.

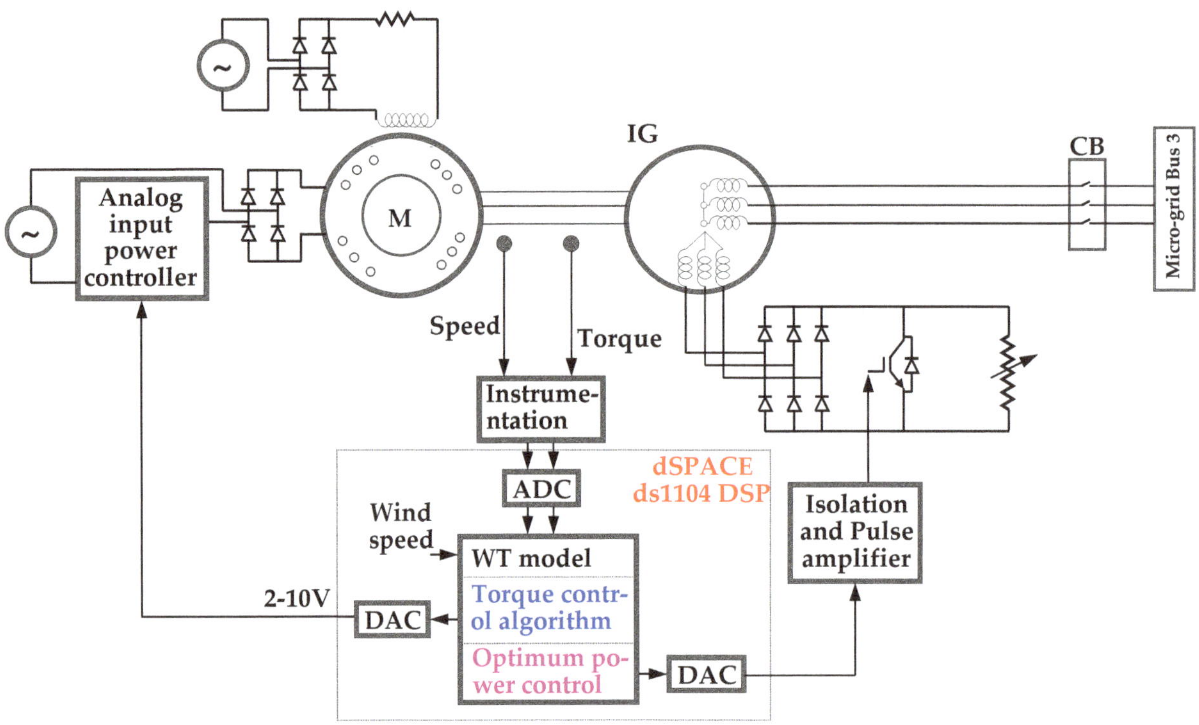

Figure 4. Laboratory arrangement for wind turbine simulator development.

output (mV) of the torque transducer. The output voltage is then sent to the DSP board through analog-to-digital converter. The torque calibration Equation in (4) is used to extract the torque information for the wind turbine model implementation.

$$N_{\text{shaft}} = 154.17 \times v_s + 1358.3 \tag{3}$$

$$T_{\text{shaft}} = \left(V_{\text{TS}} - 0.28\right) \times 1.4546 \tag{4}$$

Figure 5 shows the developed WTS test result which is obtained using four channel Tektronix TDS 3014B digital phosphor oscilloscope. Practically, wind turbine starts to rotate and produces power at the shaft, whenever wind flows through a rotor plane and as long as the flow remains. The flow may be either constant or suddenly changed. Therefore a developed WTS should be able to follow the theoretical torque of the rotor for a specific wind speed or a sudden change in the wind speed. To examine this performance, a wind pattern (**Figure 6**) follows a series of step variation was fed to the wind turbine model. The actual and the corresponding reference torque produced by the turbine shaft due to the variation in the wind speed are shown in **Figure 5**. The closeness of the DC motor actual torque and the reference torque is accurate and fast in every change in the wind speed. However, the simulator is not operating with maximum power extraction feature. **Figure 7** shows experimental performances of the WTS while maximum power extraction feature is included in the simulator. A tip speed ratio control technique is incorporated to maintain the operation of the simulator at the maximum tip speed ratio. It can be noticed from **Figure 5** that the WTS is operated at different tip speed ratios at different wind speeds that causes an error in tip speed ratio. On the contrary, in **Figure 7**, the WTS is operated at its maximum tip speed ratio which is a single value at different wind velocities that leads to close-to-zero error in the tip speed ratio.

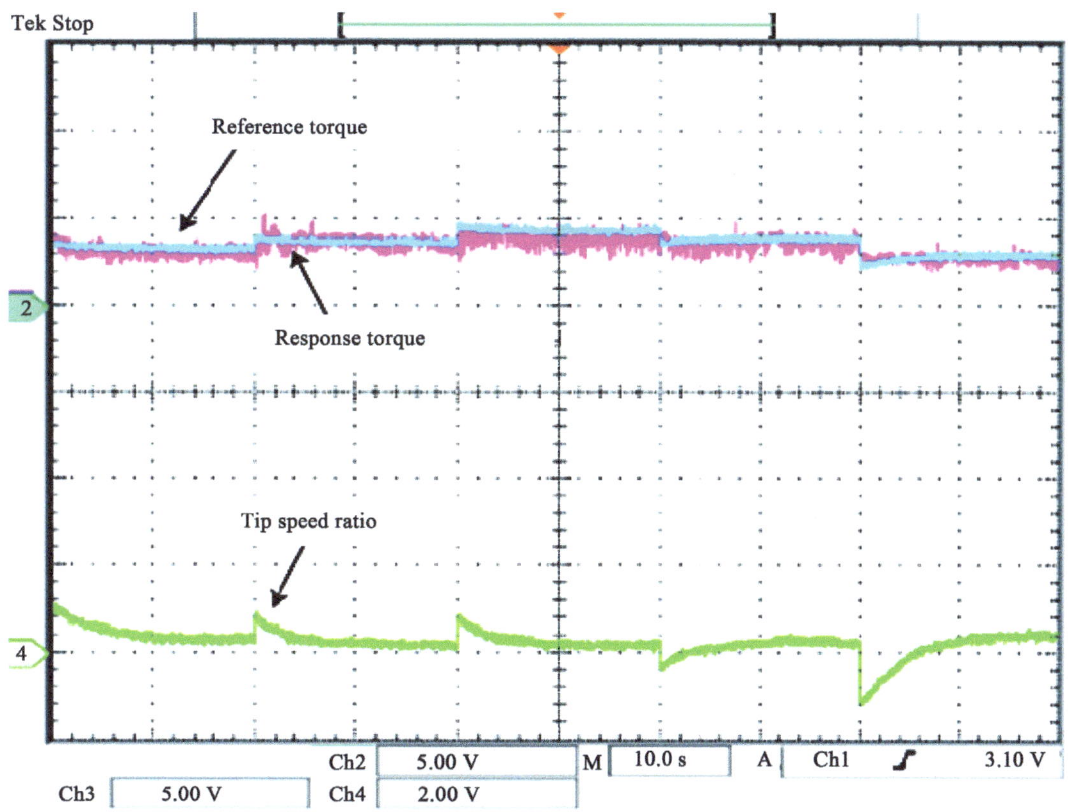

Figure 5. The experimental performances of the wind turbine simulator.

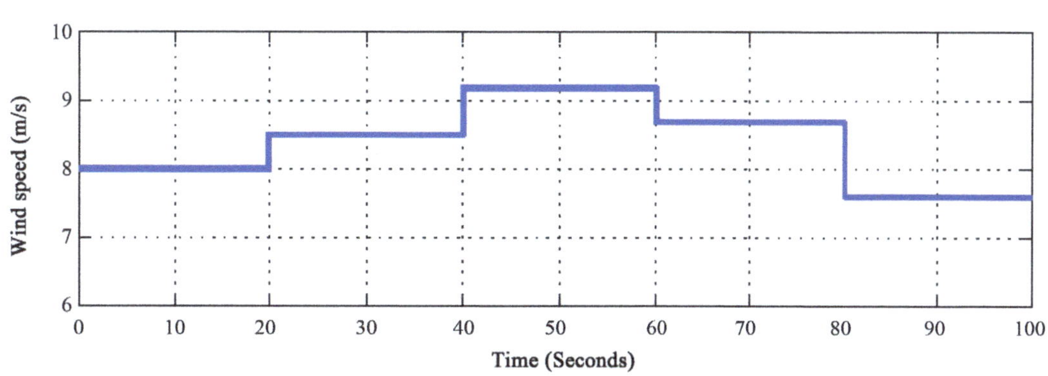

Figure 6. Wind speed variation fed into the wind turbine model.

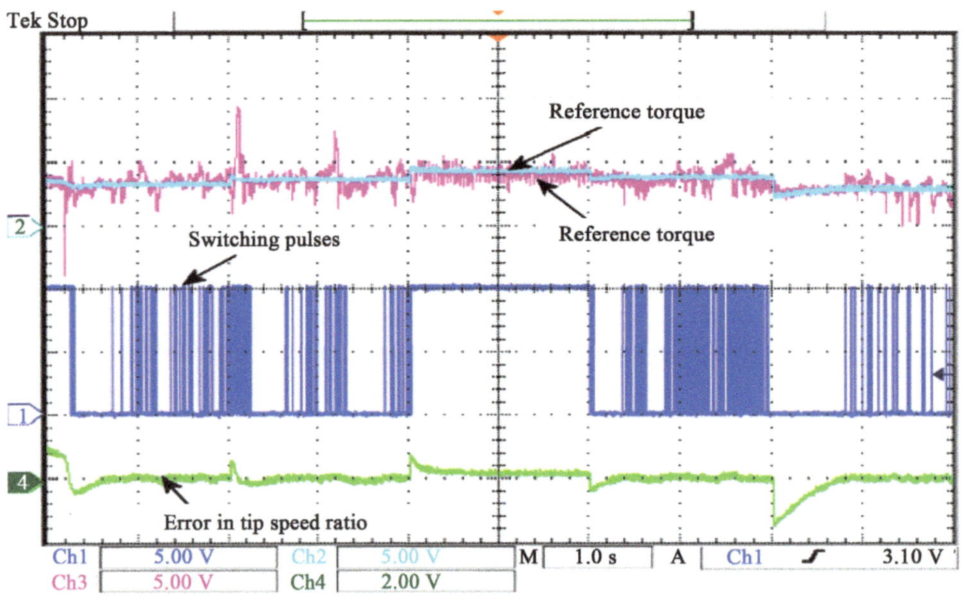

Figure 7. The experimental performances of the wind turbine simulator along with optimum tip speed ratio controller.

4. Real-Time Implementation of Active Power Controller

In order to investigate the real-time performances of the developed active power controller, an experimental test is carried out. The experimental arrangement of the active power controller along with the laboratory micro-grid prototype is shown in **Figure 8**. The controller arrangement comprises of digital signal processing (DSP) board, thyristor firing unit, thyristor switches, resistive dump load and the micro-grid test setup. The control variables and measured system information are the input to the DSP board where the controller algorithm is implemented. The detail hardware implementation of the developed controller is presented in **Figure 9**.

The micro-grid voltages at bus 3 is acquired using voltage isolator sensor which takes input ±300 V AC or DC and provide ±10 V AC or DC as the output. The output of the sensor is taken to the DSP board through the dSPACE built-in analog-to-digital converter (ADC). This signal is utilized as the input to the phase locked loop (PLL) in the active power controller algorithm shown in **Figure 10**. The output of the PLL is the system measured frequency f_{mes}, which is compared with the micro-grid system nominal frequency, f_{base}. Such frequency deviation calculates the power deviation in the system using the pre-defined droop coefficient. The power deviation in the system is utilized to derive the control signal, u_{cs}. The signal, u_{cs} is limited by the limiter to maintain the control signal in the range of input to the thyristor firing unit. Such signal is sent to the thyristor firing unit through dSPACE built-in digital-to-analog converter (DAC). The thyristor firing unit converts ±10 V to ±180 degrees. In addition, such unit also determines the synchronizing firing pulses for the thyristor switches based on the firing angle generated by the unit. In order to generate synchronizing firing pulses, the unit requires input reference voltage from the system, which is shown in **Figure 9**. An isolator and pulse amplifier is incorporated before sending pulses to the switching devices. Such pulses trigger the switches such that the current flowing into the dump load, i_{RDL} changes, and maintains the power balance between the generation and loads. Lab-volt thyristor firing unit, pulse amplifier and thyristor switch modules are used in this hardware arrangement. Furthermore, a 1.5 kW resistance box and a 0.5 kVA resistive-inductive load are emulated as resistive dump load (RDL) and load equivalent to a motor pump set, $S_{\text{motor-pump}}$, respectively.

5. Experimental Test Results

The performances of the designed active power controller for an isolated micro-grid with WPGS are tested in a laboratory prototype. The experimental results with and without the application of the developed controller for isolated micro-grid operations are presented to evaluate the performances of the controller.

Figure 11 shows the experimental results of the micro-grid system without the application of active power

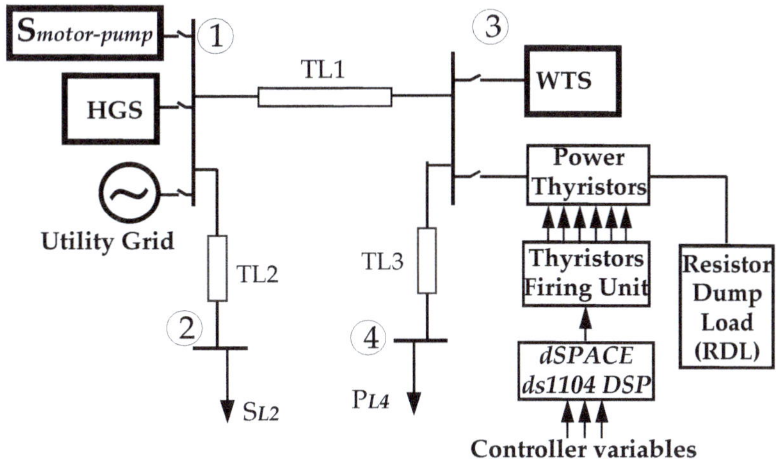

Figure 8. Block diagram of the active power controller implemented in the lab.

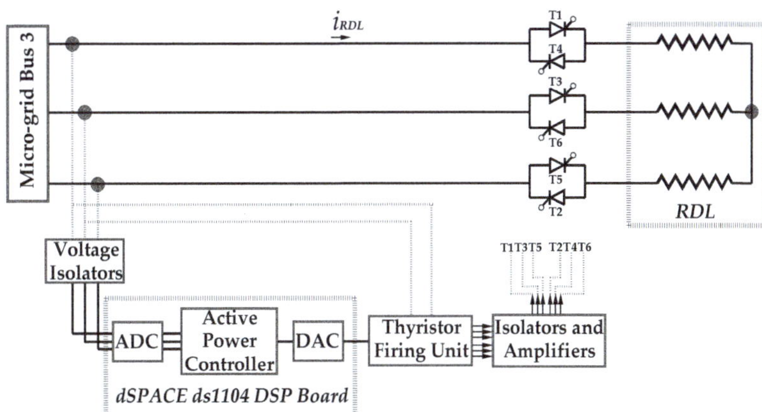

Figure 9. Real-time implementation layout of active power controller.

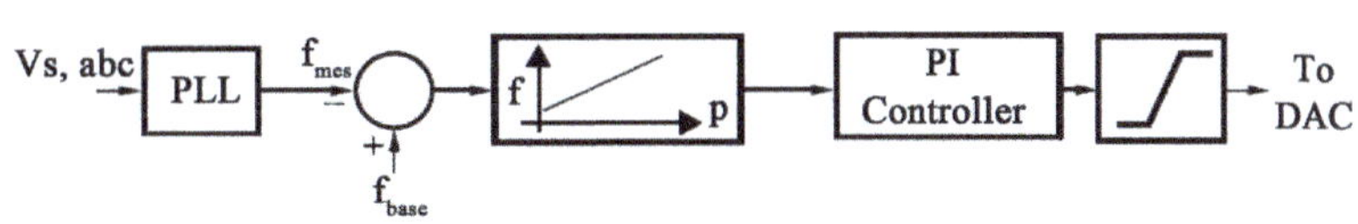

Figure 10. Active power control algorithm.

controller. **Figures 11(a)-(b)** show the micro-grid frequency, WTS simulator response, system voltage, firing angle and current flowing into the dump load. The zero value firing angle and also zero current flow into the dump load indicates the disengagement of the controller during the utility-grid connected micro-grid operation. Furthermore, the frequency and voltage of the micro-grid system are at their nominal value. However, the utility grid is disconnected and the performance results of the isolated micro-grid system are shown in **Figures 11(c)-(d)**. Since the controller is not activated yet, the micro-grid frequency is started to deviate from the nominal value because of the excess power in the micro-grid. Additionally, the excess power in the system causes the micro-grid voltage changes to higher value than the nominal value. The zero value of the firing angle and dump load current indicates the disengagement of the controller during this operation.

The experimental results of the isolated micro-grid system with the employment of the active power controller are shown in **Figure 12**. Micro-grid system frequency, firing angle, WTS response, micro-grid voltage, and current flowing into the dump load are presented in **Figures 12(a)-(b)**. At the time of utility grid disconnection, the micro-grid frequency and voltage started to deviate. However, due to the involvement of the active power controller, the micro-grid frequency and voltage become to their nominal values. The non-zero value of the firing

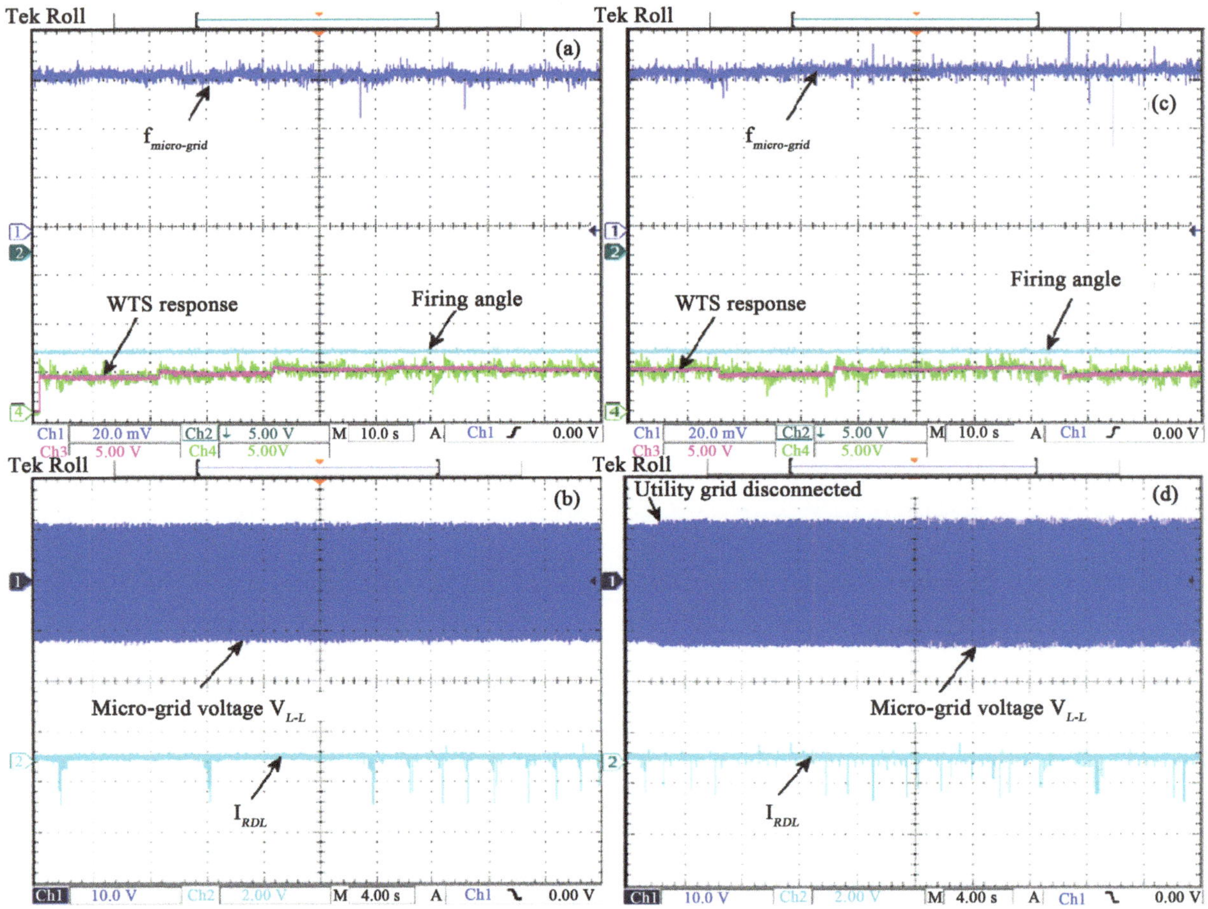

Figure 11. The experimental performances of the isolated micro-grid system without the application of the active power controller: (a) Frequency, firing angle and WTS response during grid connected operation; (b) Voltage and dump load current during grid connected operation; (c) Frequency, firing angle and WTS response at isolated condition; (d) Voltage and dump load current at isolated condition.

angle and current flowing into the dump load indicates the engagement of the controller into the system operation. Also, the WTS response indicates the variable speed operation of the WT generator in the micro-grid system.

Figures 12(c)-(d) present the performance results of the micro-grid system during a step increase in the micro-grid power demand. Since the power demand has increased in the micro-grid system, the controller adjusted the firing angle so that the current flowing into the dump load is accommodated to maintain the power balance into the micro-grid domain. Due to the accurate and fast action of the active power controller, the system frequency and voltage are maintained at their operating nominal value. Also the WT is operated at variable wind speed, which can be noticed from the WTS response.

The experimental performances of the micro-grid system are also presented in **Figures 12(e)-(f)**, while a step decrease in micro-grid load demand is applied. **Figure 12(e)** shows the micro-grid frequency, firing angle for the power semiconductor switches, and WTS response. Due to the decrease in load demand, the controller provides necessary actions to harmonize the firing angle in order to persevere the power balance between power generation and consumption in the micro-grid domain. This can be noticed from the firing angle response and the current flowing into the dump load. Because the firing angle is decreased to allow more current to flow into the dump load which lead to the fast and accurate balance in power generation and load demand. The fast and accurate balance between generation and load demand can be observed from the frequency and voltage responses shown in **Figures 12(e)-(f)**.

The power quality performances of the isolated micro-grid system is also investigated and presented in **Figure 13**. The results are presented for a short duration to observe the changes closely. **Figure 13(a)** shows the voltage and current flowing into the dump load, while the frequency spectrum for the voltage harmonics is

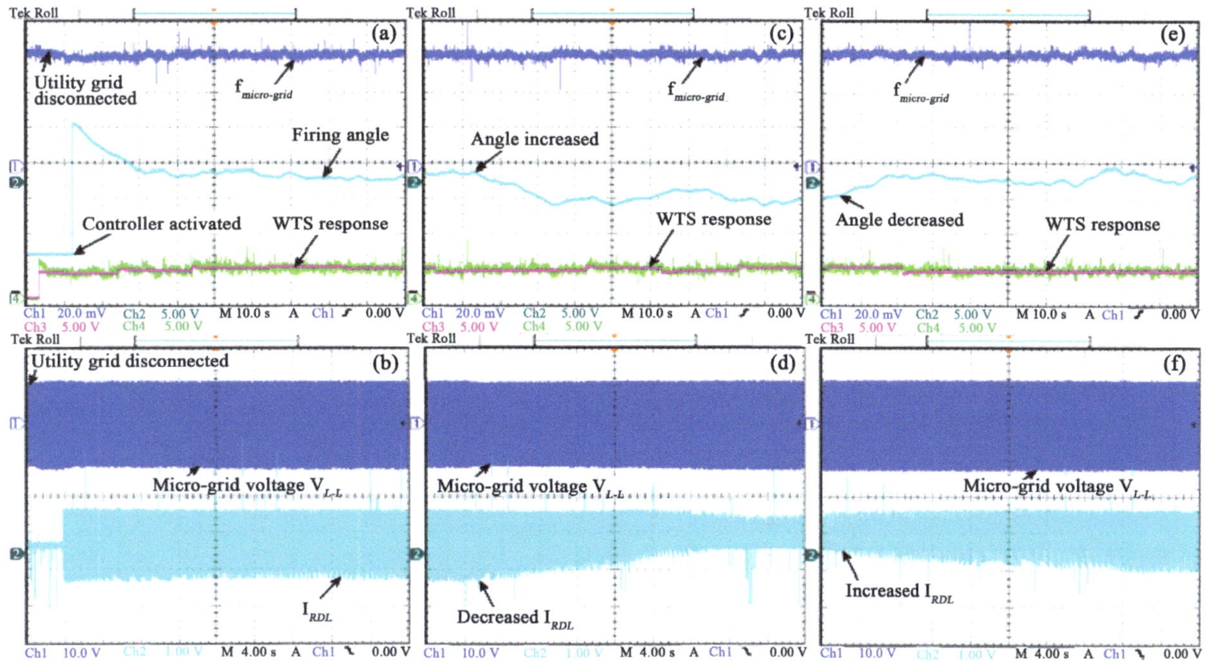

Figure 12. The experimental performances of the isolated micro-grid system with the application of the active power controller: (a) Frequency, firing angle and WTS response during isolated micro-grid operation; (b) Voltage and dump load current during isolated micro-grid operation; (c) Frequency, firing angle and WTS response at isolated condition with step increase in load demand; (d) Voltage and dump load current at isolated condition with step increase in load demand; (e) Frequency, firing angle and WTS response at isolated condition with step decrease in load demand; (f) Voltage and dump load current at isolated condition with step decrease in load demand.

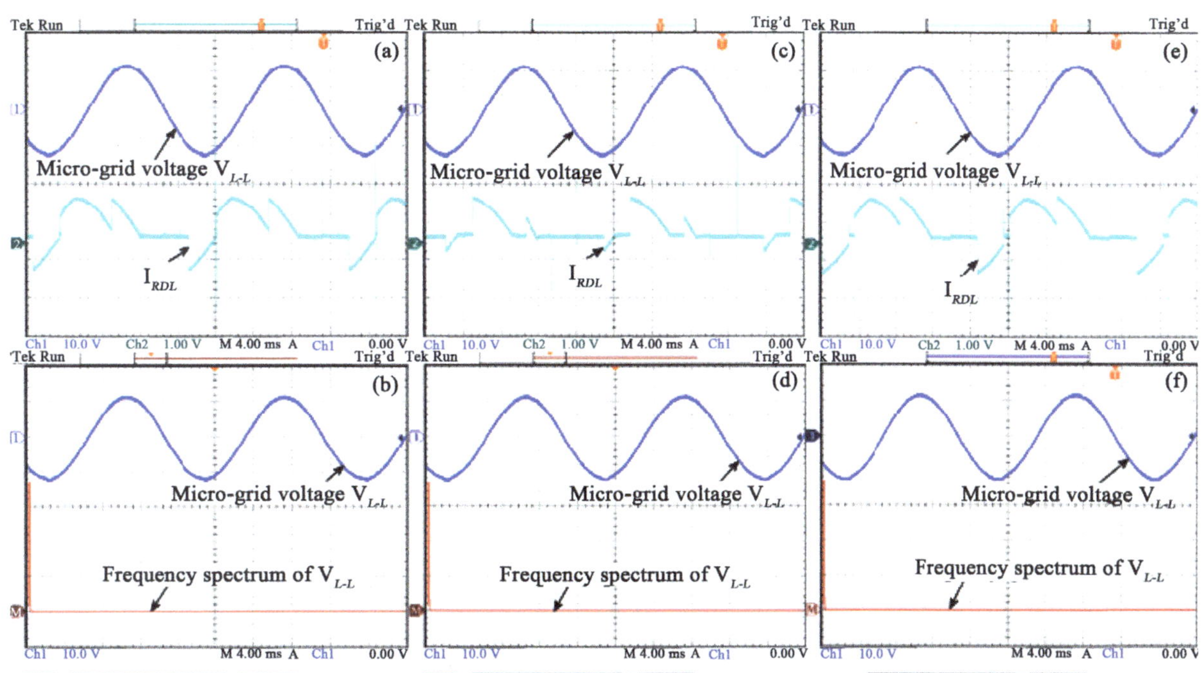

Figure 13. Power quality performances of the isolated micro-grid system with the application of the active power controller: (a) Voltage and current flowing into dump load; (b) Voltage and frequency spectrum of the voltage; (c) Voltage and current flowing into the dump load with a step increase in load demand; (d) Voltage and frequency spectrum of the voltage with a step increase in load demand; (e) Voltage and current flowing into the dump load with a step decrease in load demand; (f) Voltage and frequency spectrum of the voltage with a step decrease in load demand.

shown in **Figure 13(b)**. The frequency spectrum only shows the fundamental component of the micro-grid voltage. **Figure 13(c)** shows micro-grid voltage and current flowing into the dump load while micro-grid load demand is increased. Due to the increase in micro-grid load demand, current flowing into the dump load is also reduced. Moreover, the frequency spectrum during this operational condition also shows only the presence of fundamental component of the micro-grid voltage. The similar frequency spectrum is also noticed, while the micro-grid load demand is decreased. The presence of only fundamental frequency component in the micro-grid voltages ensures the high quality power in the isolated micro-grid system.

6. Conclusion

This paper has presented the development, implementation and performance testing of an active power controller for an isolated micro-grid system. The controller is developed based on alternating current control technique, where dump load resistor and power-frequency droop characteristics are employed. A micro-grid test bench is developed that comprises the key development of a wind turbine simulator. In addition, the other factors in the micro-grid test setup are considered based on the scope and availability of the equipments and facilities in the lab. Such test setup is utilized to test the performances of the developed active power controller. The developed controller is implemented using dSPACE ds1104 DSP board along with other equipments described in the paper in details. The performance results presented in the paper show the ability to maintain the fast and accurate power balance between the generation and load, while a stable and reliable operation was maintained during transition between the grid connected to the isolated micro-grid domain, and in the subsequent operation. Moreover, the power quality results also show the ability to generate and deliver quality power to the micro-grid load.

Acknowledgements

This work is supported by a research grant from the National Science and Engineering Research Council (NSERC) of Canada, the Atlantic Innovation Fund (AIF) Canada, and Memorial University of Newfoundland. The authors would like to acknowledge the utility company, Newfoundland Power, Canada for providing the system information and data.

References

[1] Mahat, P., Chen, Z. and Bak-Jensen, B. (2010) Under Frequency Load Shedding for an Islanded Distribution System with Distributed Generators. *IEEE Transaction on Power Delivery*, **25**, 911-918. http://dx.doi.org/10.1109/TPWRD.2009.2032327

[2] Katiraei, F. and Iravani, M. (2006) Power Management Strategies for a Micro-Grid With Multiple Distributed Generation Units. *IEEE Transaction on Power Systems*, **21**, 1821-1831. http://dx.doi.org/10.1109/TPWRS.2006.879260

[3] Lasseter, R.H. (2002) Micro-Grids. *IEEE Power Engineering Society Winter Meeting*, New York, January 2002, 305-308.

[4] Lasseter, R.H., Eto, J.H., Schenkman, B., Stevens, J., Vollkommer, H., Klapp, D., Linton, E., Hurtado, H. and Roy, J. (2011) CERTS Micro-Grid Laboratory Test Bed. *IEEE Transaction on Power Delivery*, **26**, 325-332. http://dx.doi.org/10.1109/TPWRD.2010.2051819

[5] Ahn, S., Park, J., Chung, I., Moon, S., Kang, S. and Nam, S. (2010) Power-Sharing Method of Multiple Distributed Generators Considering Control Modes and Configurations of a Micro-Grid. *IEEE Transaction on Power Delivery*, **25**, 2007-2016. http://dx.doi.org/10.1109/TPWRD.2010.2047736

[6] Ribeiro, L.A.deS., Saavedra, O.R., Lima, S.L.de. and Matos, J.G.de. (2011) Isolated Micro-Grids with Renewable Hybrid Generation: The Case of Lenis Island. *IEEE Transaction on Sustainable Energy*, **2**, 1-11.

[7] Georgakis, D., Papathanasiou, S.A., Hatziargyriou, N., Engler, A. and Hardt, Ch. (2004) Operation of a Prototype Micro-Grid System Based on Micro-Sources Equipped with Fast-Acting Power Electronics Interfaces. *Power Electronics Specialists Conference*, **4**, 2521-2526.

[8] Kojima, Y., Koshio, M., Nakamura, S., Maejima, H., Fujioka, Y. and Goda, T. (2007) A Demonstration Project in Hachinohe: Micro-Grid with Private Distribution Line. *IEEE International Conference on System of Systems Engineering*, San Antonio, 16-18 April 2007, 1-6.

[9] Katiraei, F. and Iravani, M.R. (2005) Transients of a Micro-Grid System with Multiple Distributed Energy Resources. *International Conference on Power Systems Transients* (*IPST*'05), Montreal, 19-23 June 2005, Paper No. IPST05-080.

[10] Katiraei, F., Iravani, M.R. and Lehn, P.W. (2005) Micro-Grid Autonomous Operation during and Subsequent to

Islanding Process. *IEEE Transaction on Power Delivery*, **20**, 248-257. http://dx.doi.org/10.1109/TPWRD.2004.835051

[11] Shahabi, M., Haghifam, M.R., Mohamadian, M. and Nabavi-Niaki, S.A. (**2009**) Microgrid Dynamic Performance Improvement Using a Doubly Fed Induction Wind Generator. *IEEE Transaction on Energy Conversion*, **24**, 137-145. http://dx.doi.org/10.1109/TEC.2008.2006556

[12] Majumder, R., Ghosh, A., Ledwich, G. and Zare, F. (**2009**) Load Sharing and Power Quality Enhanced Operation of a Distributed Micro-Grid. *ET Renewable Power Generation*, **3**, 109-119. http://dx.doi.org/10.1049/iet-rpg:20080001

[13] Ahshan, R., Iqbal, M.T., Mann, G.K.I. and Quaicoe, J.E. (2014) Control System Development and Test for the Operation of a Micro-Grid System—PART I. *Smart Grid and Renewable Energy Journal*, **5**, 291-301.

9

Modeling the Chain of Conversion for a PV System

Mohammed El Alami[1], Mohamed Habibi[1], Seddik Bri[2]

[1]Laboratory of Systems and Telecommunications Engineering Decision, Sciences Faculty, Ibn tofail University, Kenitra, Morocco
[2]Electrical Engineering Department, High School of Technology, Moulay Ismail University, Meknes, Morocco
Email: elalami.ensa@gmail.com

Abstract

In this work, we conduct a study of modeling and simulation of a system in the context of renewable energy in general, and solar system "photovoltaic" in particular; also, the optimizing system adapted by DC-DC converters static. Then the influence of temperature and irradiance on the optimal parameters (Power of MPP, VMPP, ...) of a solar system is analyzed in a way to operate a PV generator at its maximum power. This system includes a photovoltaic generator, "Boost & Buck", converter and a load. Modeling and simulation system (solar panel, DC-DC converter command and load) is obtained through Matlab-Simulink software.

Keywords

Photovoltaic Panel, DC-DC Converter, The Maximum Power Point, Fill Factor

1. Introduction

The greater part of the energy consumed currently comes from the use of fossil fuels like oil, coal, natural gas or nuclear power. These resources become increasingly rare while the energy demands of the world amounted continuously. Since this form of energy covers a large part of the current energy production, it is necessary to find an alternative solution to take the relay. The constraint is to use a source of economic power and low emission as the protection of the environment has become an important point. Photovoltaic solar energy is the direct conversion of part of the solar radiation into electrical energy. This energy conversion is affected throughout photovoltaic cell based on a physical phenomenon called the photovoltaic effect, which is to produce a current when the surface of the cell is exposed to light. Science is naturally interested in the "renewable" resources. Photovoltaic solar energy is increasingly used in various terrestrial applications such as lighting and telecommunications [1].

The use of solar energy will provide electricity to isolated sites of electrical networks and avoid the creation of new electrical lines which generally requires a heavy investment.

2. Modeling of the Photovoltaic System

The PV cell, also called solar cell, constitutes the basic element of the photovoltaic conversion. This is a semiconductor device that converts electric energy into light energy provided by an inexhaustible source of energy and the Sun. It exploits the properties of semiconductor materials used in the electronics industry: diodes, transistors and integrated circuits. **Figure 1** illustrates a typical PV cell where its formation is detailed. The performance of energy efficiency achieved industrially are 13% to 14% for the cells based on monocrystalline silicon, with 11% to 12% of poly crystalline silicon and finally 7% - 8% for the amorphous silicon by thin films [2]. The photopile or solar cell is the basic element of a photovoltaic generator, and the equivalent circuit of a PV cell is shown in **Figure 2**. It includes a current source, a diode, a series resistance and a shunt resistance.

The photovoltaic effect is manifested when a photon is absorbed in a semiconductor material composed of doped p (positive) and n (negative), referred to as p-n junction (or n-p). Under the effect of doping, an electric field is present in the material permanently (like a magnet which has a permanent magnetic field) [3]. When an incident photon (grain of light) interacts with the electrons in the material, it transfers its energy to the electron **hv** which finds herself liberated from its valence band and therefore suffers the intrinsic electric field. Under the influence of this field, electron migrates toward the top surface leaving a hole that migrates in the reverse direction. Electrodes placed on the upper and lower surfaces allow the electrons to collect them and to make electrical work to join the anterior surface of the hole. A cell is often modeled by the electrical equivalent diagram it is he that is shown in **Figure 2**, the photovoltaic generator is modeled by the following equations [4]:

$$I = I_{ph} - I_s\left[\exp\left(q\frac{V + I.R_s}{A.K.T}\right) - 1\right] - \frac{V + I.R_s}{R_{sh}} \quad (1)$$

$$P = V * \left[I_{ph} - I_s\left[\exp\left(q\frac{V + I.R_s}{A.K.T}\right) - 1\right] - \frac{V + I.R_s}{R_{sh}}\right] \quad (2)$$

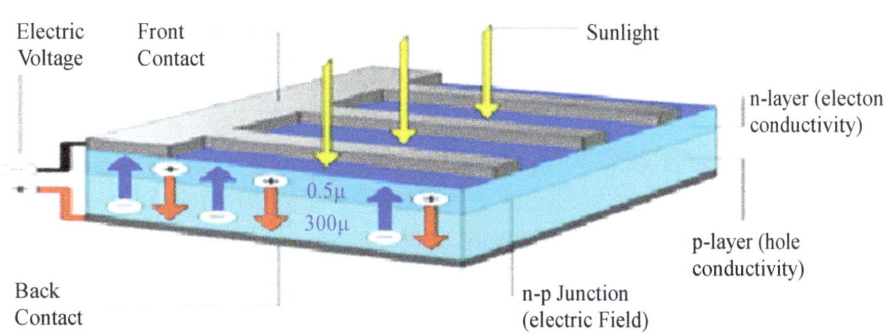

Figure 1. Schématic of a photovoltaic cell.

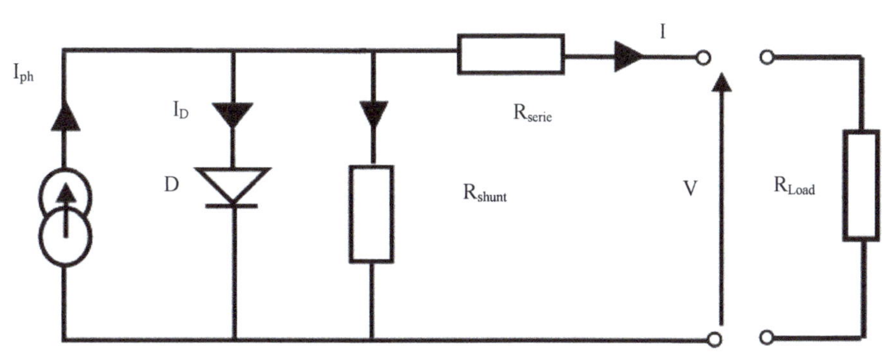

Figure 2. Equivalent circuit of a PV cell.

with $I_{ph}(\mathrm{A})$ is currently produced by the cells; $I_s(\mathrm{A})$ is the saturation current; $q = 1.6\times10^{-19}$ C load electron; $K = 1.38\times10^{-23}$ J/K Boltzmann constant; T (°C) is the operating temperature of the cell; A is a quality factor of the diode, normally between 1 and 2; $R_s(\Omega)$ and $R_{sh}(\Omega)$ are series and parallel resistors. After the determination of the various parameters of the equivalent circuit, it is possible to solve the equation of characteristic $I = f$ (V), and also to determine these characteristics (current I (A), power P (W)) by numerical method.

3. The Influence of Meteorological Conditions on the Performance of a PV System

The characteristic curve of a PV cell represents the variation of the current which it produces as a function of the voltage at its terminals, since the short circuit (zero voltage corresponding to the maximum current product) till the open circuit (no current to a maximum voltage across the cell) [5]. In this case, we chose a simple model, requiring only the parameters given by the manufacturer to evoke the influence of meteorological parameters on the maximum power point, the I-V characteristic of this model is shown in **Figure 3**, and for influences of the conditions meteorological illustrated in **Figure 4** and **Figure 5**: The electrical characteristics of the photovoltaic module that we used were obtained by numerical simulation of the model by Matlab-Simulink software.

The maximum power $P_{\max}$ suitable for MPP is very susceptible to light as illustrated in **Figure 4**: when the illumination decreases from 100 W/m^2, power $P_{\max}$ decreases by 10%. With cons, the maximum power $(P_{\max})$ decreases little with temperature particularly strong illumination. When the temperature increases from 10°C to about the ambient temperature and illumination of 1000 W/m^2 power $P_{\max}$ decreased by 4%.

The optimum voltage V_{mpp} suitable for MPP varies very little with the illumination and decreases slightly with temperature as illustrated in **Figure 5**. For illumination of 1000 W/m^2 and around ambient temperature (25°C): when the temperature varies from 10°C, the voltage ranges from 4.2%. These results show that the voltage varies relatively little in the course of the day. Also, it is considered as a first approximation that the optimal operation of PV generator substantially corresponds to operation at constant optimum voltage V_{mpp}.

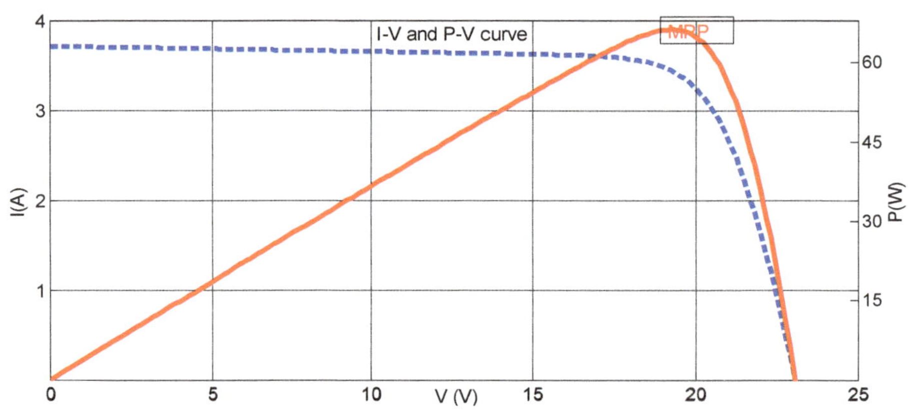

Figure 3. Characteristic current I(V) and Power P(V) of PV Cell.

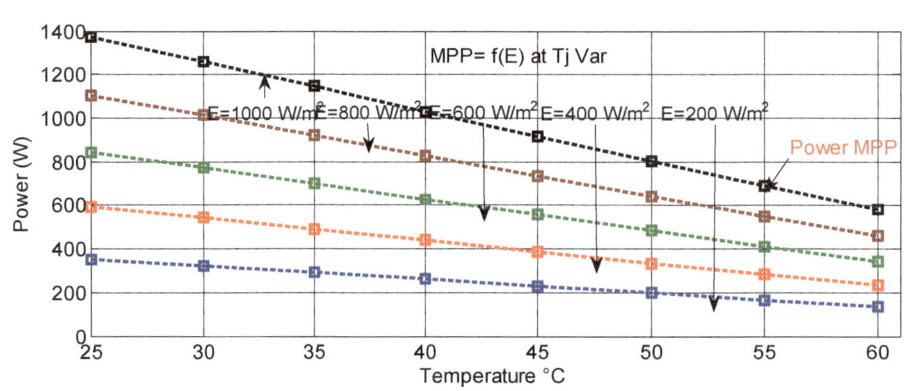

Figure 4. Influence of temperature and illumination on MPP of PV module.

The coefficient of maximum power point MPP and the optimal voltage linking the measured quantities and optimal variables is not always accurately fixed and little vary with time as a function of aging. It depends in particular on temperature and sunlight as illustrated in **Figure 4** and **Figure 5**.

4. Adaptation through of DC-DC Converter

The chopper is a DC-DC converter for converting a DC power to a given level of voltage (or current) in a continuous energy to another level of voltage (or current). Use turns out necessary to store solar energy in batteries or to feed a continuous load equivalent circuit of the chopper is represented by **Figure 6** and **Figure 7** below. Two topologies of bass conversion circuit (DC/DC), choppers are DC-DC static converters for generating a variable DC voltage source from a source of fixed voltage. The chopper consists of capacitors, inductors and switches. All of these devices in the ideal case do not consume power, this is the reason why the choppers have good yields. Usually the switch is a MOSFET transistor is a semiconductor device mode (block-saturate), for our work we throw a glance on Boost and Buck converters.

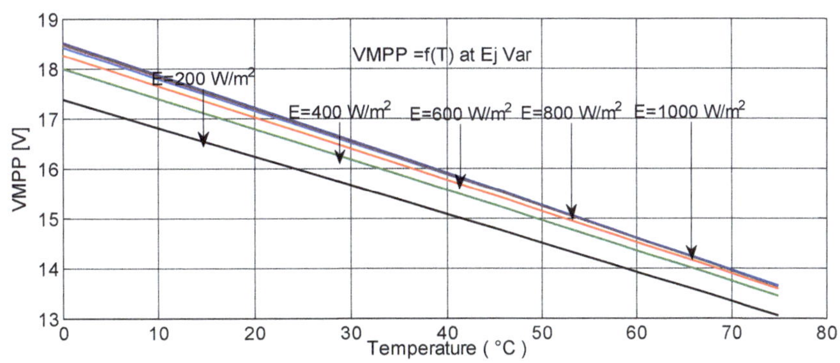

Figure 5. Variation of voltage VMPP according to on the temperature *T* and illumination E.

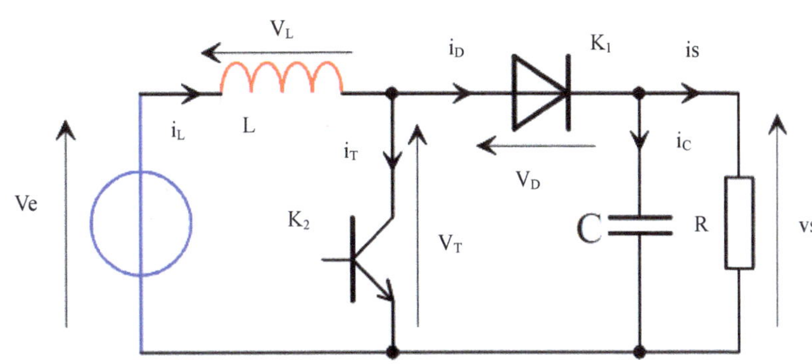

Figure 6. Boost converter.

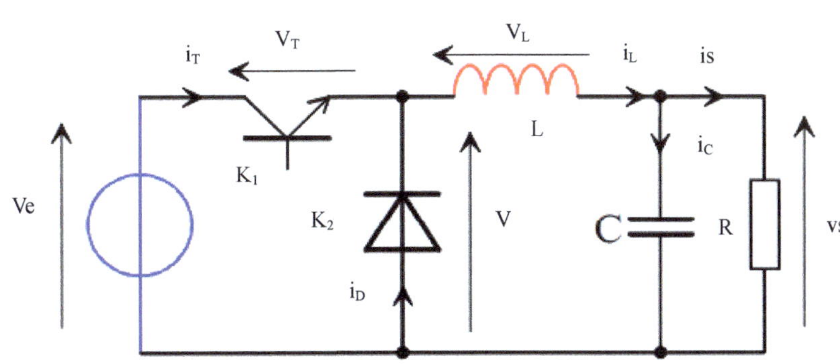

Figure 7. BUCK converter.

4.1. Buck & Boost Converters

The boost converter is known by the name of voltage booster that can be represented by the circuit of **Figure 6**. The boost converter is characterized by the fact that the output voltage is higher than the input voltage [6]. This is a direct DC-DC converter. The input source is DC-type (inductor in series with a voltage source) and output load is of the type voltage (capacitor in parallel with the resistive load). The switch K1 can be replaced by a transistor since the current is always positive and that the switching should be commanded (the Blocking and priming). A buck converter, or series chopper, is a device that converts a DC voltage into another DC voltage of lower value. The buck converter as illustrated in **Figure 7** is used in applications where you want to have an output voltage lower than the input voltage [7].

Functioning

For the BOOST, the switch K1 is closed during the αT fraction of the switching period T. The input source provides power to the load R through the inductor L. When transistor is blocked, the diode K2 ensures continuity of the current in the inductor. The energy stored in the inductor is then discharged into the capacitor and the resistance of the load. In steady state, the medium value of the terminal voltage of the inductor is zero, which imposes the following equation:

$$V_e = (1-\alpha)\cdot V_s \tag{3}$$

$$V_s = V_e \cdot \frac{1}{(1-\alpha)} \tag{4}$$

From Equation (3), the conversion ratio of boost chopper is given in the following form:

$$M(\alpha) = \frac{V_s}{V_e} = \frac{1}{(1-\alpha)} \tag{5}$$

And for the BUCK converter, the switch K1 is closed during the fraction T of the switching period T. The input source provides an power to the load R through the inductance L. When the Blocking the transistor, the diode K2 ensures continuity of the current in the inductor. The energy stored in the inductor is then discharged into the capacitor and the load resistor. In steady state, the medium value of the terminal voltage of the inductor is zero. The output voltage is given by the following equation:

$$V_s = V - V_L = \alpha \cdot V_e \tag{6}$$

From Equation (6), the conversion ratio of buck chopper is given in the following form:

$$M(\alpha) = \frac{V_s}{V_e} = \alpha \tag{7}$$

After resolution of equations by Matlab can be shown the **Figure 8** illustrates the linear relation between the conversion report and the duty ratio and also to **Figure 9** shows the evolution of conversion by the duty cycle ratio.

This is Equation (6) which shows that the buck converter is a step-down voltage because the output voltage of the converter equal to the input voltage by a coefficient which varies in the range [0,1] as illustrated in **Figure 8**. And **Figure 9**, in this case it is the Equation (3) shows that the boost converter is a voltage booster.

4.2. Simulation of the Chopper by Matlab Simulink

4.2.1. Chopper Series

Four demonstrate the role of converters Buck we used Matlab simulation software and we are taking ($E = 300$ V, $L = 1$ H, $R = 30\ \Omega$). As illustrated in the **Figure 10** with the components used and the interconnection between them, by Matlab Simulink.

After having the voltage input 300 V as shown in the **Figure 11**, we found the output value 150 V, as illustrated in **Figure 12**, different and proportional to the duty cycle alpha $\alpha = 0.5$, we conclude that the role of boost for a solar system.

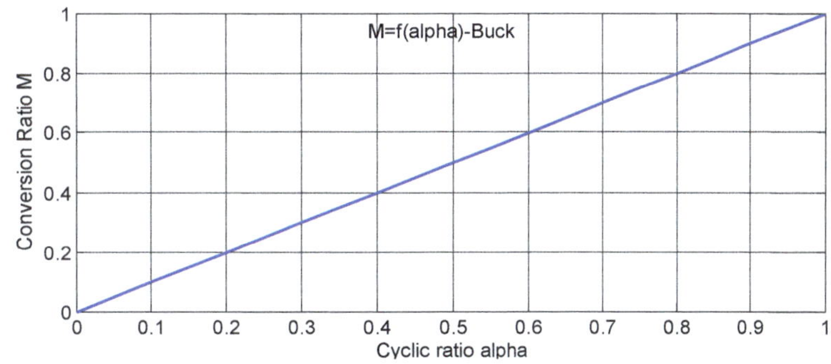

Figure 8. The conversion ratio buck according α.

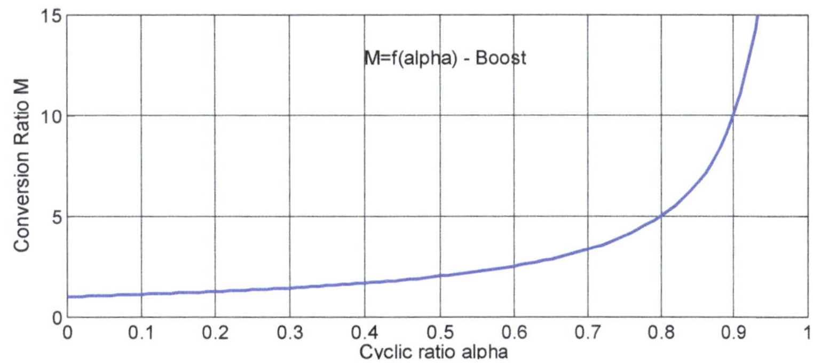

Figure 9. The conversion ratio boost according α.

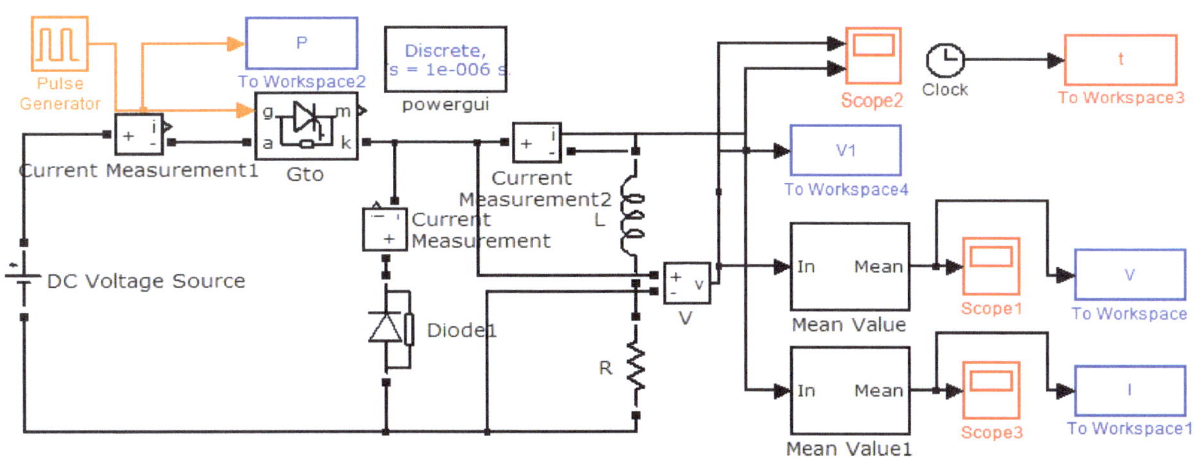

Figure 10. Block diagram of simulation of Buck converter.

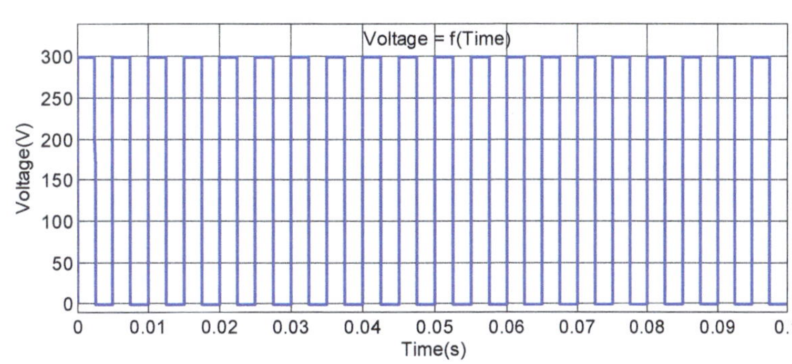

Figure 11. Input voltage to the converter as a function of time.

And to give several values of voltage as a function of duty cycle which shows the proportionality between the duty cycle and the input voltage and also to express the role of a Buck in **Table 1** below shows the variation of the average value of the voltage and current versus duty cycle.

The output voltage of the chopper is not continuous but always be positive. When the time is quite low (frequency 100 - 1000 Hz) the load does not "see" the not slots, but the medium value of the voltage. We notice whatever the nature of the charge, we have $V_s = 0 \le \alpha U \le U$ chopper series is voltage step-down (buck chopper).

4.2.2. Parallel Chopper

To show the role of converters boost we used Matlab software for the simulation and we take ($E = 24$ V, $L = 1\times10^{-3}$ H, $R = 50\ \Omega$) as illustrated in **Figure 13**.

After the results in **Figure 14** it is noted that the output voltage $V_s \sim 47$ V is not the same as the input $V_e \sim 24$ V for a duty cycle $\alpha = 0.5$ which expresses the role of boost converter. It is therefore important to play on the cyclic ratio of a share, and on the other hand, although the inductor L dimension, to get a smooth current to the capacitor C and having a desired output voltage.

5. Performance of a Chain Photovoltaic

5.1. Energy Efficiency

We recall the definitions of the different yields and measurement conditions thereof. As a result, the overall yield of the conversion chain reflects all sources of losses over the entire PV chain.

The power received by a panel of area A (m^2) is equal to $G \cdot A_{\text{eff}}$.

We will take as a definition reflecting performance the maximum capacity of a GPV and quality of the electron-photon conversion of a solar panel noted η_{pv}, the performance defined by the equation:

$$\eta_{pv} = \frac{P_{\max}}{A_{\text{eff}} \cdot G} \tag{8}$$

$P_{\max}$: The maximum power delivered by the PV generator.

G: irradiance (W/m^2) is defined as the amount of solar electromagnetic energy incident on a surface per unit of time and area.

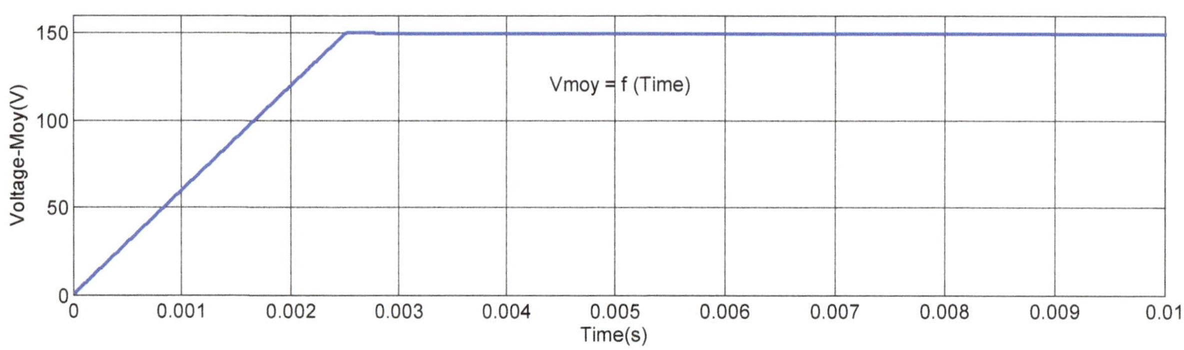

Figure 12. The medium value of the output voltage.

Table 1. The variation of the average value of the voltage and current according α.

α	Duty Cycle α.		
	0.20 0.30	**0.40**	**0.50**
U_{eMoy}	**62 89.1**	**118**	**148.5**
I_{eMoy}	**1.975 2.970**	**4.692**	**4.694**

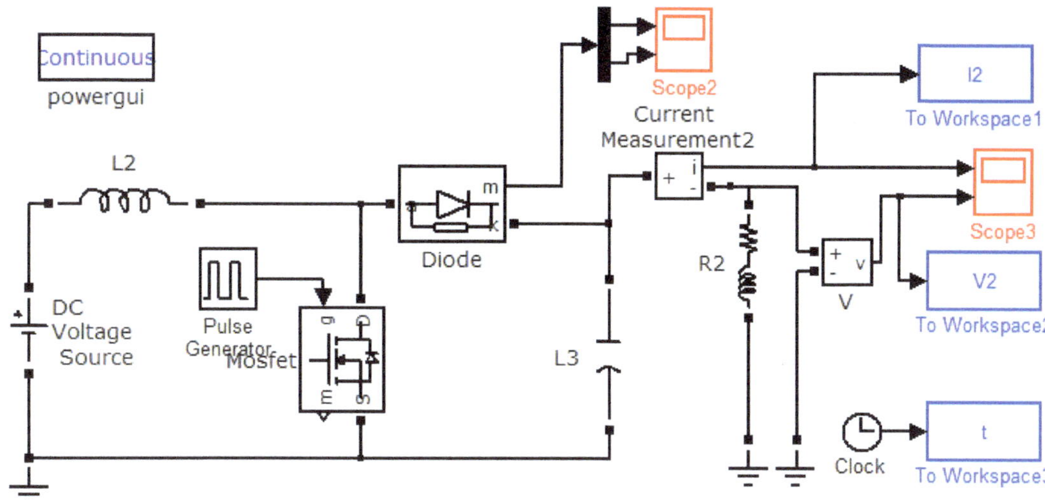

Figure 13. Block diagram of simulation of boost converter.

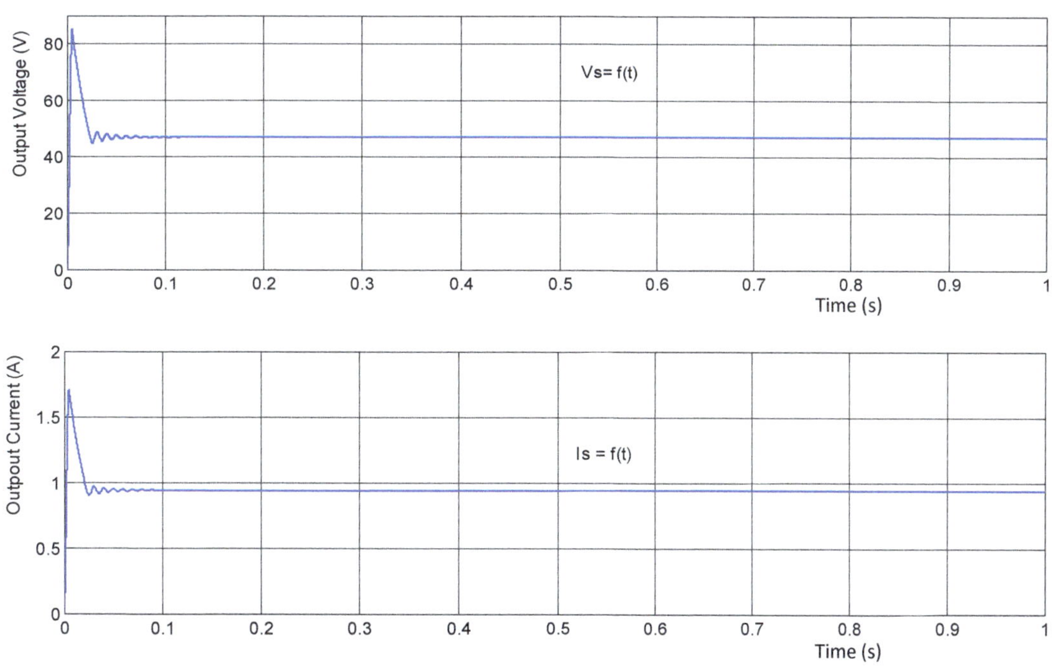

Figure 14. Curve of voltage and current output of the boost converter.

A_{eff} : Represents the area of the corresponding panel the active part and capable of able to perform of photovoltaic conversion and not the total area occupied by the solar panel.

The power P effectively delivered by an PV generator depends more on the command used in the converter (MPPT voltage control, direct connection, ...). Performance of the operating point which place we note η_{MPPT} can measure the effectiveness of the command that is responsible for the control of static converter so that the PV module furnish maximum power.

$$\eta_{MPPT} = \frac{P}{P_{max}} = \frac{P}{I_{opt} \cdot V_{opt}} \quad (9)$$

P is the power available at the terminals of GPV.

V_{opt} and I_{opt} are respectively the voltage and current optimal of generator PV.

And performance the converter noted η_{conv} , which is defined by (10), noting the power P_{out} delivered at the output the converter

$$\eta_{\text{conv}} = \frac{P_{\text{out}}}{P} \tag{10}$$

Unfortunately, more the converter is complex, plus its own performance noted η_{conv} is low. For the user, the only important overall efficiency η of the entire chain [8] [9]:

$$\eta = \eta_{pv} \cdot \eta_{\text{MPPT}} \cdot \eta_{\text{conv}} \tag{11}$$

For these variables must be optimized solar system by intelligent controls (embedded system) to increase the efficiency and effectiveness of our system.

5.2. Form Factor

Called form factor FF, also called factor curve or fill factor, the ratio between the maximum power output $P_{\max}$ cell (I_{opt}, V_{opt}) and the product of the short-circuit current I_{sc} by the open circuit voltage Voc (that is to say the maximum power an ideal cell). The shape factor indicates the quality of the cell [10]; as it approaches the unit cell is more efficient, it is of the order of 0.7 for efficient cells; and decreases with temperature. It reflects the influence of losses by both parasitic resistances and is defined by:

$$\text{FF} = P_{\max} / \left(V_{co} \cdot I_{cc}\right) \tag{12}$$

$$\text{Consequently}: \text{FF} = V_{\text{opt}} \cdot I_{\text{opt}} / \left(V_{co} \cdot I_{cc}\right) \tag{13}$$

V_{co} (V): Open circuit voltage.

I_{cc} (A): The current of a short circuit.

For our system after the simulation of the influence of meteorological parameters on FF and performance are illustrated in the following **Figure 15** and **Figure 16**.

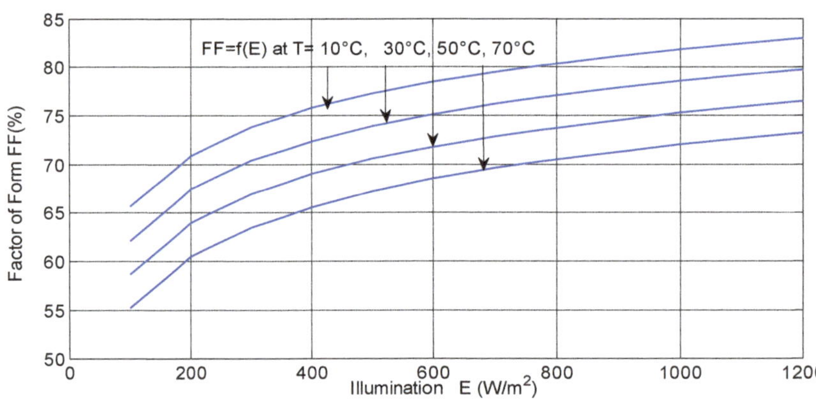

Figure 15. Variation of form factor depending of *E* and *T*.

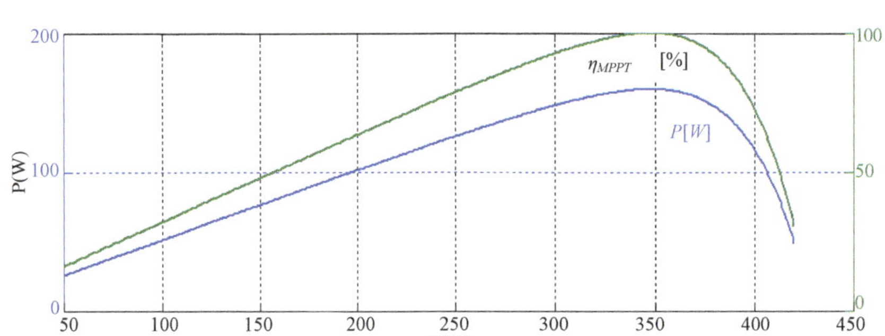

Figure 16. Power available and yield of command for different operating points.

When illumination increases the fill factor increases. The fill factor diminishes as the temperature of the device increased. The decrease in FF with the working temperature of the device outweighs the slight increase in the short circuit current. Finally for having a clearer idea of the origins of losses and yields of each part of a chain of power for a solar system, we presented the definitions of performance and form factor to extract performance against weather conditions. In all cases it is necessary to continuously monitor the output voltage of the converter, that is to say either the voltage of the battery charger or the DC bus voltage to an inverter. In addition it is also useful to monitor the battery charge, the battery current measure, then it is easier to maximize the power output of the generator set and the converter because it is this power that is useful and not the extracted panel.

Therefore to increase the efficiency of a solar system must integrate MPPT controls for example knowing that there are different types of algorithms performing the search of maximum power point MPP [11].

6. Conclusion

The performance of the PV generator is evaluated according to the standard conditions (STC): irradiance 1000 W/m^2, and Temperature 25°C. The performance of a PV solar system highly depends on the weather conditions, such as solar radiation, temperature and light, so does the performance of the PV generator degrade with increasing temperature, the decrease in intensity and the illumination load variations. To provide energy continuously throughout the year, a PV system must be properly sized and led by intelligent command to monitor MPPT for extracting the maximum energy. And for the simulation of DC-DC converters, we have shown by simulation that the average value of the output voltage can be adjusted by varying the duty cycle value. In perspective, this work will be continued with a practical realization of MPPT command based microcontroller and a follower by an Arduino board for increasing the efficiency of a solar system to extract the maximum energy.

References

[1] El Alami, M., Habibi, M. and Bri, S. (2013) Parameters Efficiency of Solar Energy. *International Journal of Emerging Trends in Engineering and Development—IJETED*, **4**, 175-186.

[2] Villalva, M.G., Gazoli, J.R. and Filho, E.R. (2009) Comprehensive Approach to Modeling and Simulation of Photovoltaic Arrays. *IEEE Transactions on Power Electronics*, **24**, 1198-1208.

[3] Hoque, A.K. and Wahid, A. (2000) New Mathematical Model of a Photovoltaic Generator. *Journal of Electrical Engineering*, **EE 28**, 8p.

[4] El Alami, M., Habibi, M. and Bri, S. (2014) Modelling of Solar System by Newton Raphson Method, MPPT Algorithm. *European Journal of Scientific Research, EJSR*, **120**, 264-276.

[5] Yu, G.J., Jung, Y.S., Choi, J.Y., Choy, I., Song, J.H. and Kim, G.S. (2002) A Novel Two-Mode MPPT Control Algorithm Based on Comparative Study of Existing Algorithms. *Conference Record of the Twenty-Ninth IEEE Photovoltaic Specialists Conference*, 19-24 May 2002, 1531-1534.

[6] Zhou, C., Ridley, R.B. and Lee, F.C. (1990) Design and Analysis of Hysteretic Boost Power Factor Circuit. *Annual IEEE Power Electronics Specialists Conference*, San Antonio, 800-801. http://dx.doi.org/10.1109/PESC.1990.131271

[7] Prasad, A.R., Ziogas, P.D. and Manias, S. (1992) A New Active Power Factor Correction Method for Single-Phase Buck-Boost AC-DC Converter. *Conference Proceedings of Seventh Annual Applied Power Electronics Conference and Exposition*, Boston, 23-27 February 1992, 814-820.

[8] Hohm, D.P. and Ropp, M.E. (2003) Comparative Study of Maximum Power Point Tracking Algorithms. *Progress in Photovoltaics: Research and Applications*, **11**, 47-62. http://dx.doi.org/10.1002/pip.459

[9] Jain, S. and Agarwal, V. (2007) Comparison of the Performance of Maximum Power Point Tracking Schemes Applied to Single-Stage Grid-Connected Photovoltaic Systems. *IET Electric Power Applications*, **1**, 753-762. http://dx.doi.org/10.1049/iet-epa:20060475

[10] Nielson, G.N., Okandan, M., Resnick, P., Cruz-Campa, J.L., Pluym, T., Clews, P.J., Steenbergen, E. and Gupta, V.P. (2009) Microscale C-Si (C)PV Cells for Low-Cost Power. 34*th IEEE Photovoltaic Specialists Conference* (*PVSC*), Philadelphia, 7-12 June 2009, 1816-1821.

[11] Femia, N., Pétrone, G., Spagnuolo, G. and Vitelli, M. (2005) Optimization of Perturb and Observe Maximum Power Point Tracking Method. *IEEE Transactions on Power Electronics*, **20**, 963-973.

10

Isolated MicroGrid's Voltage and Frequency Characteristic with Induction Generator Based Wind Turbine

Woo-Kyu Chae[1*], Hak-Ju Lee[1], Sung-Wook Hwang[1], Il-Keun Song[1], Jae-Eon Kim[2]

[1]Research Institute, Korea Electric Power Corporation, Daejeon-si, Korea
[2]Chungbuk National University, Chungbuk, Korea
Email: [*]jekim@cbnu.ac.kr

Abstract

To save on the island area's power supply cost and protect the clean environment, the Isolated MicroGrid is being duly considered. Consisting of the Wind Turbine Generator (WT), photovoltaic generator, battery system, back-up diesel generator, etc., Isolated MicroGrid, which usually uses the inverter to maintain voltage and frequency of the system, is very weak in terms of voltage and frequency stability compared to the large-scale electrical power system. If wind turbine generator is applied to this weak power system, it could experience many problems in terms of maintaining its voltage and frequency. In this paper, the measurement result of voltage and frequency is presented for MicroGrid, which consists of the Wind Turbine Generator adopting the induction generator and the battery system. MicroGrid's voltage waveform distortion and Wind Turbine Generator's output oscillation problems are analyzed using PSCAD/EMTDC. Based on the analyzed result, the importance of type and capacity choice has been suggested in case the Wind Turbine Generator is applied to the Isolated MicroGrid.

Keywords

MicroGrid, Wind Turbine, Induction Generator, Battery, Voltage, Frequency, PSCAD/EMTDC

1. Introduction

From the latter part of the 20th century, complex factors including energy-related crisis such as global warming, fossil fuel depletion, high oil prices, etc., as well as environmental issues have emerged as global issues. Energy is globally consumed in a variety of fields and methods, with the power system accounting for a large portion.

[*]Corresponding author.

Along with ways to reduce power demand, research studies on smart grid are progressing briskly to use renewable energy actively and produce systems capable of accommodating them; thus improving the power system's efficiency. Many research studies are being done to reduce power transmission loss and increase the overall energy usage efficiency. MicroGrid is introduced by USA's CERTS (Consortium for Electric Reliability Technology Solutions) to improve consumer confidence and power quality [1]-[3].

As a system featuring multiple distributed generations and loads, Grid-interconnected MicroGrid is capable of supplying energy on its own, allowing it to operate in connection with an electrical power system or to form an independent system for power supply [4] [5]. Since Grid-interconnected MicroGrid is usually operated in connection with a rugged electrical power system, there is little concern over maintaining its frequency. Thus, the main concern for the Grid-interconnected MicroGrid is electric power transaction and revenue maximization through it as well as the management of MicroGrid from the economic point of view.

Contrary to the Grid-interconnected MicroGrid, Isolated MicroGrid is not connected to the large-scale electrical power system; hence the need to maintain its optimal voltage and frequency on its own at all times. Unlike the existing simple diesel power plant, Isolated MicroGrid consists of a number of distributed generations such as wind turbine generator, photovoltaic generator, battery system, etc. [6]. Since the photovoltaic generator, battery system, etc., are connected to the MicroGrid using inverter, power does not fluctuate sharply, causing little difficulty in maintaining their frequencies. On the other hand, since wind turbine generators have diverse formats and control methods, it is important to choose and install a wind turbine generator suitable for the MicroGrid.

The Isolated MicroGrid is often confused with diesel power plant. With the diesel power plant, the diesel generator is always in operation. This means that the diesel generator should maintain its own voltage and frequency. In the diesel-renewable energy hybrid system, which connects the diesel power plant to the renewable energy system, the diesel generator often stops operating when the renewable energy output is greater than the amount of power load; hence the need for a separate device capable of parallel operation with the diesel generator while maintaining voltage and frequency. The Isolated MicroGrid typically uses a battery inverter to maintain its voltage and frequency. Surplus electric power from the renewable energy system charges the battery and gets discharged when the renewable energy system's output is insufficient. **Figure 1** shows the example diagram of isolated MicroGrid.

In this paper, we analyzed the potential problems that could occur when the SCIG (squirrel cage induction generator)-type wind turbine generator is applied to the inverter-based isolated MicroGrid using data from actual measurement. When the SCIG-type WT is started, the soft starter adopted to reduce the induction generator's inrush current generates a large amount of harmonics to distort the voltage waveform and generate overvoltage; thus rendering the wind turbine generator incapable of starting. However, not using the soft starter to avoid the overvoltage problem gives rise to serious low-voltage phenomenon. Even if the wind turbine generator is started, the induction generator, because of its characteristic, operates as induction motor even with a small fluctuation of frequency, and its output oscillates between "+" and "–". This paper presents the actual measurement data for the phenomenon, analyzes its cause and mechanism using PSCAD/EMTDC, and suggests its solution.

2. Wind Turbine Generator and MicroGrid

MicroGrid is a small-scale power supply system using renewable energy [1], whereas wind turbine generator is one of the common renewable energy sources. Since MicroGrid itself is a small-scale power supply system, however, the impact on the MicroGrid varies significantly depending on the scale and format of the wind turbine generator.

2.1. Types and Characteristics of Wind Turbine Generators

The most commonly applied wind turbine configurations are classified both by their ability to control speed and by the type of power control they use. Applying speed control as the criterion, there are four different dominating types of wind turbines, as illustrated in **Figure 2**. Wind turbine configurations can be further classified with respect to the type of power (blade) control: stall, pitch, active stall. **Table 1** indicates the different types of wind turbine configurations, taking both criteria (speed control and power control) into account [7]. There are largely 4 types of wind turbine generators in terms of mechanical composition, and the impact on the electrical power system varies by type. The characteristics by type are as follows:

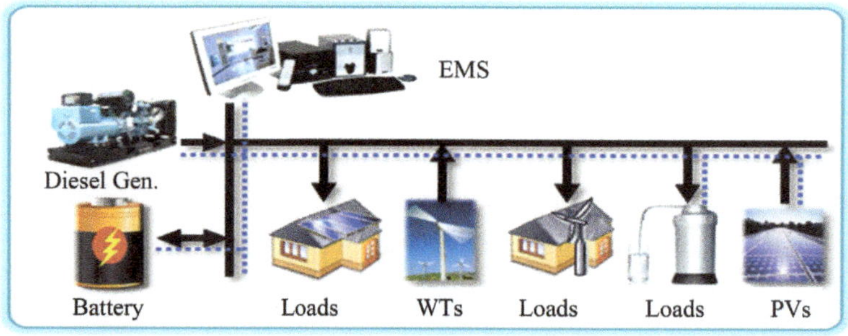

Figure 1. Example of Isolated MicroGrid.

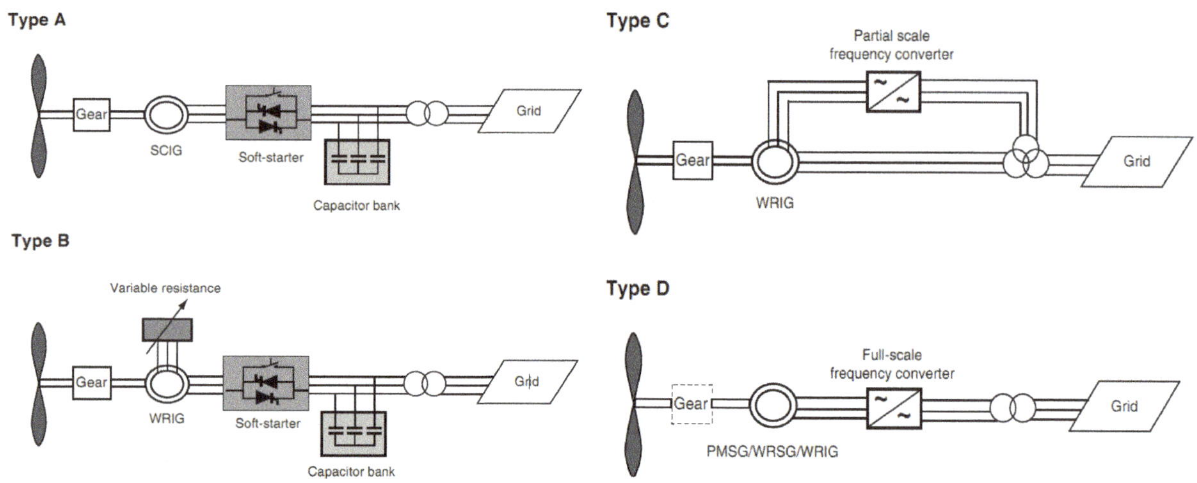

Figure 2. Typical wind turbine configurations.

Table 1. Wind turbine concepts.

Speed control		Power control		
		Stall	Pitch	Active stall
Fixed speed	Type A	Type A0	Type A1	Type A2
	Type B	Type B0	Type B1	Type B2
Variable speed	Type C	Type C0	Type C1	Type C2
	Type D	Type D0	Type D1	Type D2

2.1.1. Type A: Fixed Speed

This configuration denotes the fixed-speed wind turbine with an asynchronous squirrel cage induction generator (SCIG) directly connected to the grid via a transformer. Since the SCIG always draws reactive power from the grid, this configuration uses a capacitor bank for reactive power compensation. A smoother grid connection is achieved by using a soft-starter. Regardless of the power control principle in a fixed-speed wind turbine, the wind fluctuations are converted into mechanical fluctuations and consequently into electrical power fluctuations. In the case of a weak grid, these can yield voltage fluctuations at the point of connection. Because of these voltage fluctuations, the fixed-speed wind turbine draws varying amounts of reactive power from the utility grid (unless there is a capacitor bank), which increases both the voltage fluctuations and the line losses. Thus the main drawbacks of this concept are that it does not support any speed control, it requires a stiff grid [7].

Note: SCIG = squirrel cage induction generator; WRIG = wound rotor induction generator; PMSG = permanent magnet synchronous generator; WRSG = wound rotor synchronous generator. The broken line around the gearbox in the Type D configuration indicates that there may or may not be a gearbox.

2.1.2. Type B: Limited Variable Speed

This configuration corresponds to the limited variable speed wind turbine with variable generator rotor resistance. The generator is directly connected to the grid. A capacitor bank performs the reactive power compensation. A smoother grid connection is achieved by using a soft-starter. The total rotor resistance is controllable. This way, the power output in the system is controlled [7].

2.1.3. Type C: Variable Speed with Partial Scale Frequency Converter

This configuration, known as the doubly fed induction generator (DFIG) concept, corresponds to the limited variable speed wind turbine with a wound rotor induction generator (WRIG) and partial scale frequency converter (rated at approximately 30% of nominal generator power) on the rotor circuit The partial scale frequency converter performs the reactive power compensation and the smoother grid connection. Typically, the speed range comprises synchronous speed 40% to 30% [7].

2.1.4. Type D: Variable Speed with Full-Scale Frequency Converter

This configuration corresponds to the full variable speed wind turbine, with the generator connected to the grid through a full-scale frequency converter. The frequency converter performs the reactive power compensation and the smoother grid connection. The generator can be excited electrically [wound rotor synchronous generator (WRSG) or WRIG) or by a permanent magnet [permanent magnet synchronous generator (PMSG)] [7].

2.2. Using the Wind Turbine Generator with Isolated MicroGrid

This section summarizes the points to be considered when using the wind turbine generator in the small-scale power system.

2.2.1. Penetration Level

When incorporating renewable-based technologies into isolated power supply systems, the amount of energy that will be obtained from the renewable sources will strongly influence the technical layout, performance and economics of the system. For this reason, it is necessary to explain two new parameters—the instantaneous and average power penetration of wind—as they help define system performance [7].

The average and peak penetration of renewable generation in a hybrid power system can be defined as shown in Equations (1) and (2) [8].

$$\text{Average Penetration} = \frac{\text{Energy from Renewable Generation}(\text{kWh})}{\text{Electrical Load}(\text{kWh})} \tag{1}$$

$$\text{Peak Penetarion} = \frac{\text{Peak Power from Renewable Generation}(\text{kW})}{\text{Electrical Load}(\text{kW})} \tag{2}$$

In [9], these definitions are used to categorize hybrid power systems into three classes; low, medium and high penetration, as shown for reference in **Table 2**.

2.2.2. Hybrid Power System

The electric power production cost for the diesel power plants located in islands or remote areas-depending on their scale and distance from land can rise up to 10 times that of the large-scale electrical power system [10]. To reduce the electric power production cost, renewable energy systems such as wind turbine generators or photo

Table 2. Hybrid power system penetration classifications.

System Class	Peak Penetration	Annual Average Penetration
Low	<50%	<20%
Medium	50% - 100%	20% - 50%
High	100% - 400%	>50%

voltaic generators are sometimes installed in parallel with the diesel power plant. This kind of system is called hybrid power system [8].

Depending on its structure and the portion of renewable energy system in the whole system, the hybrid power system has its diesel generator maintain its voltage and frequency most of the time; hence the importance of choosing the wind turbine generator depending on the diesel generator's capacity and reinforcing control on the diesel generator.

2.2.3. Inverter-Based Isolated MicroGrid

Whereas the hybrid power system controls voltage and frequency through its diesel generator, the isolated MicroGrid has its battery inverter and individual distributed generations contribute to voltage/frequency control [8]. Coordination between inverters and WT is critical for performance stability. Likewise, because its system impedance is higher than that of a diesel power plant, careful attention should be paid to the harmonics generated from inverters or WT [11].

2.2.4. Electrical Characteristics of SCIG-Type WT

SCIG-type WT has been widely used in the large-scale electrical power system because of its low production cost and ease of control. However, factors that do not cause any problem in the large-scale electrical power system become big issues in the isolated MicroGrid.

SCIG-type WT uses an induction generator; the induction generator's inrush current, depending on its format, requires around 5 ~ 10 times the rated current [12]. To reduce such high inrush current, the SCIG-type WT typically adopts soft starter using thyristor. With soft starter, the inrush current can be limited to within 2 times the rated current. However, it generates much harmonic current [13], and this in turn causes serious distortion of voltage waveform, flicker, sagging, etc. in small-scale electrical power system [14]. Another drawback is that one has to overdesign the capacity of the battery inverter or diesel generator to provide the reactive power necessary for WT's startup and power generation.

3. Squirrel Cage WT's Impact on Voltage/Frequency

This section explains the structure of the currently configured isolated MicroGrid. It presents the actual measurement result, showing what kind of impact WT has on the system's voltage and frequency during startup and operation.

3.1. Structure and Test Condition of the Isolated MicroGrid

The structure of the MicroGrid tested in this paper can be seen in **Figure 3**. It consists of 1 MWh lead battery, 500 kVa inverter, 100 kW dummy load, two 250 kW WTs, and three 150 kW diesel generators as summarized in **Table 3**.

The battery inverter controls the system's voltage and frequency. To ensure the test's safety and to conduct the test under the same condition as that of the actual site, the test was conducted with 100 kW artificial load deployed. Since the diesel generator was supplying power to actual load at the time, the test was done with the

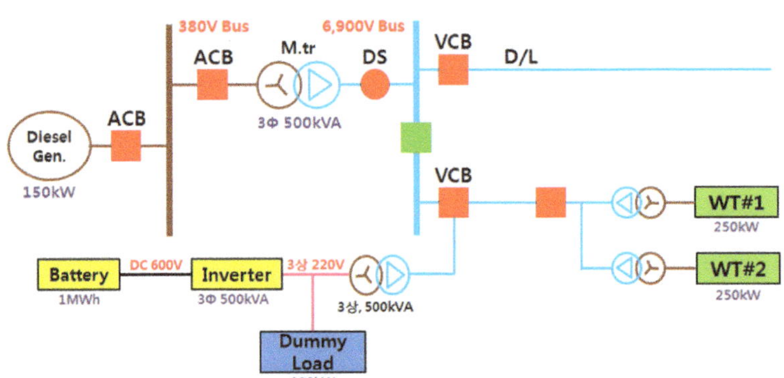

Figure 3. Schematic diagram of Island MicroGrid.

diesel generator and the actual load excluded. Simultaneous measurement was done both at the inverter's output (low voltage) and WT's output (low voltage) using Dewetron Corporation's DEWE-2520 (**Figure 4**). However, all the waveforms presented in this paper are those measured at WT's output. The average wind speed recorded at the time of test was about 8 - 10 [m/s].

3.2. Measurement Result at the Time of WT's Startup

Figure 5 presents the waveforms of voltage and current during WT's startup using soft starter. It shows that the amount of current gradually increases according to the soft starter's firing angle. It also illustrates that voltage waveform becomes greatly distorted in proportion to the current distortion level. Reasonable voltage was not applied to WT, with the soft starter malfunctioning; thus resulting in WT's failure to start up.

Figure 6 shows the waveform generated when WT was started without operating the soft starter. Because the current is not limited by the soft starter, very large starting current (approximately 700 A, at 380 V) is required. Although the voltage dropped to around 145 V (with reference voltage of 380 V) because of this, voltage waveform distortion as in **Figure 5** did not occur. In this case, WT successfully started with around 80% probability. This means that the inverter acted as if it was a motor drive to start WT.

3.3. Measurement Result during WT's Power Generation

Similar to the result in Section 3.1.2, under the test condition in this study, startup succeeded only when the soft

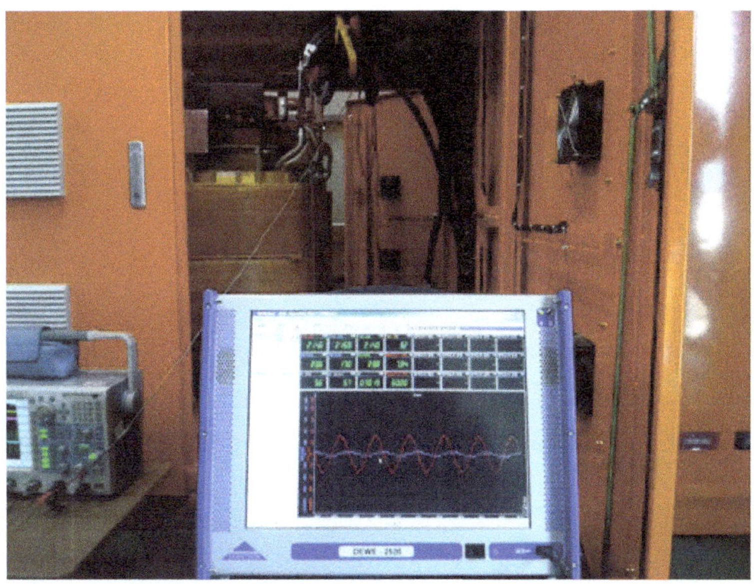

Figure 4. Measurement as shown on the inverter output side.

Table 3. MicroGrid's building blocks.

Equipment	Capacity	Number	Format
Battery	1 MWh	1	Lead-Acid
Inverter	500 kVA	1	Grid Forming
Wind Turbine	250 kW	2	SCIG Type
Dummy Load	100 kW	1	Resistor
Diesel Generator	150 kW	3	Synchronous Machine
Distribution Line	1.1 km	1	6.9 kV, TR CNCE-W 60 mm^2

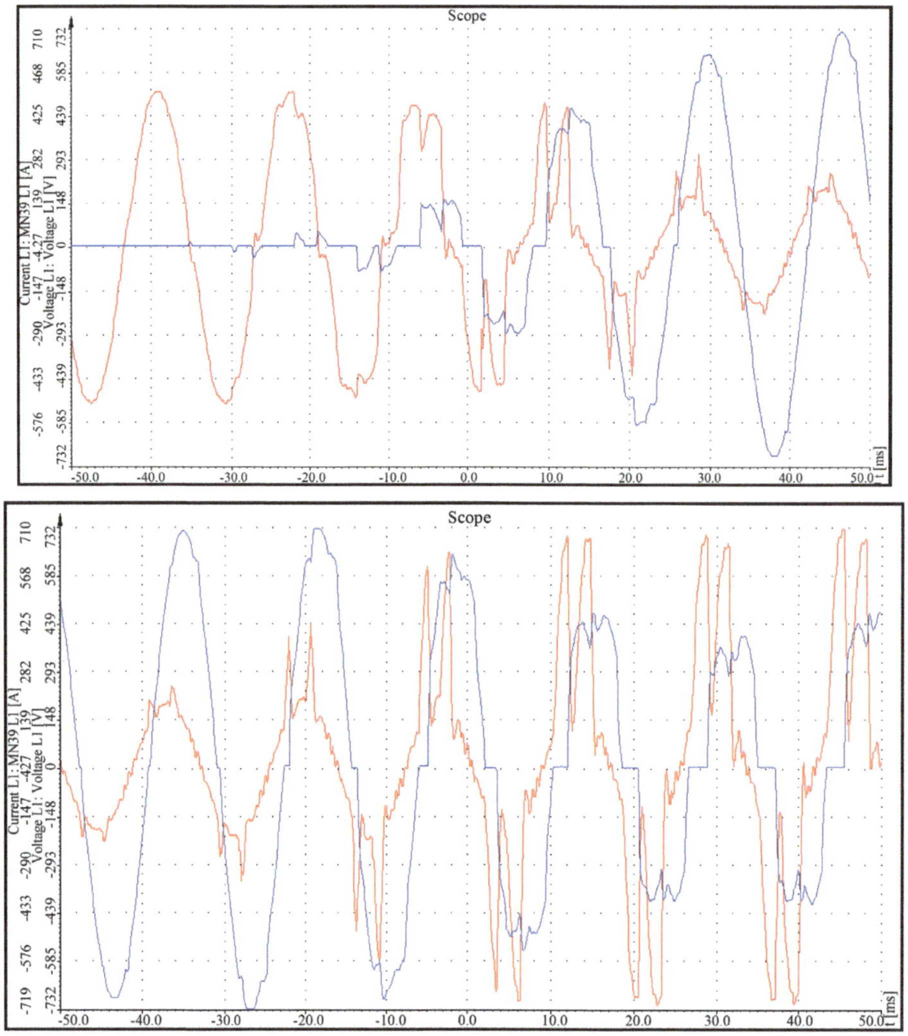

Figure 5. Voltage/current waveforms during WT's startup (soft starter in operation, Red: Voltage, Blue: Current).

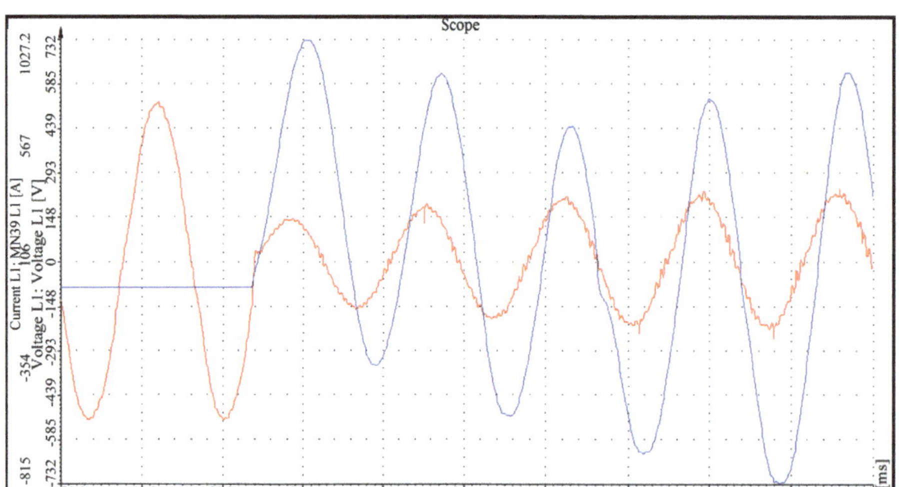

Figure 6. Voltage/current waveforms during WT's startup (soft starter not in operation, Red: Voltage, Blue: Current).

starter was not used. **Figure 7** shows the waveforms of voltage and current after WT successfully started and shifted to power generation mode. It can be seen that, even though startup succeeded, the current periodically fluctuates and oscillates.

Representing **Figure 7** with active power and reactive power produces **Figure 8**. Note that active power and reactive power oscillate with around 1 [Hz]. The oscillation range of active power is approximately −120 [kW]~ + 100 [kW], where "–(minus)" means that it operated as motor (*i.e.*, consumed energy). This suggests that WT, after shifting to power generation mode, did not work as generator only but alternately worked as both generator and motor.

The oscillation range of the reactive power is approximately –70 [kVAr] - −470 [kVAr], where "–(minus)" means that WT continuously absorbed the reactive power. This is an inevitable result considering the fact that

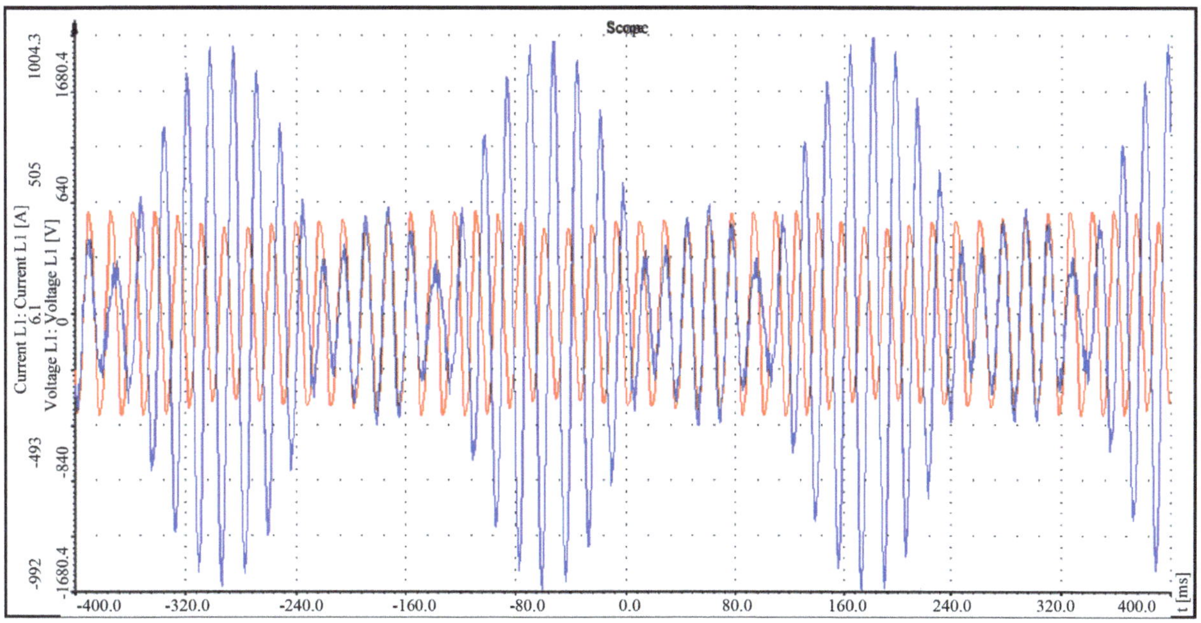

Figure 7. Voltage/current waveforms during WT's power generation (Red: Voltage, Blue: Current).

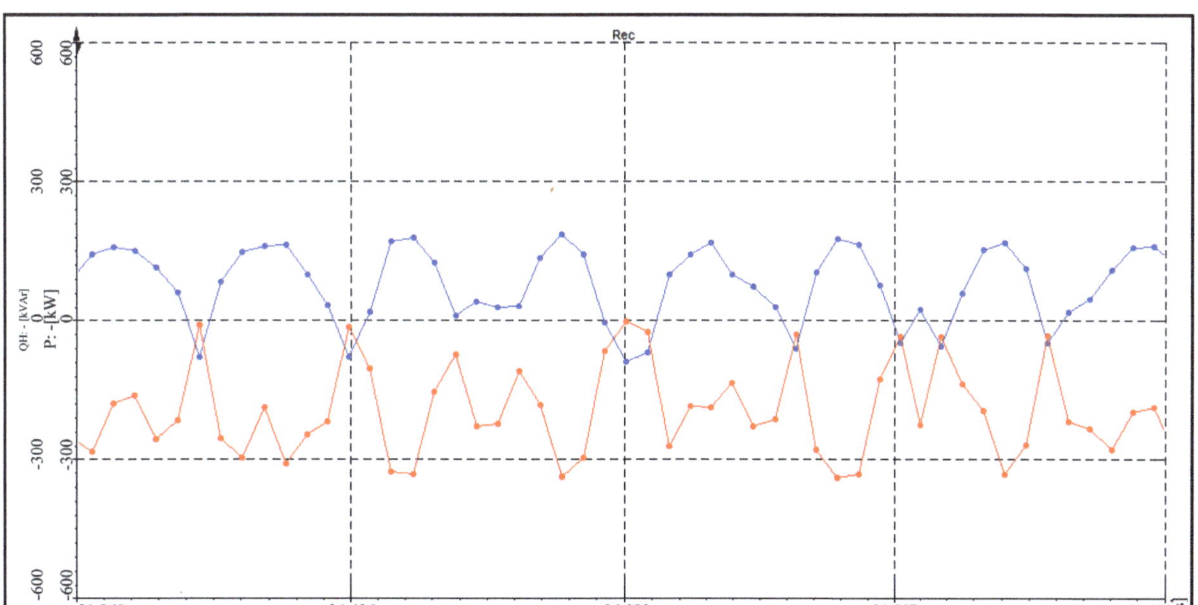

Figure 8. Voltage/current waveforms during WT's power generation (Blue: Active Power, Red: Reactive Power).

WT adopted an induction machine. As active power oscillates by 1 Hz unit, however, reactive power is continuously oscillating proportionately as well.

4. Analysis through PSCAD/EMTDC Simulation

In this section, to analyze Section 3.1's phenomenon, the isolated MicroGrid's building block was modeled using PSCAD/EMTDC. First, modeling was done under the same environment as the condition used in the test. To find out the causal analysis and improvement method, simulations were done by alternating various conditions.

4.1. Modeling Using PSCAD/EMTDC

WT, soft starter, battery, inverter, distribution line, transformer, etc., were modeled using PSCAD/EMTDC. The result of modeling can be seen in **Figure 9**. At this time, the modeling parameter used **Table 3** and actual data, and modeling was done based on the condition wherein voltage waveform distortion and output oscillation occurred as in the actual situation. The inverter was modeled as PWM inverter capable of CVCF operation. Dummy load was modeled into the inverter.

The simulation process was modeled similar to the actual test conditions. In the actual test procedure, the inverter is first started, and the line and transformer are then supplied with power. Subsequently, if optimal wind speed is reached, the WT blades begin to rotate; when their rotation speed reaches around 1000 rpm, the soft starter operates. It takes around 0.2 seconds (10 - 12 cycles) for the soft starter's rotation speed to reach the rated value (around 1200 rpm). Afterward, the soft starter is bypassed, and WT shifts to power generation mode. **Table 4** summarizes this together with the simulation processes.

4.2. Simulation Result during WT Startup

To analyze the voltage distortion phenomenon during WT startup, the soft starter was modeled to simulate the phenomenon during startup. **Figure 10** shows the waveform generated when WT was started using soft starter, whereas **Figure 11** illustrates the waveform generated when WT was started without using soft starter. **Figure 10** shows that the voltage waveform is significantly distorted, and that overvoltage is being generated. On the contrary, **Figure 11** illustrates that the situation is alleviated considerably compared to the case in **Figure 10**.

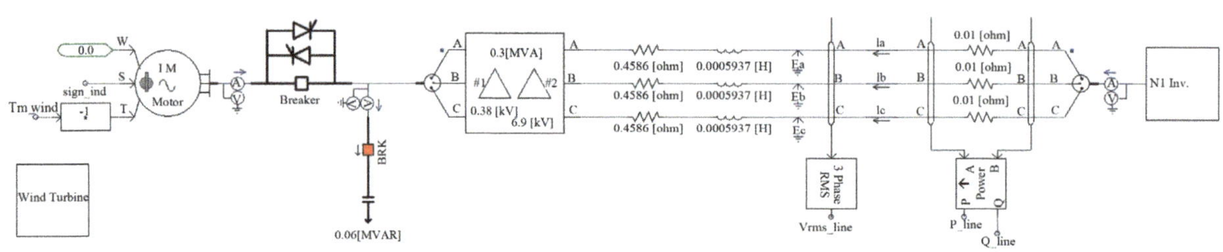

Figure 9. Simulation model using PSCAD/EMTDC.

Table 4. Simulation process.

Time	Process
0	Simulation starts.
0.2	Inverter starts.
0.5	WT begins to rotate. (RPM increases.)
0.72	WT RPM: 0.84 pu (1000 rpm) Soft Starter begins to operate.
0.92	WT RPM: 1.0 pu (1200 rpm) Soft Starter stops and gets bypassed.
1.1	WT's startup is complete, and power generation begins.

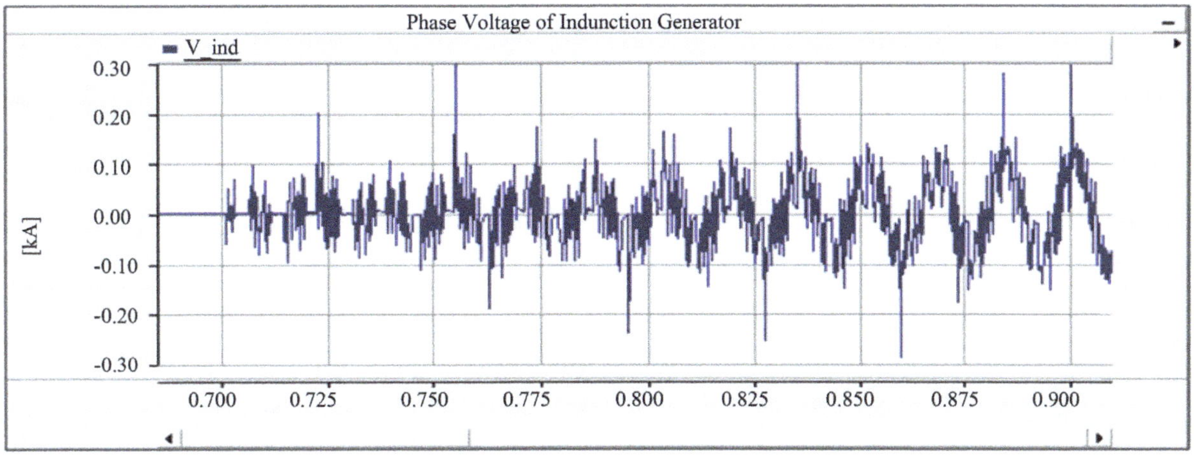

Figure 10. Voltage waveform with soft starter.

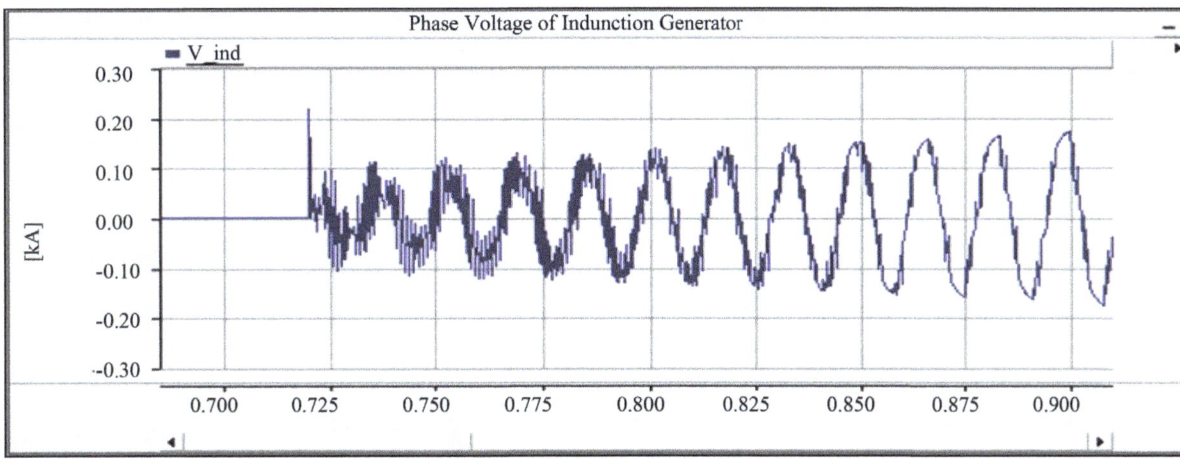

Figure 11. Voltage waveform without soft starter.

This means that the voltage waveform distortion during WT's initial startup is caused by the soft starter.

4.3. Simulation Result during WT's Power Generation

To analyze the cause of the output oscillation phenomenon during WT's power generation, simulation was first done to get the same result as the existing one. **Figure 12(b)** is the current status of WT output, and we can see that it is similar to **Figure** 7's result. To find out the condition for WT's normal operation, simulations were done by alternating various conditions. As a result, the oscillation phenomenon was confirmed to have been alleviated or aggravated according to the inverter controller's gain. The oscillation phenomenon can be seen to have been alleviated especially according to the extent of P (proportional) gain; this means that the inverter's frequency tracking performance is currently set low. **Table 5** shows the controller's gain in current state and improved state.

Figure 12 confirms that the controller's low gain caused the frequency oscillation, and that this in turn oscillated WT's rpm and voltage, too. **Figure 13** verifies that the controller tuning improved its gain, and that the oscillation phenomenon was removed. Frequency can also be confirmed to be maintained in stabilized condition at 60 Hz, with WT's rpm maintained well at the rated value.

Based on the simulation result above, the system frequency oscillation and WT's output oscillation (motoring ↔ generation) phenomena can be analyzed. Specifically, the WT considered in this paper is configured as squirrel cage induction generator (**Figure 2**'s A type), and the induction machine can work either as generator or motor depending on its rotation speed and torque in contrast to the synchronous speed [12]. It is working as a motor if it rotates slower than the synchronous speed and as generator if faster. The WT considered in this paper

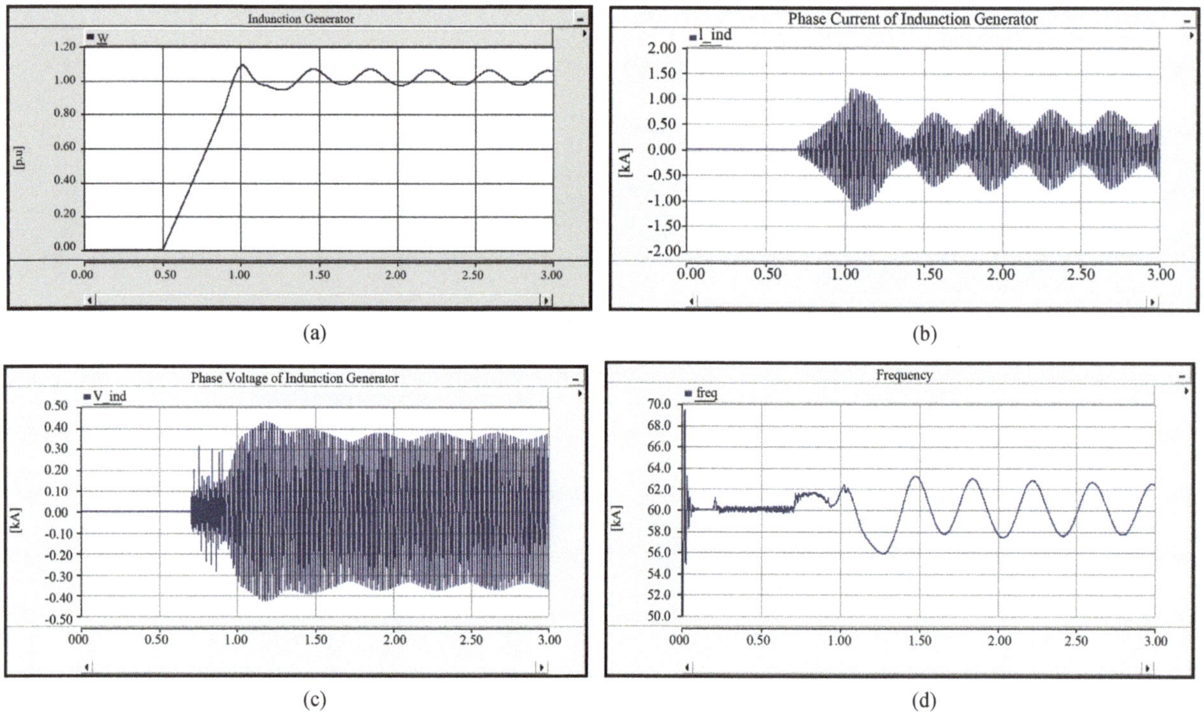

Figure 12. Simulation results for the current state. (a) WT's RPM; (b) WT's phase current waveform; (c) WT's phase current waveform; (d) Frequency of MicroGrid.

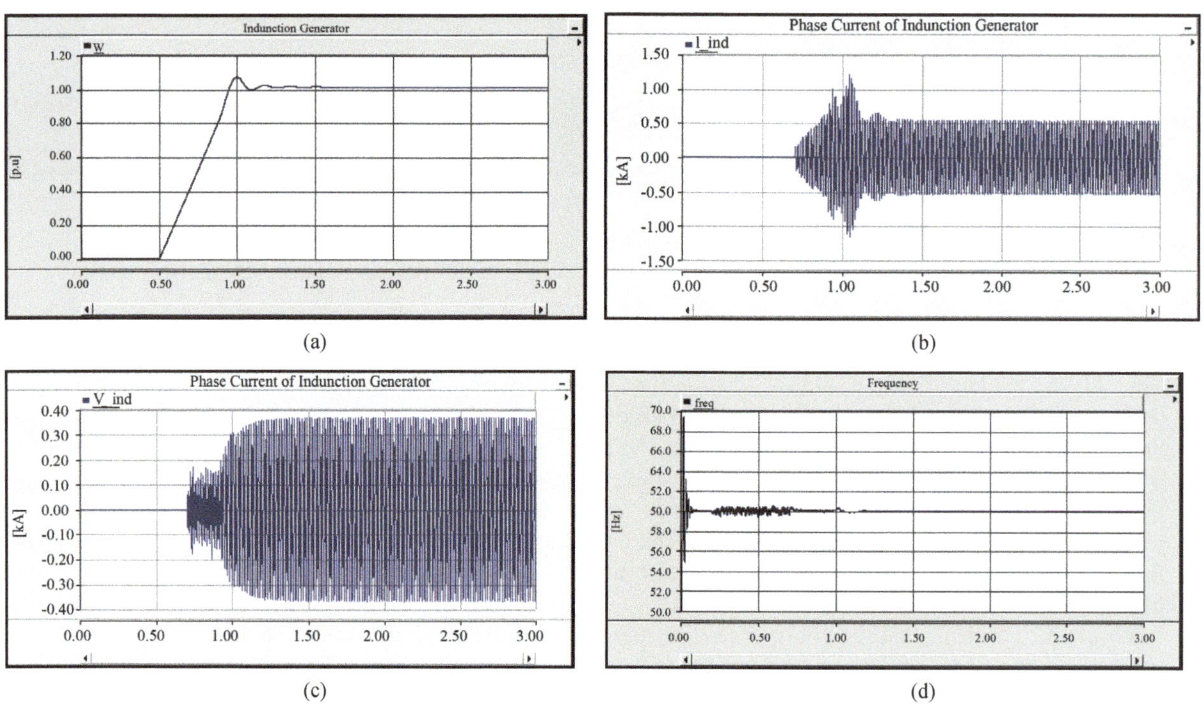

Figure 13. Simulation results for the inverter's gain change. (a) WT's RPM; (b) WT's phase current waveform; (c) WT's phase current waveform; (d) WT's phase current waveform.

has rated speed of 1211 rpm and synchronous speed of 1200 rpm during power generation. In case WT's system frequency increases to 61 Hz for some reason (ex. frequency fluctuation at the inverter), the synchronous speed at this time becomes 1220 rpm, but the wind turbine's rotation speed remains at 1211 rpm, making WT work as

Table 5. Inverter controller's gain.

		Current state	Gain change
d axis	P gain	4.0	100
	I gain	0.005	0.0001
q axis	P gain	1.0	100
	I gain	0.05	0.0001

a motor. If the synchronous speed increases from 1200 rpm to 1220 rpm according to the system frequency's increase, the generator's slip (difference between synchronous speed and operation speed) is reduced, resulting in decreased torque. Since the generator's torque acts as a brake against the wind turbine's torque, the reduction of such torque accelerates the wind turbine's rotation speed and increases the generator's rpm, which in turn increases the slip again and results in the generator's torque being increased to the original state. In other words, it gets to act as generator again.

5. Conclusions

This study presented the survey data for the voltage and frequency maintenance problem in Isolated MicroGrid composed of WT and battery system and analyzed the causes using PSCAD/EMTDC. The isolated MicroGrid considered in this paper has its voltage waveform distorted during WT's startup; this in turn generates low voltage and over voltage. Likewise, after shifting to generation mode, the output oscillation phenomenon makes it difficult to maintain system frequency, and normal operation becomes difficult.

As a result of simulation using PSCAD/EMTDC, the voltage waveform distortion during WT's startup was found to be due to the soft starter used at the time. Its solutions could include improving the battery inverter's harmonics handling performance or installing APF (Active Power Filter). WT's output oscillation phenomenon was analyzed to be due to the improper setting of inverter controller's gain, which caused WT to alternate between generation mode and motor mode.

Due to its characteristics, the SCIG-type WT can operate only at high starting current and within a very narrow frequency domain. This means that using the SCIG-type WT in isolated MicroGrid requires inverters with very large capacity and high performance. Thus, it is important to choose WT with proper type and capacity when designing isolated MicroGrid.

Finally, we will change the grid forming inverter as more proper type for isolated MicroGrid.

Acknowledgements

This work was supported by Korea Institute of Energy Technology Evaluation and Planning (KETEP) grant funded by Korea Government Ministry of Knowledge Economy (No. 20123010020080).

References

[1] Hatziargyriou, N.D., Asano, H., Iravani, R. and Marnay, C. (2007) Microgrids. *IEEE Power & Energy*, **5**, 78-94. http://dx.doi.org/10.1109/MPAE.2007.376583

[2] Hatziargyriou, N.D. (2008) Microgrids. *IEEE Power & Energy*, **6**, 26-29.

[3] Venkataramanan, G. and Marnay, C. (2008) A Larger Role for Microgrids. *IEEE Power & Energy*, **6**, 78-82. http://dx.doi.org/10.1109/MPE.2008.918720

[4] Asmus, P. (2009) Microgrids, Islanded Power Grids and Distributed Generation for Community, Commercial, and Institutional Applications. Navigant Research, Boulder. http://www.navigantresearch.com

[5] Lasseter, R.H. (2002) Microgrids. *IEEE PES Winter Meeting*, New York City, 27-31 January 2002, 305-306.

[6] Kojima, T. and Fukuya, Y. (2011) Microgrid System for Isolated Islands. *Fuji Electric Review*, **57**, 125-130.

[7] Ackermann, T. (2005) Wind Power in Power Systems. Wiley, Hoboken, 55-59. http://dx.doi.org/10.1002/0470012684

[8] Prull, D.S. (2008) Design and Integration of an Isolated MicroGrid with a High Penetration of Renewable Generation.

Doctor's Thesis, University of California, Berkeley.

[9] Baring-Gould, I.E., *et al.* (2004) Worldwide Status of Wind-Diesel Applications. 2004 Doe/Awea/Canwea Wind-Diesel Workshop, Technical Report, National Renewable Energy Laboratory, Cole Blvd.

[10] KEPCO (2012) Report for the Prime Cost Report of Diesel Power Plant. KEPCO, Seoul.

[11] Pingree, B. (2011) A New Isolated Grid Paradigm. 2011 *International Wind-Diesel Workshop*. Girdwood Alaska, 8-11 March 2011, 8-10.

[12] Chapman, S.J. (2001) Electric Machinery Fundamentals. McGraw-Hill, New York.

[13] Freris, L. and Infield, D. (2011) Renewable Energy in Power Systems. John Wiley & Sons, Hoboken.

[14] Chae, W. (2013) SCIG Type Wind Turbine's Characteristic in Island MicroGrid. ISGC & E, Jeju.

11

Ventilation of Apartment Buildings and Nursing Homes

Alo Mikola, Teet-Andrus Koiv, Hendrik Voll

Department of Environmental Engineering, Tallinn University of Technology, Tallinn, Estonia
Email: teet.koiv@ttu.ee, alo.mikola@ttu.ee

Abstract

The paper presents the results of the investigations into the indoor climate and energy efficiency of ventilation devices in residential buildings and nursing homes. The indoor climate studies discussed in the article were conducted in a nursing home where room ventilation is based on Meltem Air Handling Units (AHU). Similar studies have been carried out in apartment buildings where CO_2 levels in bedrooms were compared before and after renovating the ventilation with Meltem AHU and installing exhaust fans in the bathroom/WC and kitchen. Tenants evaluate the use of Meltem AHU in apartments very positively. The article presents the efficiency results of Meltem AHU with different external air temperatures conducted in apartments in real-life situations. The study shows that skilful renovation of ventilation in old apartment buildings enables to achieve good indoor climate and energy saving at the same time.

Keywords

Indoor Climate, Apartment Buildings, Nursing Home, AHU, Energy Efficiency of AHU

1. Introduction

The paper presents the results of the investigations into indoor climate in nursing homes and residential buildings. In the review of ventilation in European dwellings, it was pointed out that the ventilation of residential spaces in Nordic countries is often poor [1]. As a result of refurbishing, the air tightness of the building envelope increases and as many buildings have natural ventilation, the air change rate is reduced. The performances of the ventilation systems in low-energy residential houses have been presented by Maier *et al.* [2]. The energy performance of heat recovery units has been investigated by Jaber *et al.* [3] and Laverge *et al.* [4]. Ventilation heat recovery in residential buildings has been investigated by Dodoo *et al.* [5]. A number of studies on the ventilation and indoor air quality in apartment buildings have been carried out at Tallinn University of Technology by Mikola and Koiv [6]-[9].

2. Methodology

2.1. Indoor Air Quality

In accordance with the requirements [10] in Estonia, the living area must have natural or mechanical ventilation, which guarantees the air change necessary for human activity. According to the requirements in Estonia [11], air velocity in living spaces, the volume of space per person and the content of harmful substances in the indoor air must not exceed the permitted values.

Values for assessing room temperatures are taken from Estonian standard EVS-EN 15251:2007 [10]. This standard describes indoor environmental input parameters for designing and assessing the energy performance of buildings. The temperature values in case of different classes of thermal comfort are described in **Table 1**. The values to assess the relative humidity of indoor air are taken from Estonian indoor climate standard EVS 839:2003 [12]. The parameters are described in **Table 1**.

On the valid regulative basis, the CO_2 concentration in the indoor air in Estonia is set in the standard of indoor environmental input parameters [10] and designing criteria CR 752 [13]. Danish University of Technology has studied the standard in detail and gives the air flow rate of the classes of indoor climate and the maximum norms of the CO_2 content in the indoor air [14]. These norms also correspond to the values that are given in the designing criteria of indoor climate. Due to that the present study uses the maximum values given in designing criteria CR 752 for accessing the CO_2 content in the indoor air, whereby the CO_2 content in the external air is considered to be 350 ppm. The values to assess the CO_2 concentration of indoor air are described in **Table 2**. The standard for determining the initial parameters of energy efficiency [10] allows short-time deviations from the required parameters of indoor climate. The permitted deviation is up to 5%.

According to the standard for determining the initial parameters of energy efficiency (EVS-EN 15251:2007), it is possible to deduce the airflow according to the measurements of CO_2 concentrations in rooms where the main source of indoor air pollution is people. This method is called the metabolic CO_2 method. To assess the air change of a room by the metabolic CO_2 method it is possible to use Equation (1):

$$C = C_v + \frac{m}{L} - \left(C_v + \frac{m}{L} - C_0\right) \cdot \left(e^{-\frac{L}{V}\tau}\right) \quad (1)$$

where C is the indoor concentration of CO_2 at time t (g/m^3), C_v the concentration of CO_2 in the external air (g/m^3), C_0 the concentration of CO_2 in the indoor air at time 0 (g/m^3), L the volume flow rate of air entering the space (m^3/h), m the volumetric indoor emission rate of CO_2 (g/h), V the volume of the indoor space (m^3) and τ is the interval since $t = 0$.

Using literature [15], it is possible to express the relationship between the metabolic heat rate, body surface area and oxygen demand by Equation (2).

Table 1. Thermal classes of indoor climate based on Estonian standards EVS-EN 15251:2007 and EVS 839:2003.

Class of thermal comfort	Summer		Winter	
	Relative humidity, %	Room temperature, °C	Relative humidity, %	Room temperature, °C
I		23.5 - 25.5		21 - 25
II	30 - 70	23.0 - 26.0	25 - 45	20 - 25
III		22.0 - 27.0		18 - 25

Table 2. Class of the indoor climate for rooms with human activity (CR 1752) [13].

Category	Expected percentage dissatisfied, %	CO_2 concentration at outdoor air level 350 ppm, ppm	Indoor air CO_2 concentration, ppm
I (A)	15	460	810
II (B)	20	660	1010
III (C)	30	1190	1540

$$Q_{CO_2} = \frac{0.727 \cdot M \cdot RQ \cdot m^{0.425} \cdot l^{0.725}}{4.83 \cdot RQ + 16.17} \quad (2)$$

where Q_{CO_2} is the human indoor emission rate of CO_2 (l/h), RQ the ratio of M O_2 and exhaled CO_2, Q_{O_2} human oxygen demand (l/h), m the weight of person (kg), l the height of person (m), M the metabolic heat rate (W/m^2).

Human CO_2 emission rates are mainly related to the surface area of the body and the metabolic heat rate. The ratio of inhaled O_2 and exhaled CO_2 is taken as a constant value in the present study (0.7 to 1.2 met). If the value of RG is not exactly determined, it can be considered 0.83 [15]. In this case the accuracy of CO_2 emissions is approximately 3%, which is a sufficient result.

The metabolic CO_2 emissions are often observed as the average value of a 24 h period. In this case the average metabolic CO_2 emission is calculated using the weighted average method. In some studies the metabolic CO_2 emissions are shown during the sleeping period. As the CO_2 emissions vary in a wide range in the day time, it would often be much more punctual to use the night time CO_2 emissions in the calculations [10]. In the present study night time CO_2 emissions are used in the case of apartment buildings. In this study adult CO_2 emission is 13 l/h and children's CO_2 emission is 6.5 l/h during the night time. Metabolic CO_2 emissions in different studies are shown in **Table 3**.

2.2. Ventilation Heat Recovery

Member States of the European Union are required to implement energy efficiency measures for buildings under the Energy Performance of Building Directive [16]. With new building codes, EU member states intend to reduce the total energy consumption in buildings by making the buildings well-insulated and tighter. Due to the tighter and well-insulated constructions, the ventilation loads constitute a growing part of the heating demand between 30% - 60% for new and retrofitted buildings [17]. As the ventilation heating demand plays an important role in the building's total heating demand, the heat exhaust air heat recovery is inevitable.

From the exhaust air it is possible to recover only sensible heat or both sensible and latent heat [18] [19]. Energy is used to cover the heat losses due to ventilation air and to move the ventilation air for mechanical ventilation. The ventilation system also influences the air infiltration through the building envelope.

The temperature efficiency, which does not include latent heat transfer, was used to quantify the effect of heat recovery. The temperature ratio (efficiency) η_{temp} is defined as [18] [20]:

$$\eta_{temp} = \frac{\dot{m}_{in} \cdot \left(\overline{t_{supply}} - t_{outdoor}\right)}{\dot{m}_{min} \cdot \left(t_{exhausst} - t_{outdoor}\right)} \quad (3)$$

$t_{outdoor}$ is the outside air temperature, $t_{exhaust}$ is the temperature of the extract air, $\dot{m}_{in}$ is the inlet mass flow rateand $\dot{m}_{min}$ is the minimum of inlet and outlet mass flows. In the case of the regenerator, a time-averaged value of the supply air temperature $\overline{t_{supply}}$ has to be used which is given by [20]:

$$\overline{t_{supply}} = \frac{1}{\tau}\int_{t=0}^{t=\tau} t_{supply}(t) \cdot dt \quad (4)$$

Table 3. Metabolic CO_2 emissions in different studies [21]-[26].

Country, study	Object of the study	Adult CO_2 emission, l/h	Infant CO_2 emission, l/h
USA, Dietz & Goodrich 1995	Schools, residential buildings	19	12
Thailand, Leephakpreeda *et al.* 2000	Schools	16.2	16.2
Ireland, Guo and ja Lewis 2007	Residential buildings	18	18
Sweden, Pavlovas 2003	Apartment buildings	12*	12*
Czech, Jokl 2000	Schools, office buildings, residential buildings	19 (1 - 1.2 met)	18 (3 - 6 yrs and 2.7 met); 19 (14 - 16 yrs and 1 - 1.2 met)
Estonia, Kalamees *et al.* 2012	Apartment buildings	13*	6.5* (until 14 yrs)

*-night-time.

t denotes the time and τ is the semi-period, which means the duration of a supply or extract process. In the case of the recuperator, the process is in a steady state [20]:

$$t_{supply}(t) = \text{constant} \tag{5}$$

2.3. Studied Buildings

Studies have been carried out for the nursing home at Alutaguse, **Figure 1**. This nursing home has 39 rooms for 1 - 2 persons, the ventilation of which is solved with Meltem AHU.

The nursing home has hotel type rooms, the area is 25 m^2. The rooms are designed for 2 people (**Figures 2-4**) but in reality 1 or 2 people live in a room. Meltem Komfort AHU with network control has been installed in the rooms. **Figure 3** also shows that with the Meltem AHU polluted air is extracted from the bathroom and the supply air is given to the living room.

Meltem AHU can be installed in the wall construction, **Figure 5** or on the wall.

Figure 1. Outside view of the nursing home.

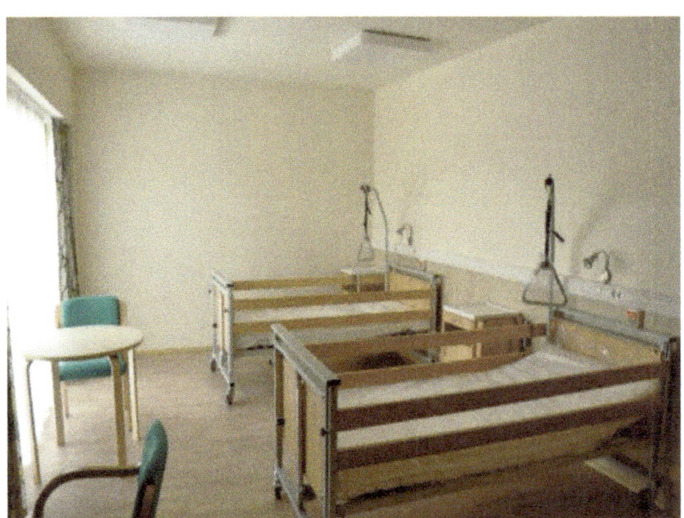

Figure 2. A two-person room.

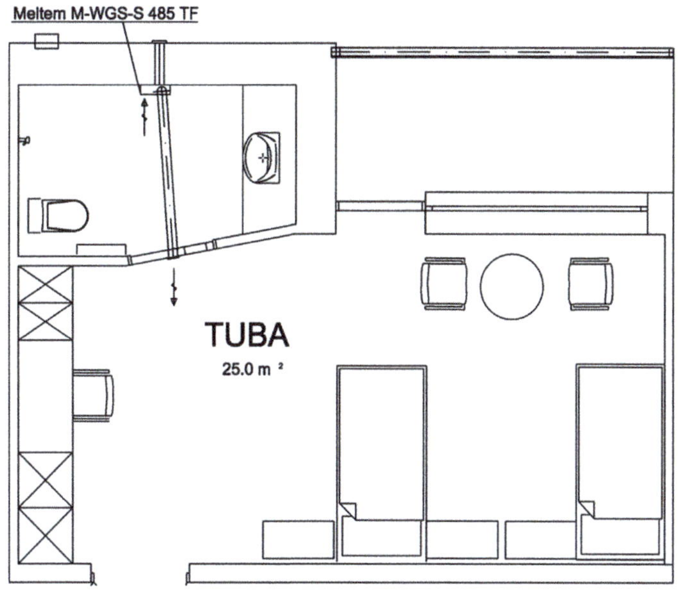

Figure 3. 2-person room solution with Meltem AHU.

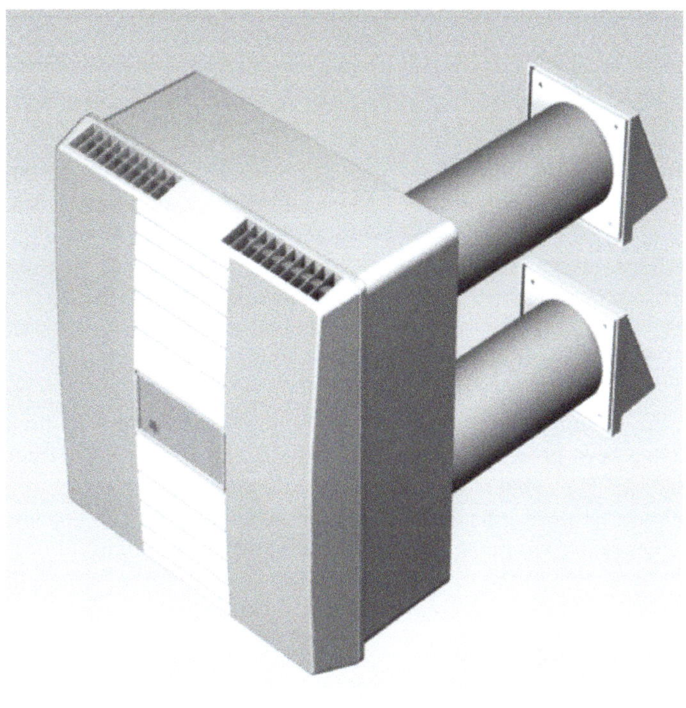

Figure 4. Meltem AHU is installed at the height of 2 m.

Meltem device has an aluminium recuperative heat exchanger, two ventilators, two filter cassettes and a control device.

Meltem room AHU has the same working principles as a standard air handling unit.

A specific feature is that the device serves up to 2 rooms, so the necessary length of the duct is minimal, typically up to 2 m, and the service is located in the AHU room. Thanks to short ducts and EC motors the electrical power of the devices is up to 15 W maximum, and hence the electric energy consumption is about 120 kWh per year, which is significantly lower than with the central balancing system.

The height of the device is 409 mm, the width is 388 mm and the depth is 196 mm. It is possible to use: stan-

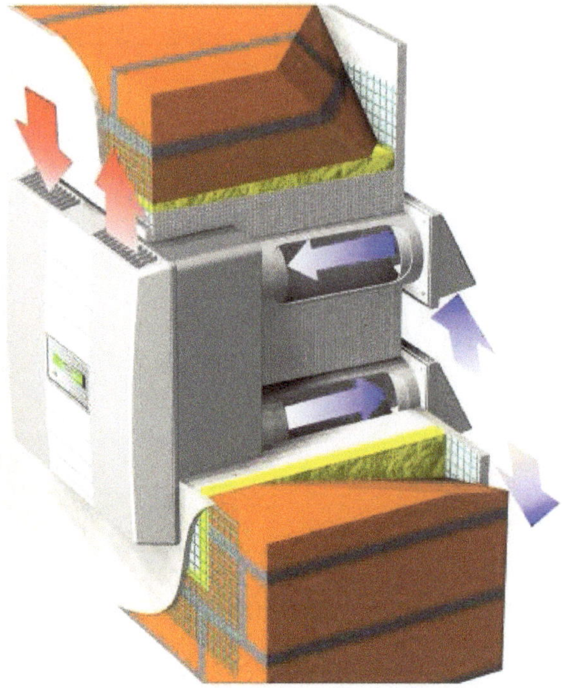

Figure 5. The operation principle of the unit Meltem.

dard filter G4, allergy F7 or absorbent carbon filter F6.

Meltem AHU is available in three modifications: Standart, Standart with network control and Komfort.

According to technical documentation Standart AHU enables three flow rates: 15/30/60 m^3/h, power consumption from 3.8 to 12.5 W and noise level from 15.5 to 36 dB(A).

Research of Meltem AHU in apartment buildings (**Figure 6**) was mostly carried out in a 2-room apartment, **Figure 7** and in 3-room apartments. The apartments were fitted with Komfort devices.

As the HVAC requirements of a nursing home and residential building are close, we can also apply the studies of Meltem AHU conducted in apartments to a nursing home.

Figure 6 shows that one Meltem AHU is used in a 2-room apartment. In addition to AHU, an exhaust ventilator has been installed in the toilet/bathroom, which works when the lighting is switched on, and in the kitchen when required. The fans have been programmed to run up to 10 minutes after the light is turned off. After that the ventilation works as natural. The exhaust air ventilator is shown in **Figure 8**.

As the ventilator capacity is 64 m^3/h, this ensures good indoor climate in the room while being used. If the ventilator is not in use, the necessary air change is ensured by natural ventilation.

3. Results

3.1. Measurements of Indoor Climate

3.1.1. The Nursing Home

Indoor climate studies in the nursing home were carried out in 4 rooms, 2 of them were single rooms and 2 double rooms. The recording results of CO_2 concentration and relative humidity in different rooms of the nursing home are presented in **Figures 9-11**.

Figure 9 shows that the relative humidity does not exceed the permissible limits of residential premises.

Economic calculations indicate that the construction costs of the ventilation on the bases of Meltem AHU and a centralised system in the nursing home are almost equal. At the same time the power consumption is lower and Meltem device allows flexible use, so the maintenance costs are significantly lower. Due to the lack of metallic ducts the number of light negative ions in the room air is also higher.

The results of the measurements conducted in different rooms show us that room air handling unit Meltem with heat recovery ensures good indoor climate in the nursing home. At the same time rational organization of

Figure 6. Facade of the house.

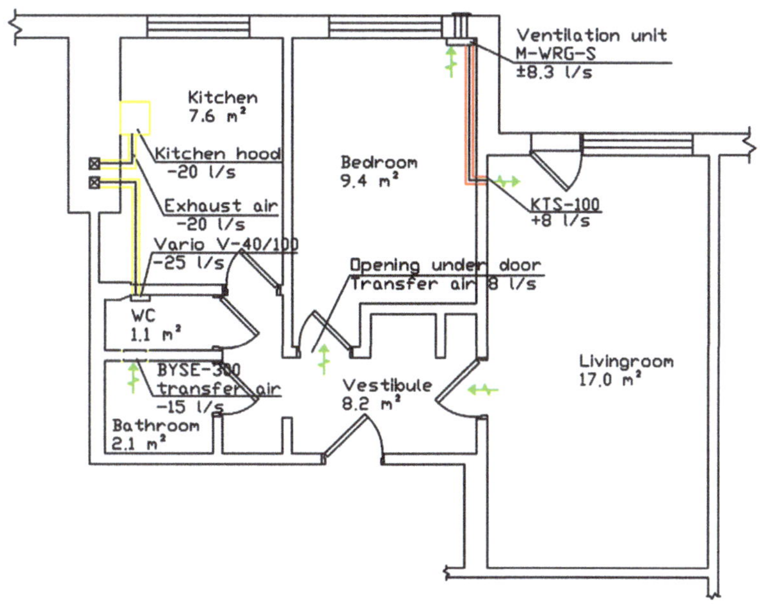

Figure 7. Meltem AHU in a 2-room apartment.

Figure 8. Ventilator in the WC/bathroom, Vario V-40/60.

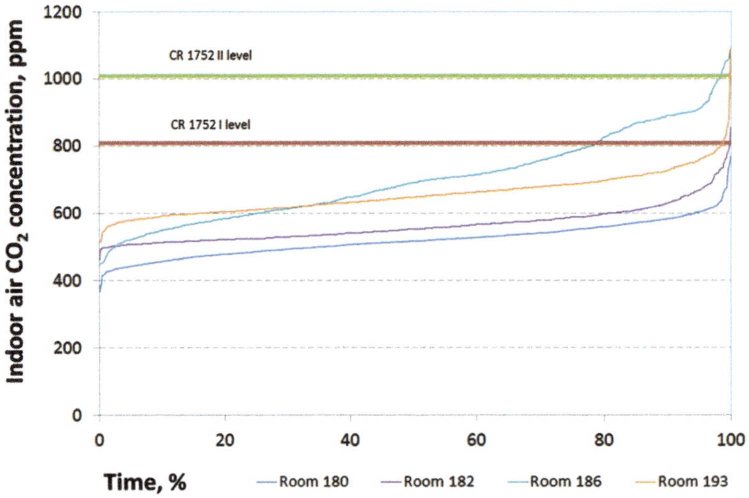

Figure 9. Cumulative distribution of CO_2 concentration in different rooms of the nursing home.

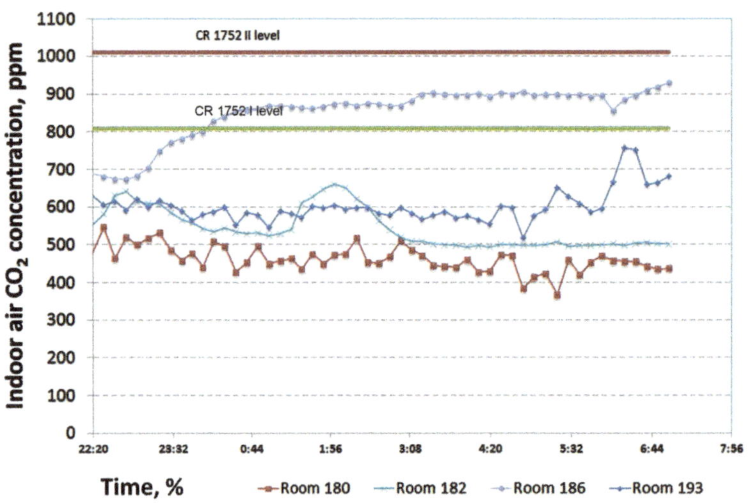

Figure 10. CO_2 concentration in different rooms, air flow rate 25 m^3/h.

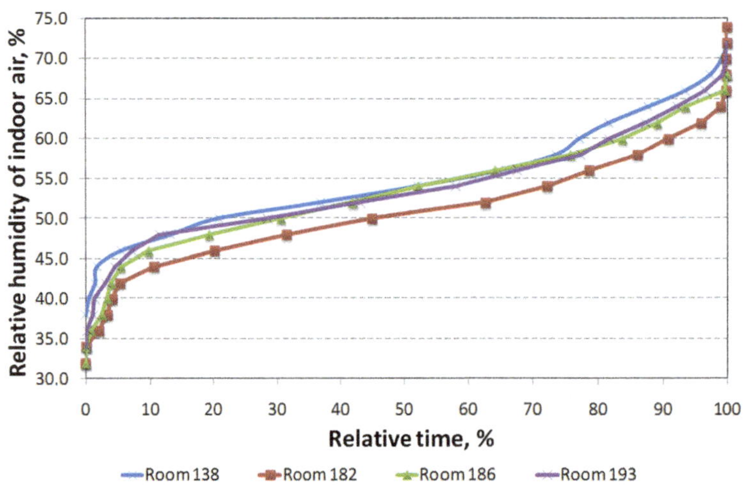

Figure 11. Relative humidity graphs in 4 rooms of the nursing home.

ventilation by Meltem significantly contributes to energy saving compared with balance ventilation.

We see that the carbon dioxide concentration in rooms was less than 1000 ppm, which can be considered a good result, and in the majority of the rooms less than 700 ppm, which can be considered a very good result.

3.1.2. Apartment Buildings

In apartment buildings studies of CO_2 concentration, relative humidity and temperature were carried out after renovating the ventilation. CO_2 level after the installation of Meltem AHU is shown in **Figure 12** and **Figure 13**.

In **Figure 12** we see that the concentration of CO_2 in the bedroom after ventilation renovation was less than 1200 ppm, with 70% of the time less than 1000 ppm, which can be considered a good result. The living room had somewhat higher CO_2 levels but more than 90% of the time the concentration was below 1500 ppm.

Tenants' assessment of the use of Meltem AHU in apartments has been very positive. Carbon dioxide levels in the bedroom have dropped nearly 1000 ppm.

An important study in the meaning of energy consumption was the determination of the temperature ratio of Meltem AHU. The measurement results are shown in **Figures 14-16**.

The temperature ratios for Meltem AHU with a wide range of external temperature are shown in **Figure 14**.

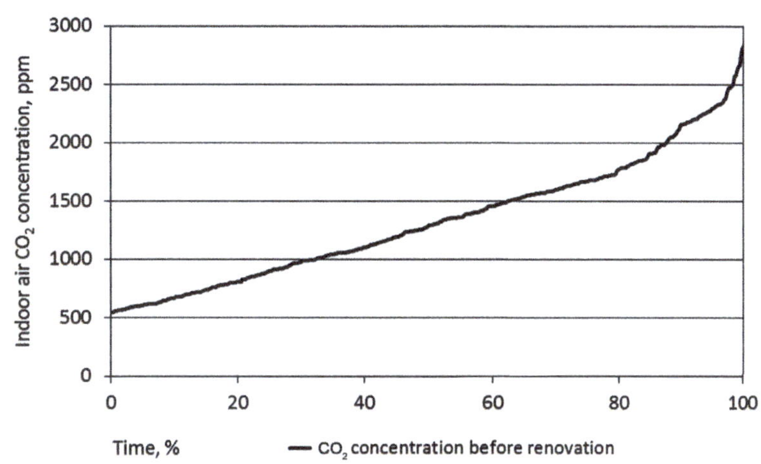

Figure 12. CO_2 concentration in the indoor air before renovating the ventilation (apartement building).

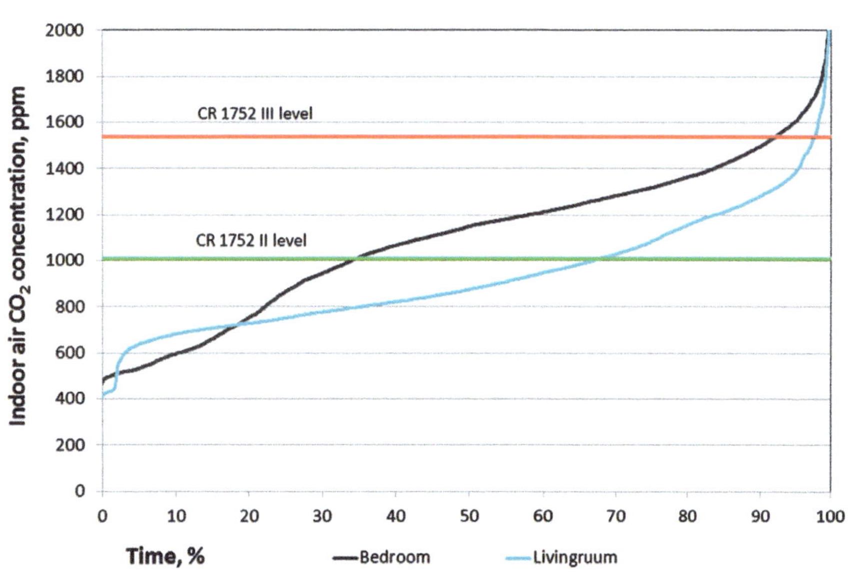

Figure13. CO_2 concentration after the installation of Meltem AHU.

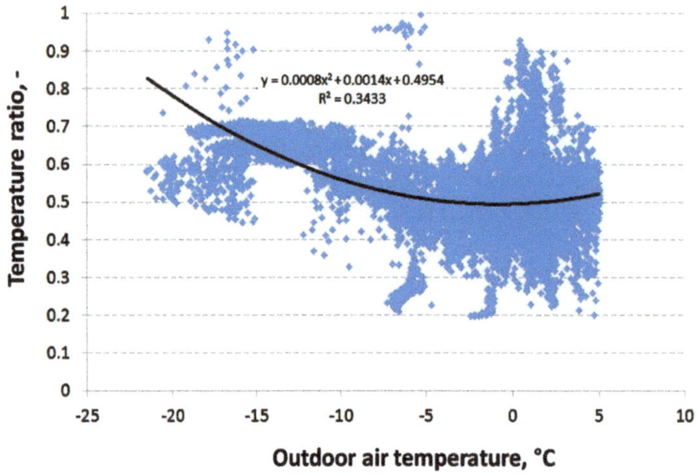

Figure 14. Temperature ratios for Meltem AHU with a wide range of external temperature.

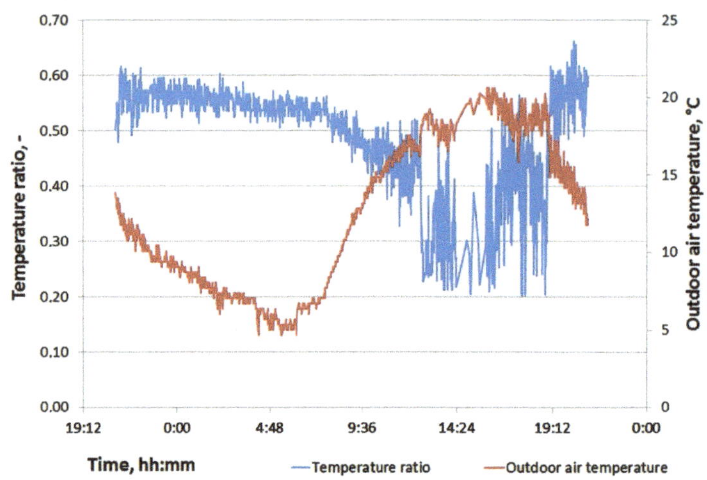

Figure 15. Temperature ratios and outdoor air temperatures for Meltem AHU at the 5th speed.

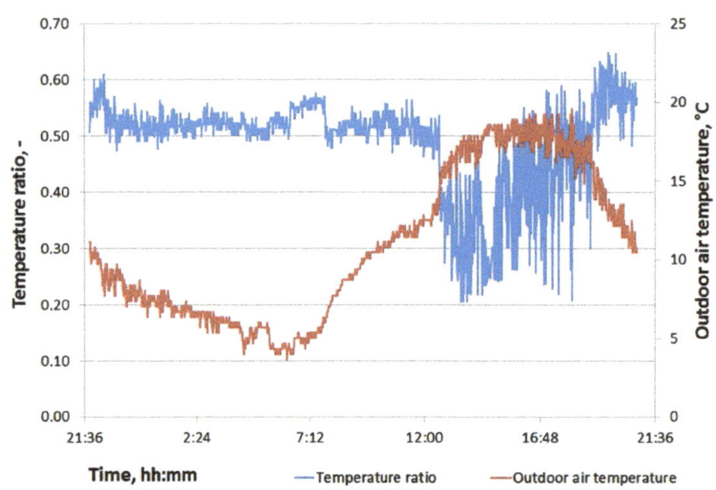

Figure 16. Temperature ratios and outdoor air temperatures for Meltem AHU at the 3rd speed.

Table 4. Noise levels of Meltem AHU-the average of 30 s (Komfort device).

Speed	Air flow rate m^3/h	Noise level in the centre of living room dB (A)	Noise level in the centre of children room dB (A)	Noise level in the centre of bedroom dB (A)
Backround noise	0	22	24.6	23.7
1	15	22.9	25.2	26
2	30	24.1	25.8	26.4
3	60	32.3	34.8	34.8

In testing Meltem AHU in the range of outdoor air temperature from +5˚C to −20˚C, the average temperature ratio (efficiency) was between 40% to 65%, with a medium size of 60%. In **Figure 15** and **Figure 16** we can see temperature ratios of Meltem AHU with the 3rd and 5th speed in positive outdoor temperatures. We can see that both temperature ratios are close, with minimum superiority to the fifth speed.

We can see that in real operating conditions the temperature ratio of Meltem AHU is on average close to 0.6, which can be considered a good result.

It should be noted that in old apartment buildings with natural ventilation the use of Meltem AHU in renovating the ventilation is one of the few suitable solutions.

3.2. Control of the Noise Level

Test measurements of the noise level of Meltem AHU have been carried out in 2-room apartments. The measurements were carried out at 5 flow rates next to the device, specifically 0.3 m from the device. Meltem AHU devices were installed in the premises of 10 m^2, where people sleep.

According to existing normative documents, the noise level should not exceed 30 dB(A) in the bedroom at night and 40 dB(A) in the living room during the day. For measuring the device Delta OHM HD 2010 was used.

The measurement results can be evaluated on the basis of the data given in **Table 4**. It is seen that Komfort AHU can be used at night at the air flow rate of 15 m^3/h, and with certain reservations at the air flow rate of 30 m^3/h. If someone sleeps in the next room, the device can be used at the air flow rate of 30 m^3/h.

4. Conclusions

The study shows that the new ventilation solution based on Meltem AHU is perfectly suitable for ventilating rooms in nursing homes and apartments. In the nursing home, Meltem AHU could provide good indoor air conditions in the rooms where two persons live and very good results in the rooms with one person. Measuring results conducted in different rooms show that the room air handling unit with heat recovery ensures good indoor climate in the nursing home. In apartment buildings, people's assessment of the use of Meltem AHU in apartments was very positive. The same studies showed that the temperature ratio of AHU was on average 60% over the season and the quality of indoor environment was satisfactory. Investigation proved that with rational and reasonable organization of ventilation we can save energy in renovating the ventilation in old apartment buildings.

Investments on ventilation with room air handling unit Meltem are close to those with the central balanced ventilation system, for example, in the nursing home.

The running costs of ventilation with the room air handling unit are lower than that with the central balanced ventilation system. The study shows that skilful renovation of ventilation in old apartment buildings enables to get satisfactory indoor climate and energy saving at the same time.

Acknowledgements

The research was supported by the Estonian Research Council, with Institutional research funding grant IUT1-15 and with the project "Development of efficient technologies for the air change and ventilation necessary for the increase in the energy efficiency of buildings, AR12045", financed by SA Archimedes. The publication of this article was supported by the ESF measure 1.2.4 Development of cooperation and innovation in universities, the sub-measure "Doctoral Schools" that finances the project "Construction and Environmental Engineering PhD School"—project code 1.2.0401.09-0080.

References

[1] Frontczak, M. and Wargocki, P. (2011) Literature Survey on How Different Factors Influence Human Comfort in Indoor Environments. *Building and Environment*, **46**, 922-937. http://dx.doi.org/10.1016/j.buildenv.2010.10.021

[2] Maier, T., Krzaczek, M. and Tejchman, J. (2009) Comparison of Physical Performances of the Ventilation Systems in Low-Energy Residential Houses. *Energy and Buildings*, **41**, 337-353. http://dx.doi.org/10.1016/j.enbuild.2008.10.007

[3] Jaber, S. and Ajib, S. (2012) Energy Recovery System in Mediterranean Region. *Sustainable Cities and Society*, **3**, 1-66. http://dx.doi.org/10.1016/j.scs.2012.01.002

[4] Laverge, J. and Janssens, A. (2012) Heat Recovery Ventilation Operation Traded off against Natural and Simple Exhaust Ventilation in Europe by Primary Energy Factor, Carbon Dioxide Emission, Household Consumer Price and Exergy. *Energy and Buildings*, **50**, 315-323. http://dx.doi.org/10.1016/j.enbuild.2012.04.005

[5] Dodoo, A., Gustavsson, L. and Sathre, R. (2000) Primary Energy Implications of Ventilation Heat Recovery in Residential Buildings. *Energy and Buildings*, **31**, 37-47.

[6] Koiv, T.-A., Mikola, A. and Kuusk, K. (2012) Energy Efficiency and Indoor Climate of Apartment Buildings in Estonia. *International Journal of Energy Science*, **2**, 94-99.

[7] Koiv, T.-A., Voll, H., Mikola, A., Kuusk, K. and Maivel, M. (2010) Indoor Climate and Energy Consumption in Residential Buildings in Estonian Climatic Condition. *Wseas Transactions on Environment and Development*, **6**, 247-256.

[8] Mikola, A. and Koiv, T.-A. (2011) Indoor Air Quality in Apartment Buildings of Estonia. *Computers and Simulation in Modern Science*: *Selected Papers from WSEAS Conferences*, **V**, 257-261.

[9] Koiv, T.-A. (2007) Indoor Climate and Ventilation in Tallinn School Buildings. *Proceedings of the Estonian Academy of Sciences*: *Engineering*, **13**, 17-25.

[10] EVS-EN 15251:2007 (2010) Indoor Environmental Input Parameters for Design and Assessment of Energy Performance of Buildings Addressing Indoor Air Quality, Thermal Environment, Lighting and Acoustics. Estonian Centre for Standardisation.

[11] Regulation of the Government of Estonia No. 38. Dwelling Requirements. 26.01.1999 (RT I 1999, 9, 38) In Estonian.

[12] EVS 839:2003 (2003) Indoor Climate. *Estonian Center of Standards* (*In Estonian*).

[13] CR 1752 (1998) Ventilation for Buildings: Design Criteria for the Indoor Environment. European Committee for Standardization, Brussels.

[14] Olesen, B.W. (2007) The Philosophy behind EN15251: Indoor Environmental Criteria for Design and Calculation of Energy Performance of Buildings. *Energy and Buildings*, **39**, 740-749. http://dx.doi.org/10.1016/j.enbuild.2007.02.011

[15] ASHRAE (1993) Handbook of Fundamentals. American Society of Refrigerating and Air Conditioning Engineers, Inc., Atlanta.

[16] European Commission (2007) 2020 Vision: Saving Our Energy. Office for Official Publications of the European Communities. http://books.google.ee/books/about/2020_vision.html?id=ujsoAQAAMAAJ&redir_esc=y

[17] Gustavsson, L., Dodoo, A. and Sathre, R. (2011) Impact of Ventilation Heat Recovery on Primary Energy Use of Apartment Buildings Built to Conventional and Passive House Standard. *World Renewable Energy Congress*, Linköping, May 2011, 8-11.

[18] Diemanu, J., Roth, K.W. and Brodrick, J. (2003) Air-to-Air Energy Recovery Heat Exchangers. *ASHRAE Journal*, **45**, 57-58.

[19] Fouih, Y.E., Stabat, P., Riviere, P., Hoang, P. and Archambault, V. (2012) Adequacy of Air-to-Air Heat Recovery Ventilation System Applied in Low Energy Buildings. *Energy and Buildings*, **54**, 29-39. http://dx.doi.org/10.1016/j.enbuild.2012.08.008

[20] Manz, H., Huber, H., Schälin, A., Weber, A., Ferrazzini, M. and Studer, M. (2000) Performance of Single Room Ventilation Units with Recuperative or Regenerative Heat Recovery. *Energy and Buildings*, **31**, 37-47. http://dx.doi.org/10.1016/S0378-7788(98)00077-2

[21] Dietz, R.N. and Goodrich, R.W. (1995) Measurement of HVAC System Performance and Local Ventilation Using Passive Perfluorocarbon Tracer Technology. *Informal Report*, BNL-61990, State University of New York, Farmingdale.

[22] Guo, L. and Lewis, O.J. (2007) Carbon Dioxide Concentration and Its Application on Estimating the Air Change Rate in Typical Irish Houses. *The International Journal of Ventilation*, **6**, 235-245.

[23] Leephakpreeda, T., Thitipatanapong, R., Grittiyachot, T. and Yungchareon, V. (2001) Occupancy-Based Control of Indoor Air Ventilation: A Theoretical and Experimental Study. *ScienceAsia*, **27**, 279-284.

http://dx.doi.org/10.2306/scienceasia1513-1874.2001.27.279

[24] Pavlovas, V. (2006) Energy Savings in Existing Swedish Apartment Buildings. Chamlers University of Technology, Göteborg.

[25] Jokl, M.V. (1998) Evaluation of Indoor Air Quality Using the Decibel Concept Based on Carbon Dioxide and TVOC. *Building and Environment*, **35**, 677-697. http://dx.doi.org/10.1016/S0360-1323(99)00042-6

[26] Kalamees, T., Ilomets, S., Arumägi, E., Alev, Ü., Kõiv, T.-A., Mikola, A., Kuusk, K. and Maivel, M. (2011) Indoor Hygrothermal Conditions in Estonian Old Multi-storey Brick Apartment Buildings. *The 12th International Conference on Indoor Air Quality and Climate*, Austin, 5-10 June 2011, 6 p.

12

Effect of Exhaust Gas Recirculation on Performance and Emission Characteristics of a Diesel-Piloted Biogas Engine

Meshack Hawi[1*], Robert Kiplimo[2], Hiram Ndiritu[1]

[1]Department of Mechanical Engineering, JKUAT, Nairobi, Kenya
[2]Department of Marine Engineering, JKUAT, Nairobi, Kenya
Email: [*]mhawi@jkuat.ac.ke

Abstract

In this research, a Direct Injection Compression Ignition (DICI) engine was modified into a dual-fuel engine that used biogas as the primary fuel and diesel as pilot fuel, with the focus on reduction of harmful exhaust emissions while maintaining high thermal efficiency. The effect of exhaust gas recirculation (EGR) on engine performance and emission characteristics was studied. The EGR system was developed and tested with different EGR percentages, *i.e.* 0%, 10%, 20% and 30%. The effect of EGR on exhaust gas temperature and performance parameters like brake specific fuel consumption, brake power and brake thermal efficiency was studied. The performance and emission characteristics of the modified engine were compared with those of the conventional diesel engine. The results showed that EGR led to a decrease in specific fuel consumption and an increase in brake thermal efficiency. With increase in percent (%) of EGR, the percentage increase in brake thermal efficiency was up to 10.3% at quarter load and up to 14.5% at full load for single fuel operation while for dual-fuel operation an increase up to 9.5% at quarter load and up to 11.2% at full load was observed. The results also showed that EGR caused a decrease in exhaust gas temperature; hence it's potential to reduce NO_X emission. However, emissions of HC and CO increased slightly with EGR.

Keywords

DICI Engine, Dual-Fuel Engine, EGR, NO_X

1. Introduction

Regulations for particulate matter (PM) and oxides of nitrogen (NO_X) emissions from diesel engines have been

[*]Corresponding author.

strengthened and therefore there is need to find alternative fuels for the engines and ways of reducing harmful exhaust emissions while maintaining high thermal efficiency. The concept of using alternative gaseous fuel in diesel engines has gained worldwide attention. Increasing petroleum fuel prices and their deterioration to environment have led to the search for alternative fuels in the past several years [1]. Biogas is one such fuel, which is produced easily by anaerobic fermentation of organic material extracted from varied sources like animal wastes, vegetable wastes, wastes from households, wastes from the food and fodder industry and waste from productive livestock husbandry [2].

Diesel engines are typically characterized by low fuel consumption and very low CO emissions. However, the NO_X emissions from diesel engines still remain high. Hence, in order to meet the environmental regulations, it is highly desirable to reduce the amount of NO_X in the exhaust gas. Diesel engines are predominantly used in heavy duty machinery such as tractors, heavy lorries, trucks and electrical generator sets. Owing to their low fuel consumption, they have become increasingly attractive for smaller lorries and passenger cars as well. But higher NO_X emissions from diesel engines remain an environmental concern. In order to reduce emission levels, some external engine features can be applied, such as EGR or after treatment systems. EGR systems have been used to reduce emissions of nitrogen oxides (NO_X) from diesel engines. Depending on the engine operating conditions, these systems divert 5% - 30% of an engine exhaust stream back to the combustion chamber [3].

In diesel engines, NO_X formation is a highly temperature dependent phenomenon and takes place when the temperature in the combustion chamber exceeds 2000 K [4]. Therefore, in order to reduce NO_X emissions in the exhaust, it is necessary to keep peak combustion temperatures under control. One simple way of reducing the NO_X emission of a diesel engine is by late injection of fuel into the combustion chamber. This technique is effective but increases fuel consumption by 10% - 15%, which necessitates the use of more effective NO_X reduction techniques like exhaust gas recirculation (EGR) [4]. Recirculating part of the exhaust gas helps reduce NO_X. When exhaust gas is recirculated, it is important to investigate the effect on engine performance characteristics as well.

Exhaust gas consists of CO_2, N_2 and water vapor mainly. When a part of this exhaust gas is recirculated to the cylinder, it acts as diluent to the combusting mixture. This also reduces the O_2 concentration in the combustion chamber. The specific heat of the EGR is much higher than fresh air; hence EGR increases the heat capacity (specific heat) of the intake charge, thus decreasing the temperature rise for the same heat release in the combustion chamber [4]. Diesel engines can be modified to operate with gaseous fuel in dual-fuel mode. Dual-fuel engines at part load inevitably suffer from lower thermal efficiency and higher emission of carbon monoxide and unburned fuel. Hosseinzadeh *et al.* [5] used a quasi-two-zone combustion model developed for studying the second-law analysis of a dual-fuel (diesel-gas) engine operating under part load conditions. They made an attempt to investigate the combustion phenomenon from second-law point of view at part load and use exhaust gas recirculation (EGR) to improve the aforementioned problems. Therefore, the availability analysis was applied to the engine from inlet valve closing (IVC) until exhaust valve opening (EVO). Various availability components were identified and calculated separately with crank position. The different cases of EGR (chemical, radical and thermal cases) were then applied to the availability analysis in dual-fuel engines at part loads. They found that the chemical case of EGR had negative effect and in this case the unburned chemical availability was increased and the work availability decreased in comparison with baseline engine (without EGR). The thermal and radical cases had positive effects on the availability terms especially on the unburned chemical availability and work availability. The results showed that second-law efficiency was increased by using low amount of radical and thermal cases of EGR.

Their results showed that radical, thermal and combined cases of EGR had positive effects on availability terms. With employment of these cases of EGR, the combustion process improved due to enhancement of the oxidation process in lean mixtures. Also in these cases, the presence of very active radicals and a higher intake charge temperature could promote the combustion process and therefore reduced the chemical availability of unburned fuel in the exhaust gases. Their results also indicated that the dual-fuel engine can obtain maximum work availability, minimum destruction availability and minimum chemical availability of unburned fuel when 2% of thermal or combined cases of EGR was introduced to the dual-fuel engine under part load operating conditions.

2. Description of the Experimental Procedure

The engine typically used for this study is a single cylinder four stroke Direct Injection (DI) diesel engine. It is water cooled, naturally aspirated constant speed compression ignition engine whose major specifications are shown in **Table 1**. The engine was coupled to a hydraulic dynamometer through which load was applied by increasing the water supply to rotor blades through a centrifugal pump. The engine was tested at 0, 25, 50, 75 and 100 percent brake load conditions. The engine had capability to run either on pure diesel or dual-fuel mode. The engine was modified to run on biogas by introducing it in the intake manifold pipe via a mixing device. The biogas flow rate was kept fixed for a given speed and load, and a variable pilot injection quantity controlled by a manual regulator fixed at the injection pump. The experimental apparatus is shown in **Figure 1**.

Biogas generated from water hyacinth, in a digester was collected in a flexible bag (3 m^3) and used for tests in the laboratory (engine room). The properties of the gas are shown in **Table 2**. The gas was then supplied to the engine through a biogas transfer pump. The volume flow rate of the gas was controlled by a valve and measured using a gas flow meter before being admitted into the intake manifold. For every test, the initial and final meter reading was recorded as well as time taken to consume the gas. The volume of gas consumed by the engine was calculated by subtracting the initial from the final meter reading. Mass flow rate was calculated using volume of gas consumed, density of the gas and the time. The power output of the engine was measured using a Fuchino SF-3.5 hydraulic dynamometer by recording the dial reading (in kgf) and then converting into brake power (hose power) by applying Equation (1). The EGR circuit consisted of a stainless steel pipe and stainless steel butterfly valves fitted to control the quantity of exhaust gas being re-circulated. The volume of pilot (liquid) fuel consumed was measured by a calibrated glass tube (burette) by measuring the time required for the consumption of 10 ml of fuel. During exhaust gas recirculation, the EGR control valves were used to regulate the amount of exhaust allowed back into the combustion chamber. During the experiments, dynamometer reading (load), engine speed, fuel consumption and exhaust gas temperature were recorded. Exhaust gases were analyzed on line by Horiba MEXA-544GF emission analyzer in which HC, CO and CO_2 were measured. A portable Laser Type digital tachometer-RS 445-9557 was used to measure the engine speed at all operating conditions. Thermocouples were fixed at the exhaust manifold and at engine coolant inlet and outlet to the engine to measure the temperature of exhaust gas, temperature of cooling water at entry to the engine and temperature of cooling water at exit from the engine respectively.

Table 1. Engine specifications.

S/No.	Specification	Value
1	Number of cylinders	1
2	Number of strokes	4
4	Rated power (kW)	7.5
5	Speed (rpm)	1500
6	Compression ratio	16.5:1
7	Fuel consumption (g/kWh)	251
8	Injection pressure (bar)	200
9	Injection timing (^{0}btdc)	23
10	Cylinder capacity (cc)	950
11	Combustion system	Direct injection

Table 2. Properties of biogas.

S/No.	Quantity	Value
1	Methane (%)	49 - 53
2	Carbon dioxide (%)	21 - 29
3	Nitrogen (%)	17 - 19
4	Density (kg/m^3)	1.22
5	LHV (kJ/kg)	19,100
6	Octane number	130
7	Auto-ignition temperature (˚C)	632 - 813

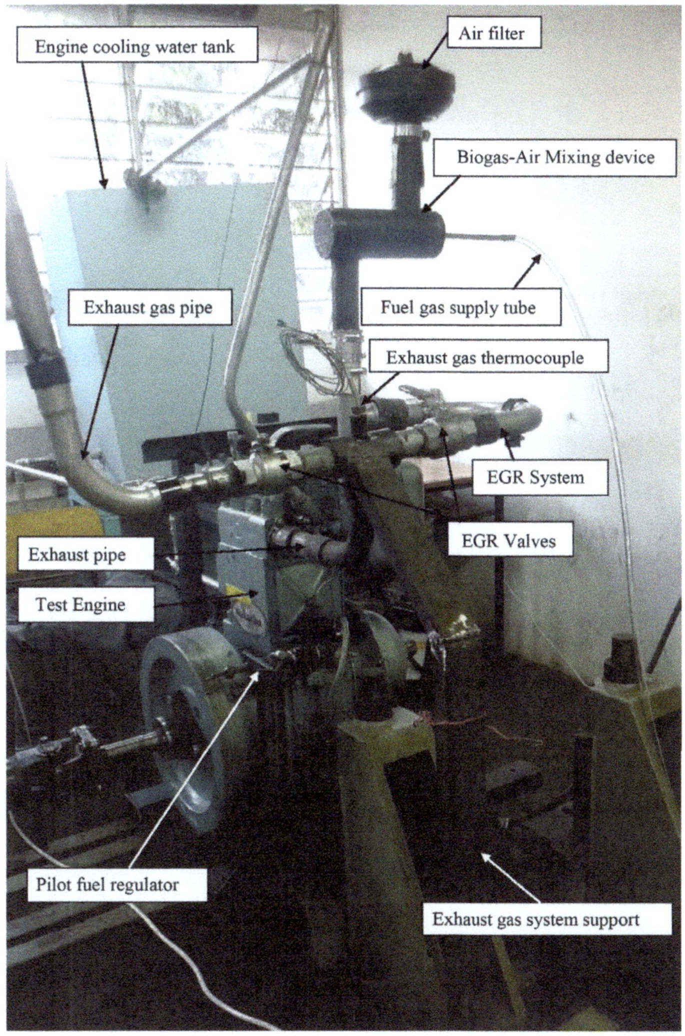

Figure 1. Photograph of the experimental set up.

3. Performance Analysis of the Engine

The following equations were used in analysis of the collected data.

Engine brake horse power:

$$P.S = \frac{W \times N}{1000} \tag{1}$$

where $P.S$ = brake horse power, W = torque indication in kg, N = revolution per minute.

Specific fuel consumption:

$$SFC = \frac{\dot{m}_f}{\dot{W}} \tag{2}$$

where SFC = specific fuel consumption, $\dot{m}_f$ = rate of fuel flow into engine and W = engine power.

Brake specific fuel consumption:

$$BSFC = \frac{\dot{m}_f}{\dot{W}_b} \tag{3}$$

where $BSFC$ = brake specific fuel consumption, $\dot{m}_f$ = rate of fuel flow into engine and W_b = brake engine power.

Thermal efficiency:

$$\eta_{th} = \frac{\dot{W}}{\dot{m}_f \times CV} \tag{4}$$

where η_{th} = thermal efficiency, W = engine power, $\dot{m}_f$ = rate of fuel flow into engine and CV = calorific value of fuel.

Brake thermal efficiency:

$$\eta_{bth} = \frac{\dot{W}_b}{\dot{m}_f \times CV} \tag{5}$$

where η_{bth} = brake thermal efficiency, W_b = brake power, $\dot{m}_f$ = rate of fuel flow into engine and CV = calorific value of fuel.

Brake thermal efficiency of engine in dual-fuel mode:

$$\eta_{bth} = \frac{\dot{W}_b}{\dot{m}_{fd} \times CV_d + \dot{m}_{fg} \times CV_g} \tag{6}$$

where BTE = η_{bth} = brake thermal efficiency, W_b = brake power, $\dot{m}_{fd}$ = rate of flow of diesel fuel into engine, $\dot{m}_{fg}$ = rate of flow of biogas fuel into engine, CV_d = calorific value of diesel fuel and CV_g = calorific value of biogas fuel.

Diesel Substitution:

$$ds = \frac{D_d - D_{dg}}{D_d} \times 100 \tag{7}$$

where ds = diesel substitution, per cent, D_d = diesel consumption by the engine on single fuel mode in kg/h, D_{dg} = diesel consumption by the engine on dual-fuel mode in kg/h.

Percentage of exhaust gas recirculation [3]:

$$\%\text{EGR} = \frac{\text{Volume of EGR}}{\text{Total Intake Charge to the Cylinder}} \times 100 \tag{8}$$

4. Results and Discussions

4.1. Engine Performance

4.1.1. Brake Specific Fuel Consumption (BSFC)

Figure 2 shows the variation of BSFC with engine load for various percentages of EGR; by varying the percentage of exhaust gas which is recirculated under 0%, 25%, 50%, 75% and 100% load. It was found that brake specific fuel consumption decreased with increase in EGR and reached a minimum value at about 20% EGR and then increased slightly up to 30% EGR from 50% load to full load. The decrease in BSFC with EGR is due to increase in intake charge temperature which increases the rate of combustion of the fuel, hence causing a decrease in BSFC.

4.1.2. Brake Power

Figure 3 shows the variation of engine power output with load. It shows that EGR has no significant effect on the power output of the engine up to 30% EGR. This is because the recirculated gases do not displace too much of the air required for combustion which could cause a decrease in power output.

4.1.3. Brake Thermal Efficiency

From **Figure 4**, showing variation of BTE with engine load, the brake thermal efficiency increases with increase in engine load for all operating modes. The BTE increases also with EGR up to about 20% EGR and then starts to fall. The reason for increase in BTE with EGR is due to re-burning of HC that enters combustion chamber with the recirculation of exhaust gases and also EGR increases intake charge temperature which increases the rate of combustion [6]. Recirculation of too much exhaust gases, however, displaces much of the necessary air for combustion, hence leading to decrease in thermal efficiency.

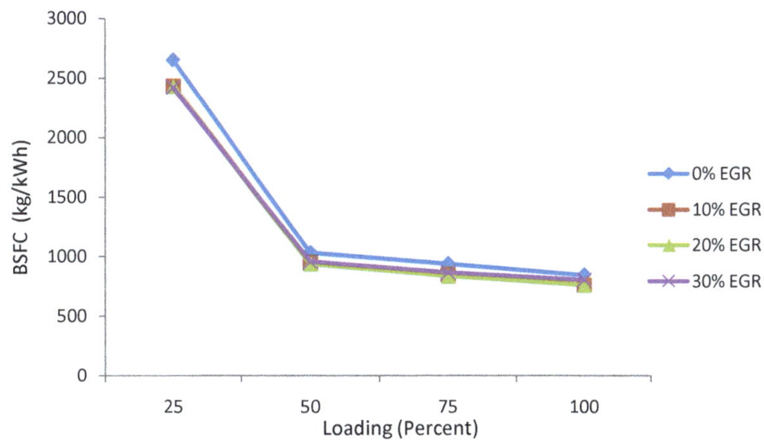

Figure 2. Variation of brake specific fuel consumption with engine load.

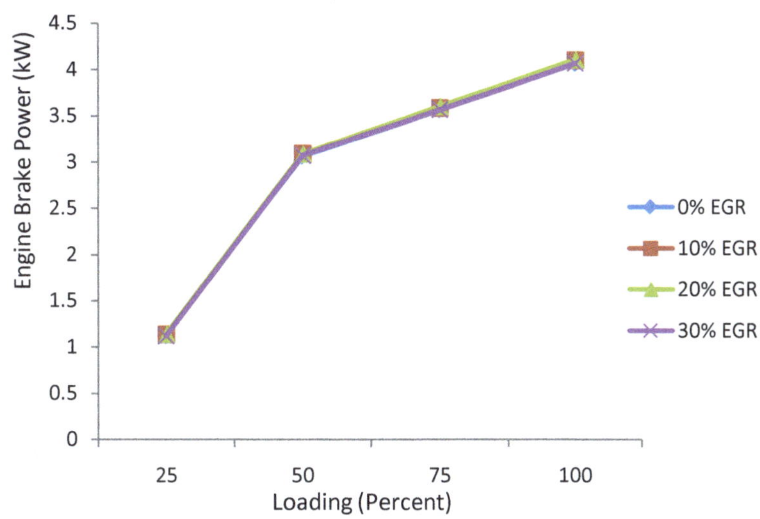

Figure 3. Brake power vs engine load.

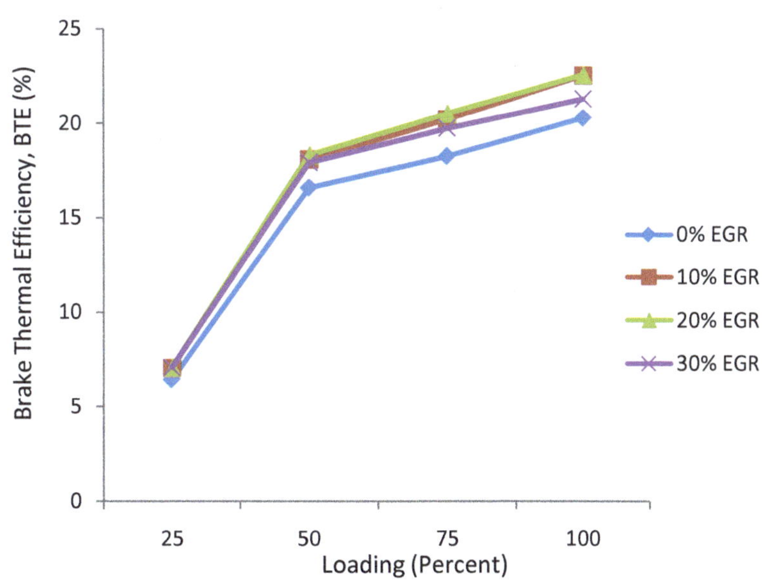

Figure 4. Brake thermal efficiency vs engine load.

4.1.4. Exhaust Gas Temperature (EGT)

Figure 5 shows the variation of the exhaust gas temperature with engine load for various conditions of EGR. The exhaust gas temperature was found to increase with increase in engine load for all operating modes. This is due to the increase of total energy input at high load following higher fuel consumption. It was also found that with increase in percentage of EGR, the exhaust gas temperatures reduced. This can be attributed to the oxygen deficient operation under EGR which results in lower combustion temperatures and furthermore specific heat of exhaust gas is more than that of intake air which also contributes to the lower combustion temperatures [6]. However, the exhaust gas temperatures were found to decrease with increase in percent EGR only up to about 20% EGR, beyond which it started to increase. This shows that beyond 20% the exhaust gas starts to add heat in the combustion chamber rather than absorbing it.

4.2. Emissions

4.2.1. Variation of Carbon Monoxide Emission with Load

Figure 6 shows the variation of CO emissions with increase in percentage of EGR at 0%, 25%, 50%, 75% and

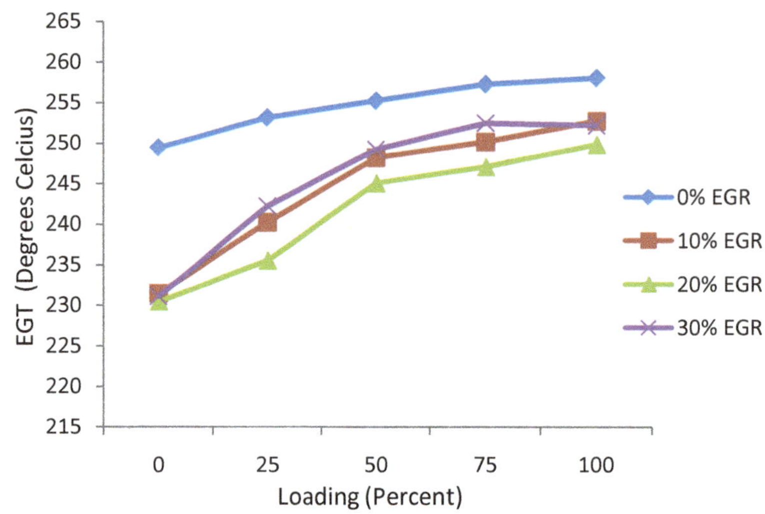

Figure 5. Exhaust gas temperature vs engine load.

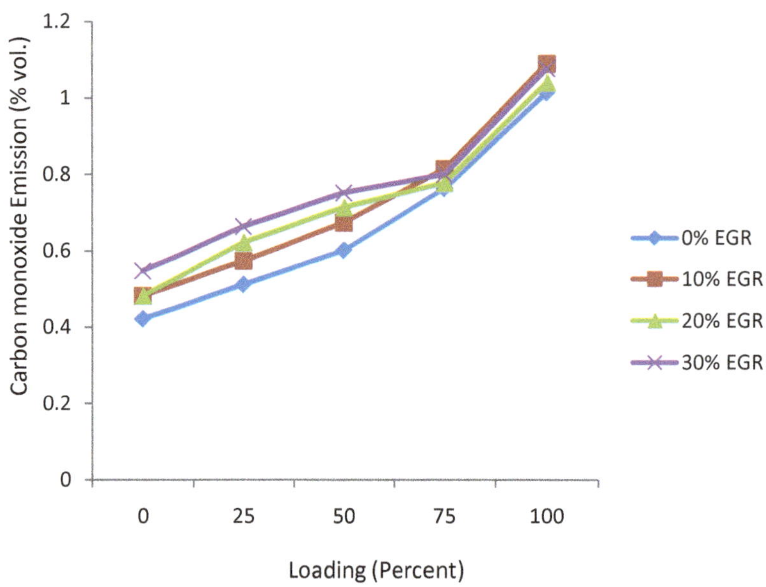

Figure 6. Carbon monoxide emission vs engine load.

100% loads. With increase in the percentage of EGR, CO increases; however, the effect of CO emission is found to be less at higher engine loads of 75% and 100%. The emissions are found rapidly increasing with EGR up to 75% load and beyond which there is only an insignificant rise. The deficiency of oxygen with the increase in EGR percentage can be attributed to the increase in CO emission for EGR.

4.2.2. Unburned Hydrocarbon Emission

Figure 7 shows the variation of HC emission with increase in percentage of EGR at 0%, 25%, 50%, 75% and 100% loads. It can be seen from the figure that for all loading conditions, the HC emission increases with increase in EGR. The increase in HC emission is as a result of increase in the CO_2 content of the inducted mixture instead of fresh-air. The figure also shows that unburned hydrocarbon emission decreases with increase in load for all EGR ratios. Normally dual-fuel operation exhibits higher emission of unburned hydrocarbon at light loads. At light loads the pilot quantity being small so flame cannot propagate fast and far enough to ignite the entire mixture. As the result it causes higher HC emissions but with increase in load the hydrocarbon emission decreases. As load progresses the pilot quantity increases and burns the surrounding fuel-air mixture sufficiently [7].

4.2.3. Carbon Dioxide Emission

From **Figure 8**, emission of carbon dioxide was found to increase with engine load while the same decreased slightly with EGR from 10% to 30% EGR. This is as a result of substitution of some of the fresh air with exhaust gas leading to decrease in CO_2 which is a product of fuel combustion.

5. Conclusions

In this work, a single cylinder four stroke diesel engine was modified into a dual-fuel engine that used diesel as pilot fuel and biogas as the primary fuel. An experimental study was then conducted on the engine to investigate various effects of EGR on the performance and emission characteristics of a dual-fuel single cylinder four stroke DI diesel engine under various experimental conditions. The following conclusions were drawn from the analysis:

1) It was found that with increase in percent of EGR, the brake thermal efficiency increased and specific fuel consumption decreased. It was observed that with rise in % of EGR the percentage increase in brake thermal efficiency was up to 10.3% at quarter load and up to 14.5% at full load for single fuel operation while for dual-fuel operation an increase up to 9.5% at quarter load and up to 11.2% at full load was observed. From experimental results it was found that an EGR percentage of around 20% resulted in maximum BTE and minimum BSFC.

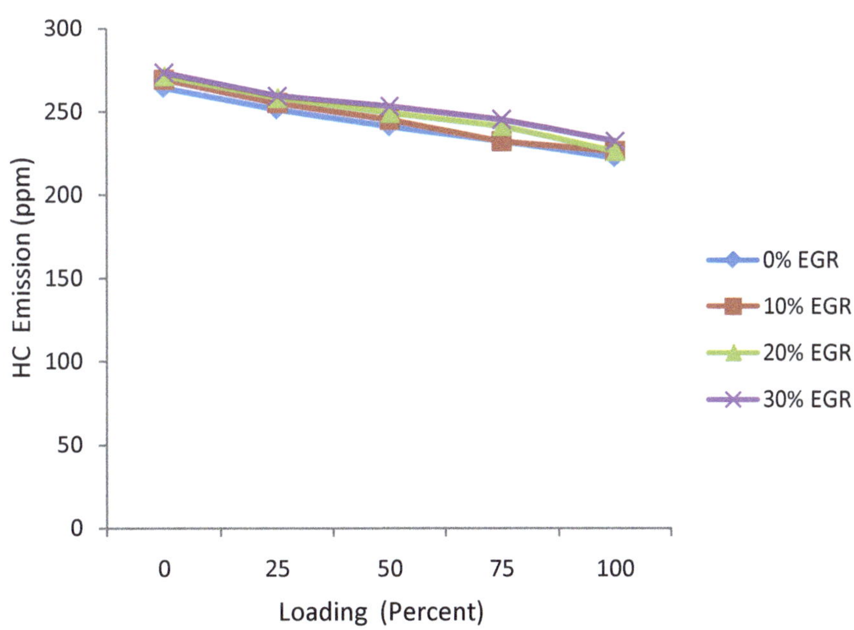

Figure 7. Unburned hydrocarbon emission vs engine load.

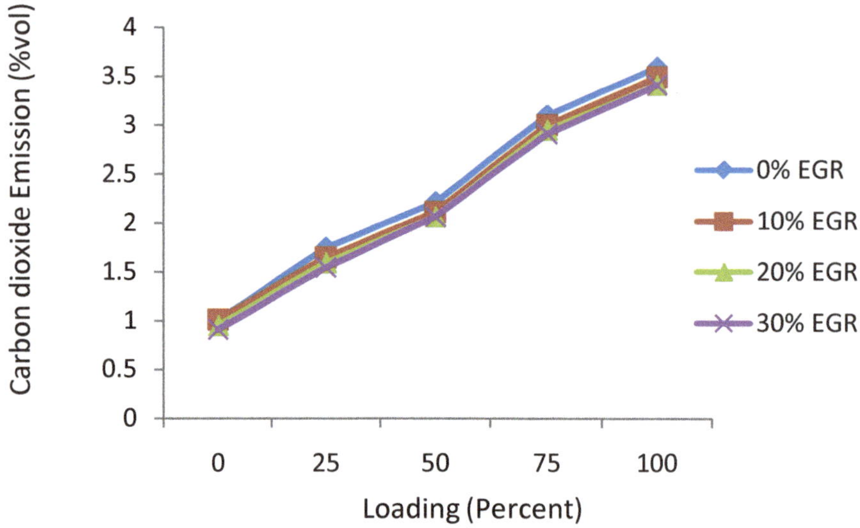

Figure 8. Carbon dioxide emission Vs Engine load.

Brake thermal efficiency increased at low EGR ratios due to the recirculation of active radicals from EGR that made the combustion process to be enhanced, so resulting in an improvement in brake thermal efficiency.

2) Results also showed that the brake power was not affected significantly by EGR.

3) Experiments were carried out using the setup to prove the efficacy of EGR as a technique for NO_X reduction. It was seen that the exhaust gas temperatures reduced drastically by employing EGR. This indirectly showed the potential for reduction of NO_X emission. This can be concluded from the fact that the most important reason for the formation of NO_X in the combustion chamber is the high temperature of about 2000 K at the site of combustion. It was found that EGR reduced the exhaust gas temperature by up to 7.6% at minimum engine loading and up to 2.3% at maximum loading.

4) EGR caused a slight increase in Carbon monoxide (CO) and Hydrocarbon (HC) emissions. Increasing EGR reduced the amount of oxygen and led to incomplete combustion and therefore increased CO and HC emission due to lower combustion temperature.

Acknowledgements

The authors gratefully recognize the support from Japanese International Corporation Agency (JICA) for providing the diesel engine, electronic diesel injection pump and electronic diesel fuel injector used in this research. The authors would also like to highly appreciate Jomo Kenyatta University of Agriculture and Technology (JKUAT) for funding this research and for providing laboratory facilities for conducting experiments.

References

[1] Karim, G.A. (2003) Combustion in Gas Fueled Compression: Ignition Engines of the Dual Fuel Type. *Journal of Engineering for Gas Turbines and Power*, **125**, 827-836. http://dx.doi.org/10.1115/1.1581894 http://gasturbinespower.asmedigitalcollection.asme.org/article.aspx?articleid=1421670

[2] Tippayawong, N., Promwungkwa, A. and Rerkkriangkrai, P. (2007) Long Term Operation of a Small Biogas/Diesel Dual-Fuel Engine for On-Farm Electricity Generation. *Bio-Systems Engineering*, **98**, 26-32. http://dx.doi.org/10.1016/j.biosystemseng.2007.06.013 http://www.authorstream.com/Presentation/ravikantghotekar-2049447-long-term-operation-small-biogasdiesel-dual-fuel-engine-fa/

[3] Kumar, N.R., Sekhar, Y.M.C. and Adinarayana, S. (2013) Effects of Compression Ratio and EGR on Performance, Combustion and Emissions of Direct Injection Diesel Engine. *International Journal of Applied Science and Engineering*, **11**, 41-49. http://www.cyut.edu.tw/~ijase/2013/11%281%29/5_023008.pdf

[4] Agrawal, A.K., Singh, S.K., Sinha, S. and Shukla, M.K. (2004) Effect of EGR on the Exhaust Gas Temperature and Exhaust Opacity in Compression Ignition Engines. *Sadhana*, **29**, 275-284. http://www.researchgate.net/publication/257925344_Effect_of_EGR_on_the_exhaust_gas_temperature_and_exhaust_

opacity_in_compression_ignition_engines

[5] Hosseinzadeh, A., Saray, R.K. and Mahmoudi, S.M.S. (2010) Comparison of Thermal, Radical and Chemical Effects of EGR Gases Using Availability Analysis in Dual-Fuel Engines at Part Loads. *Energy Conversion and Management*, **51**, 2321-2329. http://dx.doi.org/10.1016/j.enconman.2010.04.005 http://www.researchgate.net/publication/245160135_Comparison_of_thermal_radical_and_chemical_effects_of_EGR_gases_using_availability_analysis_in_dual-fuel_engines_at_part_loads

[6] Venkateswarlu, K., Murthy, B.S.R.C. and Subbarao, V.V. (2013) The Effect of Exhaust Gas Recirculation and Di-Tertiary Butyl Peroxide on Diesel-Biodiesel Blends for Performance and Emission Studies. *International Journal of Advanced Science and Technology*, **54**, 49-60. http://www.sersc.org/journals/IJAST/vol54/5.pdf

[7] Mahla, S.K., Das, L.M. and Babu, M.K.G. (2010) Effect of EGR on Performance and Emission Characteristics of Natural Gas Fueled Diesel Engine. *Jordan Journal of Mechanical and Industrial Engineering*, **4**, 523-530. http://jjmie.hu.edu.jo/files/v4n4/JJMIE-41-09_Revised%2812%29/JJMIE-41-09_modified.pdf

Nomenclature

DICI—Direct Injection Compression Ignition
DI—Direct Injection
CI—Compression Ignition
SI—Spark Ignition
BTE—Brake Thermal Efficiency
BSFC—Brake Specific Fuel Consumption
EGT—Exhaust Gas Temperature
PM—Particulate Matter
CO—Carbon Monoxide
CO_2—Carbon Dioxide
HC—Hydro Carbons
NO_X—Oxides of Nitrogen
LHV—Lower Heating Value

13

Behavior an Induction Generator without and with a Voltage Regulator

M. L. Elhafyani[1*], S. Zouggar[1], M. Benkaddour[2], A. Aziz[1]

[1]Laboratory of Electrical Engineering and Maintenance (LEEM), University Mohammed 1, Oujda, Morocco
[2]Department of Physics, Faculty of Science, University Mohammed 1, Oujda, Morocco
Email: [*]m_elhafyani@yahoo.fr

Abstract

During the isolated use of a wind system, the output voltage of the self-excited induction generator depends on the variation characteristic of its parameters: the excitation condensers, the drive speed and the load. Therefore, the regulation of the tension appears to be of great interest. We focused on the use of an analogical regulator of tension, with the aim of controlling the tension at the exit of the self-excited induction generator. So we modelled, implanted and simulated a wind system (Self-excited induction generator, converters (AC/DC, DC/DC) and load it) in the Orcad/Pspice environment. In the first time the behaviour of the asynchronous generator was analyzed when the load, the excitation capacitor and the drive speed vary in the absence of any form of regulation. This analysis was conducted with the aim of defining the limits of the machine exploitation. In the second time the functioning mode is controlled by an analogical control of tension. The results of simulation show the good performances of the system during the application of the proposed voltage regulator.

Keywords

Wind Power, Self-Excited Induction Generator, Converters (Buck-Boost), Regulation

1. Introduction

The exploitation of the renewable energy sources raises a growing interest, not only in the Third World countries, but also in the industrialized countries, particularly for the developing countries and for any isolated site generally. The use of the mechanical converters (synchronous machine, asynchronous machine), is a subject of studies and research in the range of products of the electric power [1]-[12]. As far as the production of low

[*]Corresponding author.

power electrical energy (<50 kw) is concerned, the three-phase self-excited induction generator (SEIG) with cage or wound-rotor, competes as a generator with the synchronous machine. This is due primarily to its sturdiness, robustness, its simplicity of excitation (simple connection of condensers at the stator boundaries) and its relatively modest cost.

During the isolated use of the asynchronous machine [1] [10]-[12], a variation of the capacity, load or diver speed involves an imbalance of the stator output voltage. This problem can be solved by operating, the condensers, the load or the drive speed by the intermediary of a regulator.

In this paper, we present the capacity variation effect, speed of drive and load as function of the wind system without and with analogical regulator.

2. Behavior of the Machine without the Voltage Regulator

2.1. Modelling of the Wind System

2.1.1. Equivalent Circuit of the Wind System

The **Figure 1** represents the synoptic diagram of an asynchronous machine (1.5 KW) related to on the energy converter (AC-DC) and feeds a resistive load (R_S):

The Induction Asynchronous Generator does not excite without external power given to its stator, the rotor windings are shorted. Rotor winding will have current only when there is connection of voltage from Grid/battery operated AC source.

The initiation of the generator is possible only from a speed-dependent threshold value of the capacitances and the magnetic state of the machine (linear portion of the magnetization curve) (**Figure 2**).

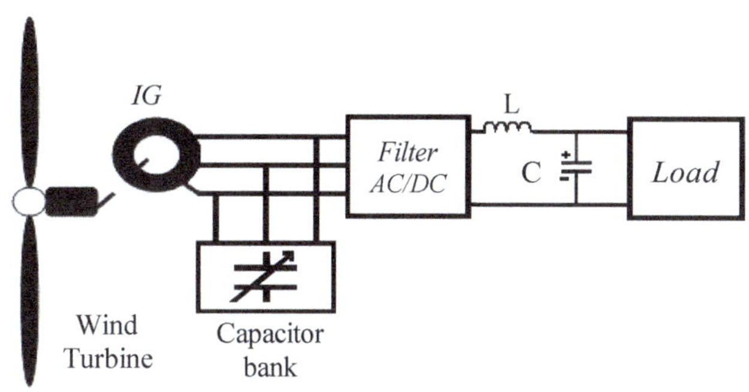

Figure 1. Synoptic diagram of a SEIG with a rectifier dependent has a resistive load.

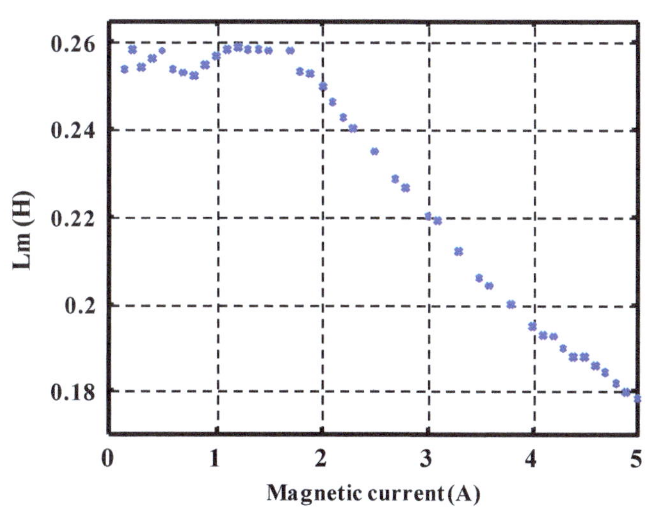

Figure 2. Experimental curve of the magnetizing inductance [4].

2.1.2. Modelling of the Wind-Turbine

By applying the theory of momentum and Bernoulli's theorem, we can determine the incident power (theoretical power) due to wind:

$$P_{inci} = \frac{1}{2} \cdot \rho \cdot A \cdot V_w^3 \tag{1}$$

The wind turbine acts like a primary strength to train the induction generator [7]. The power produced by the wind turbine part is given by the following formula

$$P_{wt} = \frac{1}{2} \cdot C_P(\lambda) \cdot \rho \cdot A \cdot V_w^3 \tag{2}$$

where:

ρ: the air density ($\rho = 1.225$ kg/m^3 at atmospheric pressure);
A: the swept area (m^2);
C_p: the coefficient of wind turbine;
V_w: the wind velocity (m/sec);

$C_p(\lambda)$ is called the power coefficient, which expresses the aerodynamic efficiency of the turbine. It depends on the ratio λ, which represents the ratio between the speed at the end of the blades and the wind speed. The ratio λ can be expressed by the following relation:

$$\lambda = \frac{\Omega \cdot R}{V_w} \tag{3}$$

2.1.3. Symbolization of the Mechanical Equation (See Figure 3)

The implementation and the simulation of the self-excited asynchronous machine were achieved by using the Orcad-Pspice tool whose libraries are rich in electrical and analogical components, and by elaborating symbolization procedures [13]-[15].

In electrical technology, there is no software for simulating mechanical parameters directly. We did this simulation on the Orcad/Pspice software-using the following analogies:

Mechanical	Electrical
Speed	Voltage
Torque	Current
Moment of inertia	Capacitance of a capacitor
Viscous friction	The inverse of resistivity

The mechanical equation of the induction machine is given by the following formula:

$$J\frac{d\omega_{mr}}{dt} = T_e - B_m \cdot \omega_{mr} - T_r \quad \text{or} \quad \omega_{mr} = \frac{2}{p}\omega_r \tag{4}$$

$$T_e = \left(\frac{2}{p}\right) J \frac{d\omega_r}{dt} + \left(\frac{2}{p}\right) B_m \cdot \omega_r + T_r \tag{5}$$

The electromagnetic torque T_e is given by:

$$T_e = \sqrt{3}\left(\frac{2}{p}\right) Lm\left(i_{as} \cdot i_{\gamma r} - i_{cs} \cdot i_{\alpha r}\right) \tag{6}$$

With:

J: Moment of inertia;
T_e: Couple électromagnétique en N·m;
T_r: Load torque (N·m);
B_m: The constant friction (kg·m^2/s);

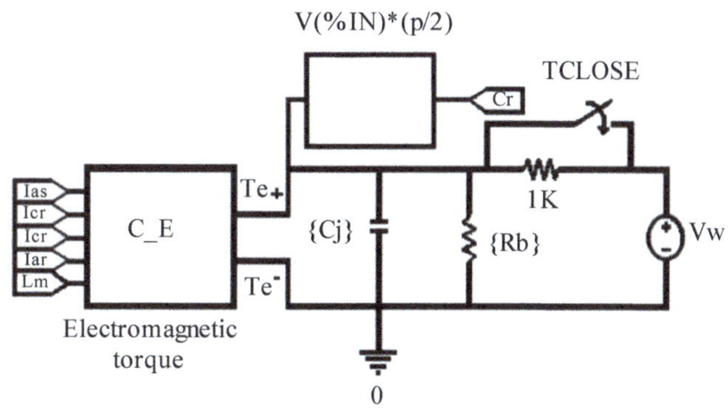

Figure 3. Electrical diagram of the mechanical equation.

ω_r *et* ω_{mr}: Electrical and mechanical speed (rad/s).

Applying the analogy between mechanical and electrical variables, we have:

$$J\frac{d\omega}{dt} = \sum_k T_k \Rightarrow C\frac{dV}{dt} = \sum_j i_j$$
$$\frac{1}{B_m} \Rightarrow R \tag{7}$$

2.2. Simulation Results without the Voltage Regulator

2.2.1. Speed Variations Effect

Figure 4 represents the stator voltage variation at the output of the rectifier according to the drive speed of the machine for a capacity of C = 80 μF and a load R_S = 70 Ω. We note that the voltage increases with the speed (the nominal voltage $220\sqrt{2}$ corresponds to a speed of the ω_r = 298 rad/s).

2.2.2. Load Variations Effect

Figure 5 illustrates the evolution of the stator tension on the outlet side of the rectifier according to the load at different values of the drive speed, and at the excitation capacitor C = 80 μF. One fixes shows that a variation speed or load involves a variation of the tension (the nominal voltage $220\sqrt{2}$ V corresponds to a load R_S = 70 Ω, a C = 80 μF capacity and a speed ω_r = 298 rad/s).

2.2.3. Capacity Variations Effect

Figure 6 represents the voltage waveform evolution of a stator phase (**Figure 1**) according it's of the self excitation capacitor C. At several values of the speed, one observes that the voltage waveform increases with the capacity value knowing well that the induction generator cannot start unless the excitation capacitor has reached a certain threshold.

The objective of these tests is to study the behavior of the asynchronous machine (SEIG) during variation of the speed, capacity and load, and to determine the exploitation limits the generator without resorting to any form of regulation.

3. Modeling and Simulation the Wind System in the Orcad-Pspice Environment

3.1. Behaviour of the Machine with the Voltage Regulator

The synoptic diagram equivalent of the system (asynchronous machine, energy converters (AC-DC, DC-DC, regulator, and load) is given in **Figure 7**. The voltage regulator is intended to control the output voltage of the reversible inverter Buck-Boost in spite of the variations of speed, load and capacity.

This diagram is implemented and simulated in the Orcad-Pspice environment.

Generally the conception and the modeling of a wind system (self-excited induction machine, static converters (AC/DC and DC/DC)) require the use of performing simulators on electronics and electrical engineering to implement each block of the wind system and in particular the model of the self-excited induction generator in

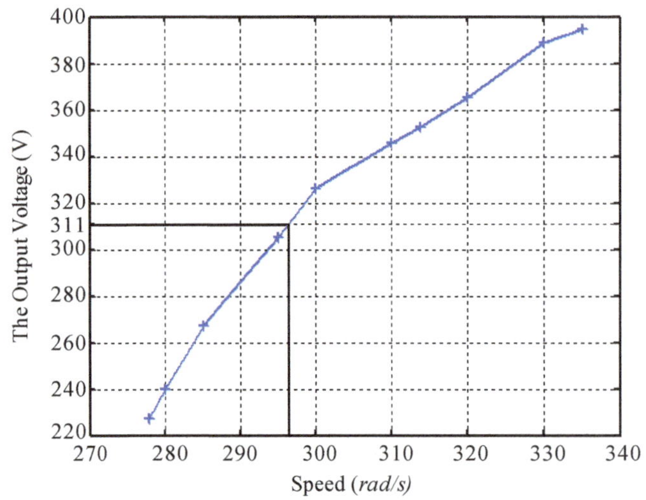

Figure 4. The output voltage evolution of the system (**Figure 1**) as a function of the drive speed at ($C = 80$ μF and $R_S = 70$ Ω).

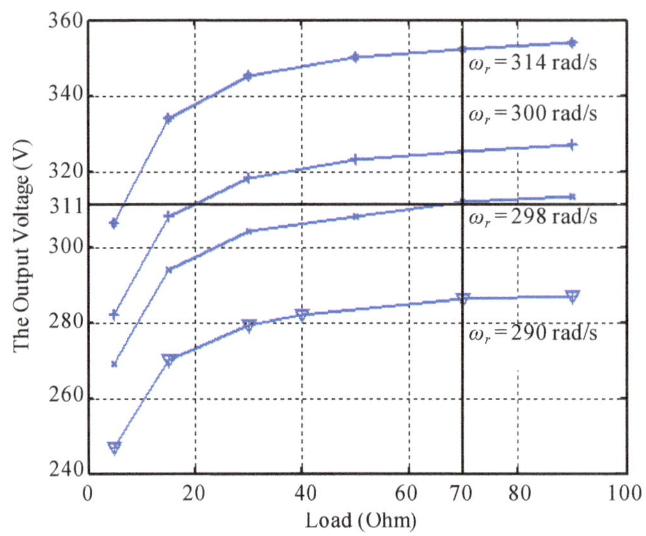

Figure 5. Variation of the output tension of the rectifier as a function of the load for different speeds ($C = 80$ μF).

transient state. In our research framework, the Orcad/Pspice software can largely meet our need in electric simulations.

The main advantages of this software are:

- Orcad/Pspice simulation software is the most widely used in the industry.
- It is based on the industry standard SPICE and therefore gives access to libraries of models developed by industrial manufacturers, and the availability of a large number of libraries on the market.
- It allows the combination of digital and analog components without any problems.
- It allows the ability to integrate Orcad electrical schematics in Matlab/Simulink tool especially with the version 16 of Orcad/Pspice.
- It is a very complete software because it allows to simulate all aspects of systems that we find in electrical engineering: power electronics, analog and digital electronic control, slaving, and mechanics.

3.2. Modelling of the Voltage Regulator

The model of the voltage regulator is given in **Figure 8**.

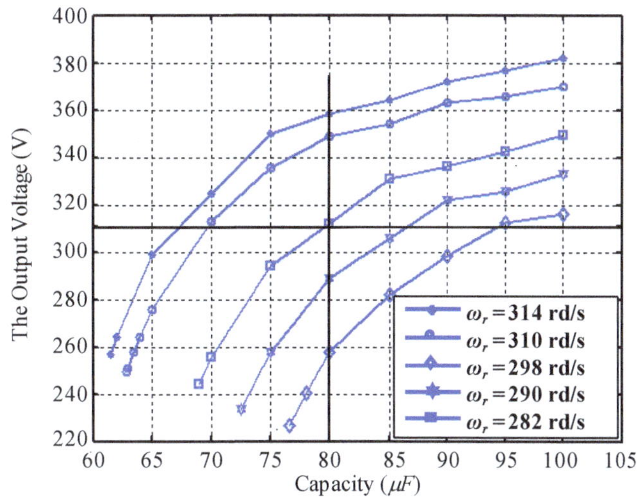

Figure 6. Evolution of the output tension of the system (**Figure 1**) according to the capacity for different speeds at (R_S = 70 Ω).

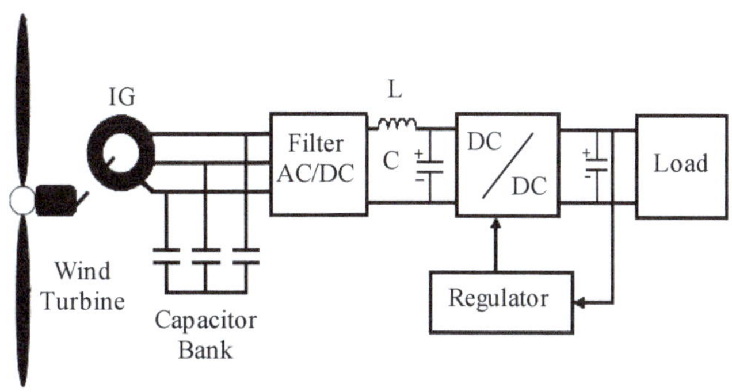

Figure 7. Descriptive diagram of the system.

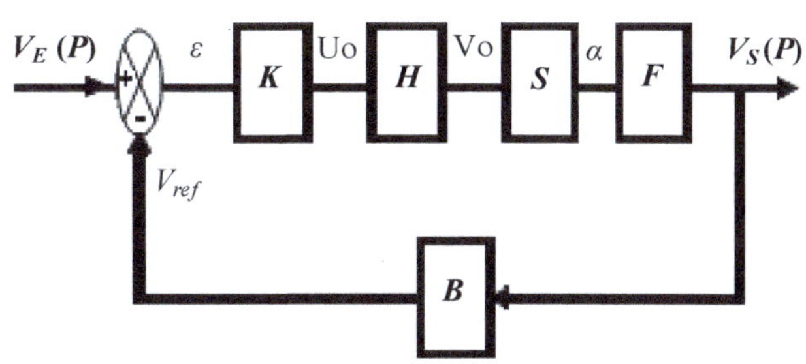

Figure 8. Equivalent circuit of the voltage regulator.

The different transfer functions of the various elements of **Figure 8** are as follows:
B: divider of return:

$$B = \frac{V_{ref}}{V_s} \tag{8}$$

K: Corrector Proportional-Integrator PI

$$K(P) = \frac{U_0(P)}{\varepsilon(P)} = \frac{A_0}{1+\tau p} \tag{9}$$

H: Corrector Integrator

$$H(P)=\frac{V_0(P)}{U_0(P)}=\frac{1}{\tau_1 p} \tag{10}$$

S: Comparator circuit Puls Wave Modulation (PWM):

$$S(P)=\frac{\alpha(P)}{V_0(P)} \tag{11}$$

The transfer functions of the inverter:

$$F(P)=\frac{V_s(P)}{\alpha(P)}=\frac{V_E}{k\cdot p(1+\tau_2 p)} \tag{12}$$

with: α = Cyclic ratio

P: Laplacian operator

The total function of transfer of the system is as follows:

$$\begin{gathered} Ht(p)=\frac{Vs}{Ve} \\ Ht(p)=K(p)\cdot S(p)\cdot F(p)\cdot H(p)\cdot B \end{gathered} \tag{13}$$

with:

$$A_o = 2.2\times10^{-3},\ \tau = 12.69,\ \tau_1 = 0.0423\ et\ S = 1.3\times10^{-3}.$$

3.3. The Analog Voltage Control Algorithm

When you submit your final version, after your paper has been accepted, prepare it in two-column format, including figures and table We have implemented in the environment Orcad/Pspice the wind system of **Figure 7** without the Analog Voltage Control (*AVC*) with taking into account the parameters of the converter given in the previous paragraph. Then we have analyzed the adaptation of the generator *SEIG* (LEROY SOMER: LA3 loop asynchronous machine 1.5 KW) and studied the effect of the variation of the resistive load (R_S), the excitation capacity, the drive speed and the duty cycle to the output system.

The *IRF*450 switch of the converter is controlled by a pulse generator outputting a square wave of high frequency 100 kHz and a variable duty ratio.

3.3.1. Influence of the Load and Duty Cycle of the Voltage

In **Figure 9**, we have shown the output voltage of the converter according to the R_S load and the duty cycle α for an excitation capacity C = 70 µF and a driving speed ω_r = 298 rad/s. From these curves of the figure below we deduced that the output voltage of the converter increases with increasing the R_S load and the duty cycle.

3.3.2. Influence of the Capacity and the Duty Cycle on the Output Voltage

Figure 10 illustrates the voltage at the output of the Buck/ Boost converter as a function of the capacity and the duty ratio α for a resistive load $R_S = 70\ \Omega$ and a driving speed ωr around 314 rad/s. From these curves we have observed that the output voltage of the converter depends on the self-excitation capacity and the duty cycle.

The voltage required by our specifications $(V_S = 311\ \text{V})$ is given by a capacitor $C = 85\ \mu\text{F}$ and a ratio α of 0.25.

3.3.3. Influence of the Speed and the Duty Cycle on the Output Voltage

Figure 11 shows the output voltage of the converter as a function of the load and the duty cycle α for an excitation capacity of 75 µF and a resistive load of 70 Ω.

From these curves we found that the output voltage of the converter also varies with the duty cycle when the speed varies.

As to the Buck/boost converter, the variations in the output voltage according to the duty ratio, obtained in the simulation Orcad/Pspice, are consistent with the relation (14) and the proportionality of the duty ratio with the voltage at the variation in the mechanical and electrical quantities of the machine.

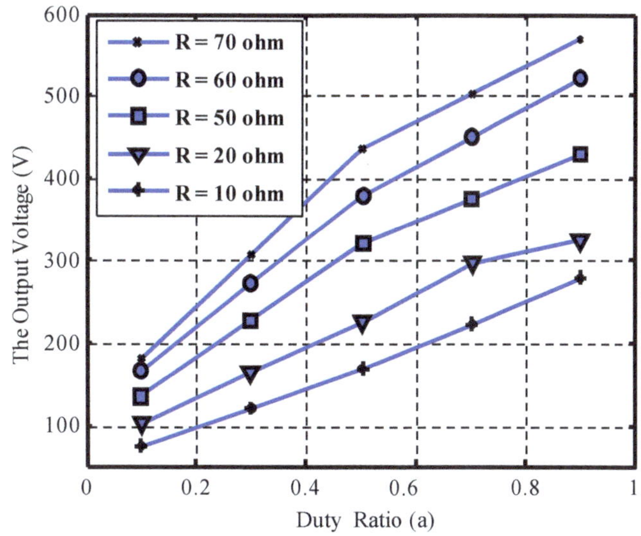

Figure 9. Typical simulation of the voltage V_S at the output of the buck/boost converter in function of the load and the duty ratio: $C = 70\ \mu\text{F}$ and $\omega_r = 298$ rad/s.

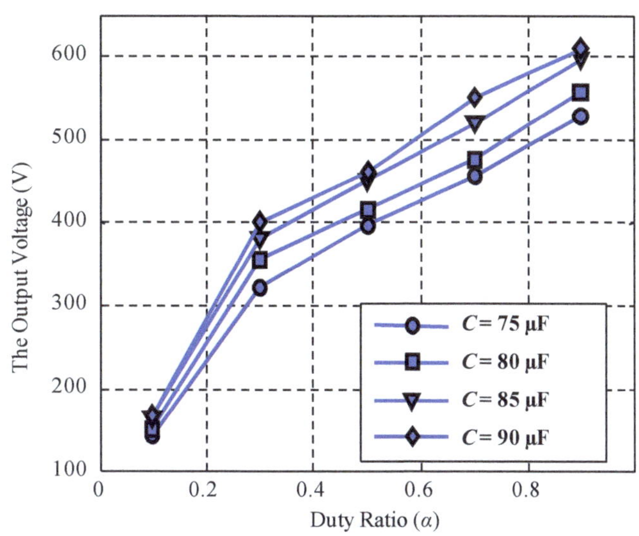

Figure 10. Typical simulation of the voltage V_S at the output of the buck/boost converter according to the capacity and the duty ratio for: $\omega_r = 314\ \text{rad/s}$ and $R_S = 70\ \Omega$.

$$|Vs| = \frac{\alpha}{1-\alpha} E \qquad (14)$$

3.3.4. Summary of Variations in the Parameters Load, Speed and Capacity

We saw in the previous section when the excitation capacity, the load and the driving speed varies, an adaptation of the system must be achieved by including between the generator (asynchronous machines SEIG + Rectifier) and the R_S load an adapter (**Figure 1**). This adapter must also include the *DC-DC* power converter, an *AVC* command to regulate the voltage in the operating conditions considered. The *AVC* command must meet our specifications so as to maintain a constant voltage on the terminals of the wind system studied.

So, for a given a wind system operating. The design of a command to adjust the duty cycle α, depending on weather changes (wind speed), changes of the capacity and of the load is needed in order to operate the wind system in optimal conditions.

Figure 12 shows the search algorithm of the duty cycle as a function of the output voltage whatever the variation of the SEIG characteristic.

In the following paragraph, we study the design and the operation in the Orcad/Pspice environment, of the Analog Voltage Control (AVC) linked to a converter (buck/boost) operating at high frequency (100 KHz).

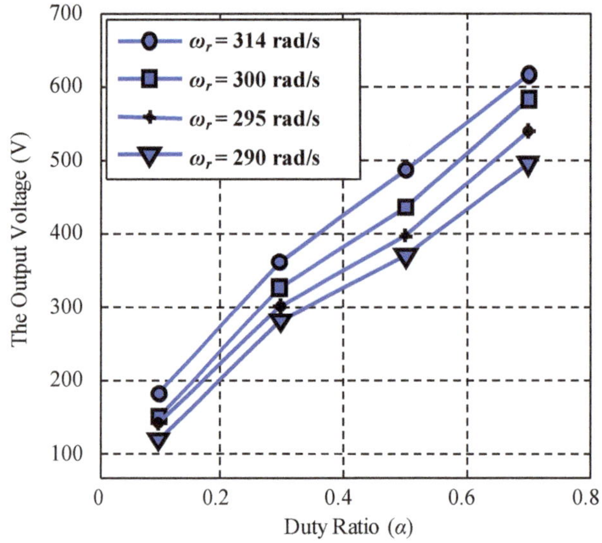

Figure 11. Typical simulation of the voltage V_S at the output of buck/boost converter according to the driving speed and the duty cycle for: $C = 75$ µF and $R_S = 70\ \Omega$.

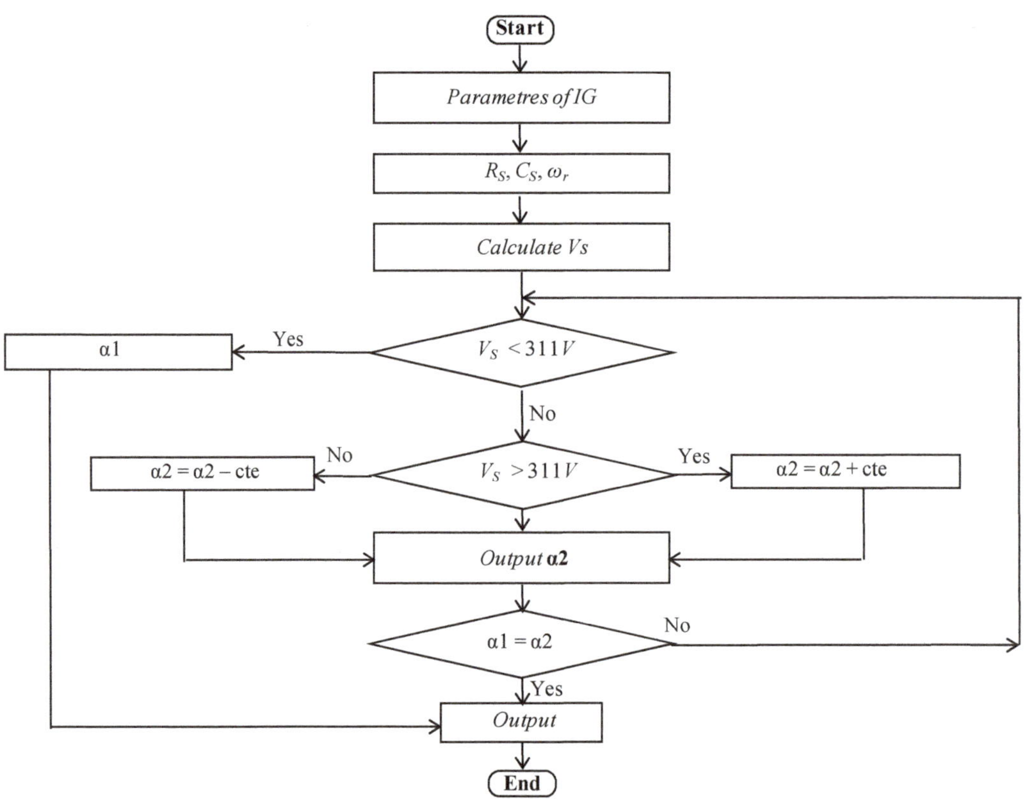

Figure 12. Algorithm for calculating the duty ratio.

4. Results of Simulation Analog Voltage Control Algorithm

Figure 13 represents the nominal output voltage of the inverter at fixed values of the capacity of self excitation, load and wind speed.

Figure 14 illustrates the output voltage evolution of the inverter when the driver speed decreases by 8%. We note as light variation of the voltage at the time of the speed variation and then it is stabilized at its initial value (The wished value).

The voltage output variation of the inverter, when the self-excitation capacitor increases by 12%, is showed in **Figure 15**. We observe that the output voltage reaches the nominal value of stabilization after as light increase at the time of the capacity variation.

Figure 16 represents the output voltage evolution of the system (**Figure 7**) when the load increases by 22%. We note a slight variation of the voltage at the time of the load variation; then is stabilized at the nominal value fixed by our schedule of conditions.

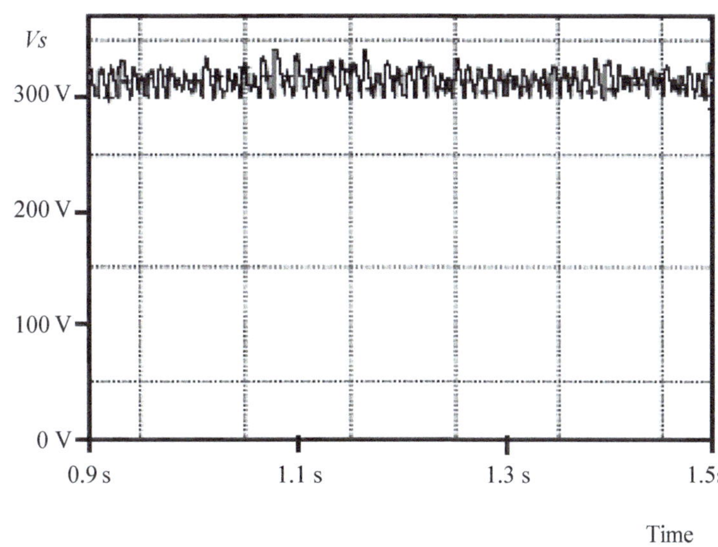

Figure 13. Variation of the output voltage of the chopper (**Figure 5**) at ω_r = 298 rad/s, C = 80 μF and R_S = 70 Ω.

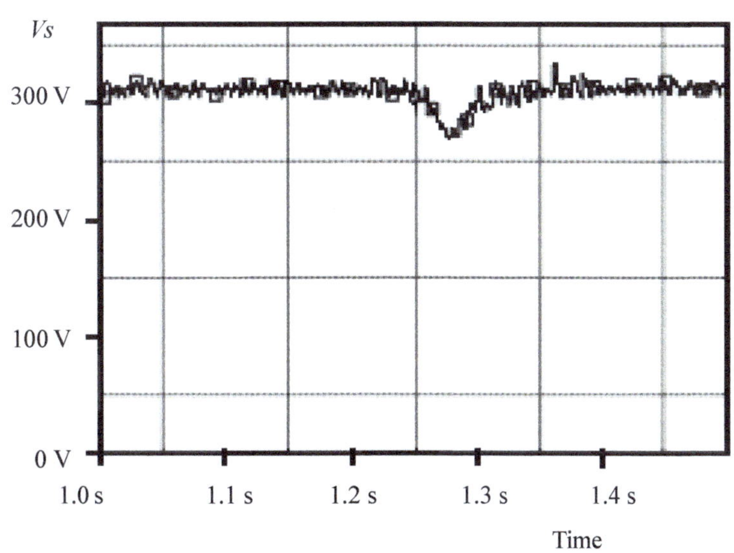

Figure 14. Variation of the output voltage of the chopper when the speed decreases by 8% at C = 80 μF and R_S = 70 Ω.

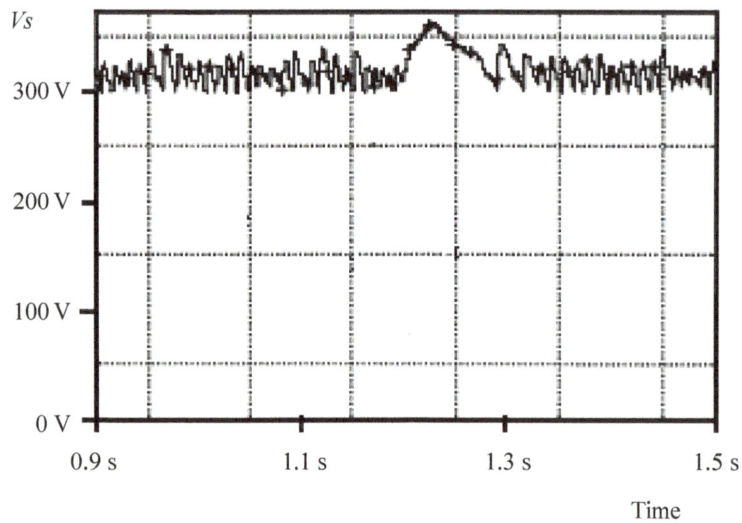

Figure 15. Variation of the output voltage of the chopper when the excitation capacitor increases by 12% at $\omega_r = 298$ rad/s and $R_S = 70\ \Omega$.

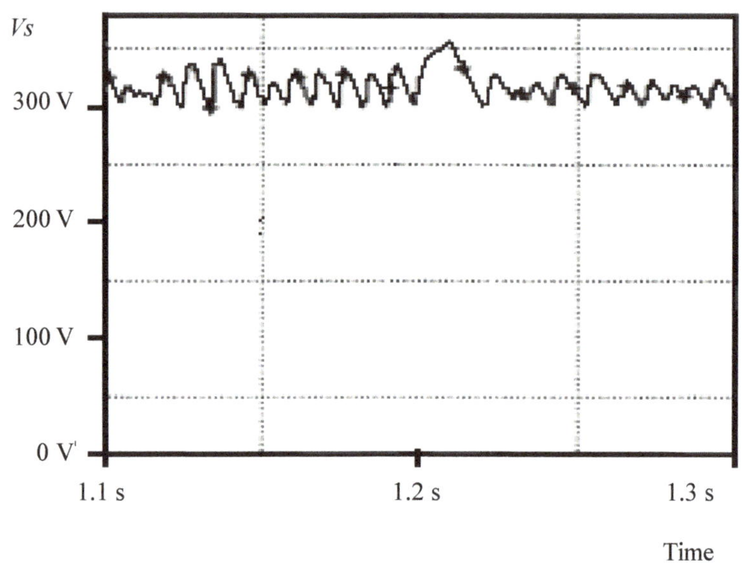

Figure 16. The output voltage evolution of the chopper when the load increases by 22% at $\omega_r = 298$ rad/s and $R_S = 70\ \Omega$.

5. Conclusions

In this paper we proposed a robust law of the control circuit applied to a wind system with the aim of finding a constant value of tension $\left(220\sqrt{2}\ \text{V}\right)$ at the exit of the system provided that the drive speed does not exceed a maximum value $\omega_r = 330$ rad/s.

From the results of simulation of the studied system (asynchronous machine, converters (AC-DC and DC-DC) and load), it appears that:

- The voltage at the exit of the inverter is almost invariant for all the curves, even if one varies the characteristic parameters of the system (speed, capacity or load).
- The control designed makes the system oscillate around the desired tension.
- The disturbances which exist during the oscillations around the value $220\sqrt{2}$ V are due to the transient state of the asynchronous machine and the losses in the components of the system (Mosfet, diode and component of the regulator).

The SEIG has the following characteristics:

$$P = 1.5\ \text{kW},\ f = 50\ \text{Hz},\quad 220/380\ \text{V},\ I_s = 4.4\ \text{A},\ p = 2,$$
$$r_s = 5.51\ \Omega,\ r_r = 2.24\ \Omega,\ Xr = 6.9\ \Omega,\ Xs = 6.9\ \Omega,$$
$$Xm = 38.81\ \Omega.$$

References

[1] Elhafyani, M.L., Zouggar, S., Zidani, Y. and Benkaddour, M. (2006) Permanent and Dynamic Behaviours of Self-Excited Induction Generator in Balanced Mode. *Journal of Physics*: *Condensed Matter*, **7**, 49-53.

[2] Premalatha, K. and Sudha, S. (2012) Self-Excitation and Voltage Control of an Induction Generator in an Independent Wind Energy Conversion System. *International Journal of Modern Engineering Research* (*IJMER*), **2**, 454-461.

[3] Marra, E.G. and Pomilio, J.A. (2000) Spice-Assisted Simulation of the $\alpha\beta\gamma$ Model of Cage-Rotor Induction Machines Including Saturation. IND2000_FDI05, 497-502.

[4] Elhafyani, M.L., Zouggar, S. and Benkaddour, M. (2005) Comportement permanent et Transitoire d'une Génératrice asynchrone auto-excitée en régime équilibré. PREMME_05, Université My Ismail, FST-Errachidia, Maroc.

[5] Zidani, Y. and Naciri, M. (2001) A Numerical Analytical Approach for the optimal Capacitor used for the Self Excited Induction Generator. *Proceedings of the* 32*nd IEEE PESC Power Electronics Specialists Conference*, Vancouver, June 2001, 216-220.

[6] Zidani, Y. and Naciri, M. (2001) A Steady State Analysis of the Self Excited Induction Generator Controlled by an Electronic Load Governor. *Proceedings of the* 4*th IEEE PEDS Power Electronics and Drive Systems*, Bali, October 2001, 903-907.

[7] Mujadi, E., Sallan, J., Sanz, M. and Butterfield, P. (1999) Investigation of Self Excited Induction Generators for Wind Turbine Application. *IEEE Industry Application Conference* 34 *IAS Annual Meeting*, Arizona, October 1999, 3-7.

[8] Seyoum, D., Grantham, C. and Rahman, F. (2001) Analysis of an Isolated Self-Excited Induction Generator Driven by a Variabe Speed Prime Mover. *AUPEC* 2001, Curtin University of Technology, Perth, September 2001, 23-26.

[9] Mujadi, E., Gregory, B. and Bord, D. (1999) Self Excited Induction Generators for Variable-Speed Wind Turbine Generation. *IEEE Industry Application Conference* 34 *IAS Annual Meeting*, Arizona, October 1999, 343-352.

[10] Barara, M., Akherraz, M., Abbou, A., Elhafyani, M.L. and Moutchou, M. (2013) Fuzzy Modeling Magnetizing Curve of a Self Excited Induction Generator and Effect of Mutual Inductance. 2013 *International Renewable and Sustainable Energy Conference* (*IRSEC*), Ouarzazate, 7-9 March 2013, 287-291.

[11] Ouchbel, T., Zouggar, S., Elhafyani, M.L., Oukili, M., Sedik, M. and Rabhi, A. (2013) Regulation of the Excitation Reactive Power of the Asynchronous Wind Turbine at Variable Speed. *Smart Grid and Renewable Energy*, **4**, 272-280. http://dx.doi.org/10.4236/sgre.2013.43033

[12] Ouchbel, T., Zouggar, S., Elhafyani, M.L., Oukili, M., Sedik, M. and Aziz, A. (2014) Power Maximization of an Asynchronous Wind Turbine with a Variable Speed Feeding a Centrifugal Pump. *Journal Energy Conversion and Management* (*EC & M*), **78**, 976-984.

[13] Elhafyani, M.L., Zouggar, S. and Benkaddour, M. (2006) Symbolization of the Asynchronous Machine in the Orcad/Pspice environment. JER06, EST-Oujda, Morocco.

[14] Aziz, A., Kassmi, K., Olivié, F. and Sarrabayrouse, G. (2005) Symbolization of the Electric Diagram of the Marketing Solar Panels in the Orcad-Pspice Environment. Rapport LAAS (05270), Toulouse, 17.

[15] Aziz, A., Kassmi, K., Maimouni, R., Olivié, F., Sarrabayrouse, G. and. Martinez, A. (2005) Integration of the New Components in the Library of the Orcad-Pspice Simulator. 2 Symbolization of the Electric Diagrams: Application to the Photovoltaic Systems. Rapport LAAS/CNRS (05271), Toulouse, 15.

Nomenclature

SEIG	:	Self-Excited Induction Generator
IG	:	Induction Generator
ω_r	:	Mechanical Rotor Speed
r_s, r_r	:	Stator and Rotor Resistances
Xs, *Xr*	:	Stator and Rotor Leakage Impedance
Xm	:	stator Mutual Impedance
C	:	Excitation Capacitor
α	:	Cyclic Ratio
AVC	:	Analog Voltage Control
AC-DC	:	The energy Converter (Rectifier)
DC-DC	:	The energy Converter (Buck-Boost)
PWM	:	Pulse-Width Modulation
R_S	:	The Resistive Load
Vs	:	The Output Voltage
E	:	The Input Voltage

Indices s and r stand for stator and rotor respectively.

14

Islanding Protection and Islanding Detection in Low Voltage CIGRE Distribution Network with Distributed Generations*

Ghullam Mustafa Bhutto[1,2#], Ehsan Ali[2,3], Jagdesh Kumar[2], Muhammad Akram Bhayo[2]

[1]Department of Energy Technology, Aalborg University, Copenhagen, Denmark

[2]Department of Electrical Engineering, Quaid-e-Awam University of Engineering, Science & Technology, Nawabshah, Pakistan

[3]IM2NP, Aix-Marseille University, Marseille, France
Email: #gmu@et.aau.dk, #gmustafabhutto@yahoo.com, ehsanaliburiro@gmail.com, bhayoakram@yahoo.com, jagdesh.k24@gmail.com

Abstract

The Power Quality (PQ), security, reliability etc., are the prime objectives of the power system. The protection is developed in such a way that it should be selective, fast, reliable and the cost effective. The study about the islanding protection in Low Voltage (LV) CIGRE distribution and networks like this has been proposed in this paper. This is achieved by developing the protection against the short circuit faults which might appear at the Medium Voltage (MV) bus. The protection of the network with significant penetration of the Distributed Generations (DGs) is a complicated process. The DG units which are directly connected to the grid such as synchronous or induction generators contribute large short power, whereas the DG units which are connected to the grid via inverters carry small amount of the short circuit power. This creates the problems in the protection of the network. If the proper protection coordination measures have not been taken, it might cause the mal-function of the protection devices which put the portion of the power network into the security threats. The selection of the islanding protection devices in this paper is made to protect the network against bi-directional currents at the time of short circuit fault. The LV CIGRE distribution network will enter into islanding if a fault is cleared at the MV bus by the proposed islanding protection devices. It is therefore, essential to detect the islanding in the CIGRE power network. The detection of the island in this network is another major objective of this paper. The detection of the island is proposed by using the technique which is based on the voltage phase angle difference. The simulations are carried out by using DIgSILENT power factory software version 15.0.

*1) Subject Classification: Smart Grids. 2) All authors are mutually agreed. 3) It's the original work of all the authors. 4) We did not submit this manuscript before. 5) The research work carried out for this paper is related to the corresponding author's Ph.D. study, so some of the results or texts are directly or indirectly copied/used in this paper.

#Corresponding author.

Keywords

Islanding Protection, Loss of Grid (LOG), Over Current Protection, Under Voltage Protection, Fuses, Relays, Island Detection (ID)

1. Introduction

The large scale integration of DG units into the distribution grid presents several problems for the network. The islanding is one of the challenging problems in the electric network. Islanding, Loss of Grid (LOG) [1] is a situation where electrical system becomes electrically isolated from the rest of the power network and yet continues to be energized by the DG units connected to it. The main cause of the LOG is the occurrences of the faults in the network. If fault is not removed or a faulty portion is not quickly isolated from the healthy system, it will propagate into healthy parts of the network and will cause possible damages to the equipments available in the network. The urgent action is necessary to be taken in order to protect power network. The protection of the power network together with DG units and its coordination is complicated because of the bidirectional flow of the current and the difference in short circuit level of DG units and main power grid. This paper describes the new method of islanding protection.

It is also necessary to detect islanding in a power network as fast as possible as it might cause injuries to line worker and may create reclosing problems. Several island detection techniques have been proposed in the literature [2]-[6]. The island detection techniques can be generally divided into two categories: active methods and passive methods. The main passive techniques used for islanding detection include under/over frequency, vector shift, Rate Of Change Of Frequency (ROCOF) [7] [8], Under/Over Voltage, rate of change of voltage and power factor [9] etc. The main active techniques include reactive power Error Export Detection (REED) [10], fault level monitoring [11], and system impedance monitoring [5].

Passive methods are familiar for detecting LOG because of their simplicity and low cost. The ROCOF and vector shift methods are the most common methods used among the passive methods. These methods especially ROCOF method have some of the disadvantages that they do not properly differentiate between the island and transient conditions. Unnecessary tripping might occur in the case of sudden load changes, faults etc., if ROCOF method is used to detect islanding.

This paper describes the detection of LOG by using new technique which is based the voltage phase angle difference between the grid and the distribution system. This method is described in [7] but further details are added in this paper.

A test distribution network set up by CIGRE comprising Wind Turbine Generator (WTG), Photovoltaic (PV) solar generation units and sensitive power electronics based equipment such as Voltage Source Converters (VSCs) at different locations has been chosen for the study. The single line diagram of this distribution system is shown in **Figure 1**. Further details of CIGRE network can be found in [12].

The LV CIGRE distribution network modeled in DIgSILENT power factory software 15.0 is shown in **Figure 2** together with appropriate protection devices.

Due to economic reasons, the protection of whole CIGRE network is mainly performed by the fuses. The detailed study about the protection of entire CIGRE network done by the author is presented in [13]. The LOG protection is performed by using fuse mainly in this study but the under voltage relay is also used to clear a fault if fuse is unable to detect faults in some conditions.

2. Simulations Results

The simulation results of the islanding protection and islanding detection are presented in the section given below.

2.1. Islanding Protection

A three phase fault with a fault impedance of zero ohm ($Z_f = 0$) is applied at time t = 5 s on the MV bus (*i.e.*, bus R0). There is a bi-directional flow of the current to the fault point in this condition. A fuse F0 as shown in

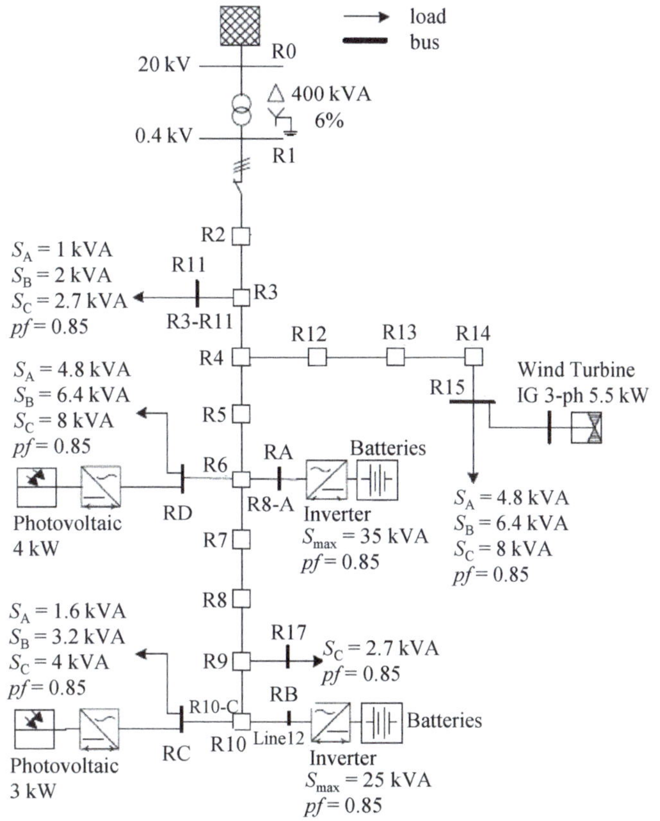

Figure 1. The single line diagram CIGRE distribution system test network [12].

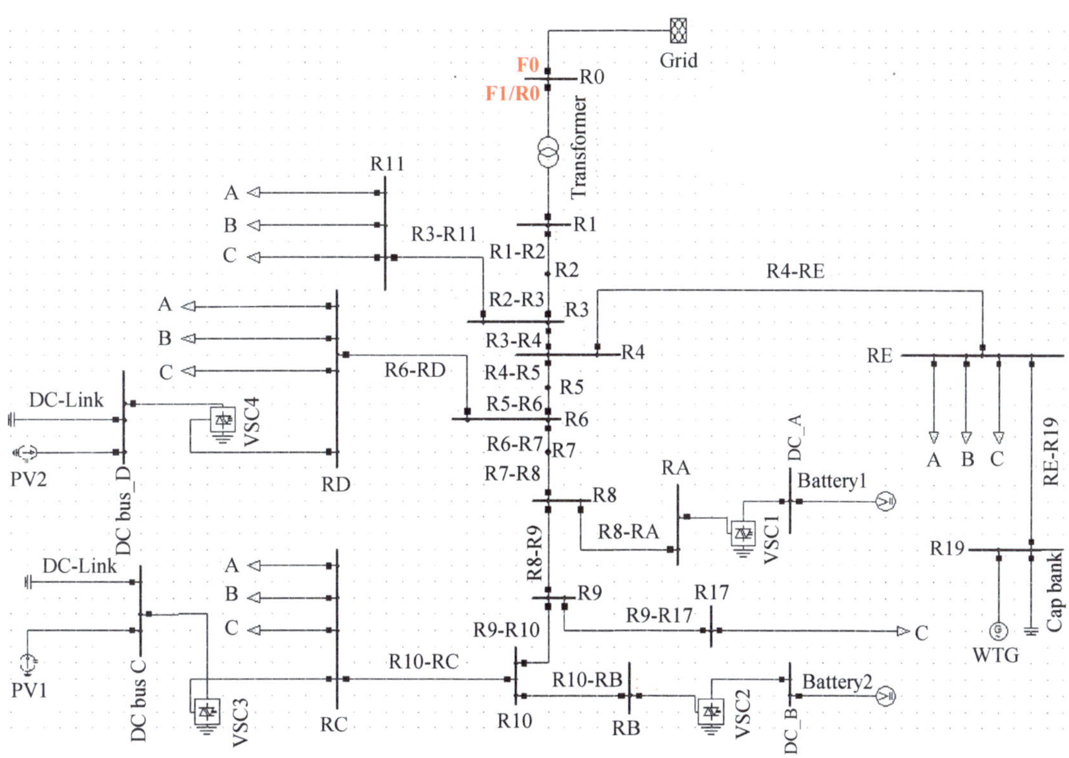

Figure 2. The CIGRE network modeled in DIgSILENT power factory together with protection devices.

Figure 2 is used to disconnect the faulty section from the grid side. This enters the distribution network into island. Fuse F0 is selected to the next higher rated available than that of a fuse F1 to ensure the protection coordination. Fuse F1 does not clear a fault because it was selected for the protection in the forward direction (*i.e.*, in the grid connected mode). The distribution network has a small amount of short circuit power when it is isolated from the grid, that's why fuse F1 does not clear a fault from the backward direction. The details about why fuse F1 do not clear a fault are described below with the help of simulation results.

A fault still exist in island portion of the network as it was only cleared from the grid side, therefore; to clear a fault in island portion of the network; a new approach of using the under voltage relay is introduced in this report. The under voltage relay R0 as shown in **Figure 2** is used to isolate a fault from the DGs side of the network in this regard.

The current delivered by external gird which flows through a fuse F0 is shown in **Figure 3**. It can be seen in **Figure 2** that grid delivers a huge amount of the current at the instant of a fault. When a fuse of 20 A (*i.e.*, F0) sees this value of the current it blows within a very short duration as shown in **Figure 3**. Hence, a fault has been cleared from the grid side and should also be cleared from the DGs side very quickly.

The current seen by fuse F1 is shown in **Figure 4**. It can be seen in **Figure 4** that the current which passes through the fusing element of fuse F1 increases at the instant of a fault and soon it decays. The peak in the current at the initiation of a fault is due to the current surge produced by DG units and it decays because of the current limiters used for these units. This current is coming from the DG side of the MG and passes through the 400 kVA transformer used between LV and MV buses. This current flowing through LV to the HV winding of the transformer (*i.e.*, to the faulted point) passes through the transformer impedance. The current seen in **Figure 4** is small because of the transformer impedance. This current in the steady state condition after the current peak does not reach to the value which melts fusing element.

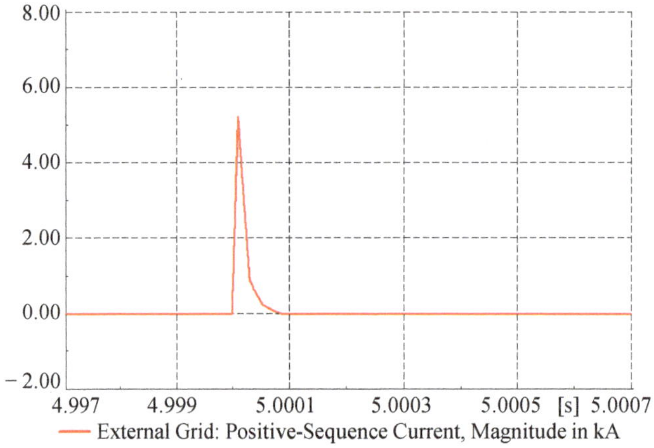

Figure 3. The external grid current.

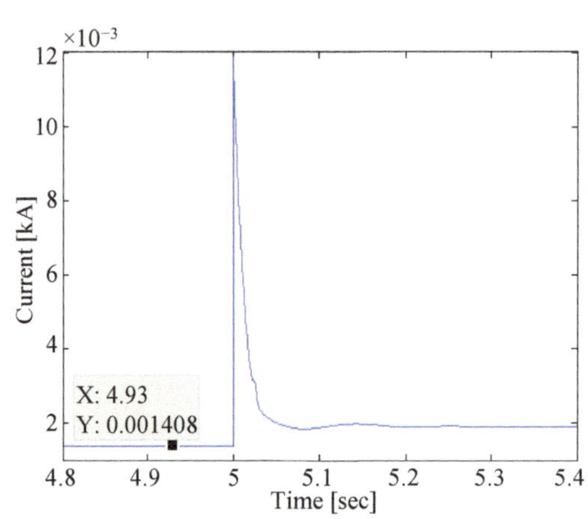

Figure 4. Current seen by fuse F1.

The circuit breaker is used to clear this kind of fault in island portion of the network. The circuit breaker receives a trip signal from the under voltage relay R0. The circuit breaker named VD4 made by ABB is used in this regard. The minimum opening time of this circuit breaker lies between 33 - 60 ms [14]. The minimum opening time of circuit breaker used for this study is selected as 33 ms. The time of the operation of a fuse F0 and circuit breaker receiving trip signal by under voltage relay R0 is shown in **Figure 5**.

The plot in **Figure 5** demonstrates the clearance of the fault from both sides of the network and islanding protection has been achieved. The 1 on the plot in **Figure 5** shows that a fuse operates in a normal operation and the contacts of the circuit breaker are in a close position whereas 0 displays that a fuse has been blown out and the breaker contacts are opened. It can be seen in **Figure 5** that fuse F0 clears a fault from the grid side faster than the circuit breaker used on the DGs side of the network. A fuse blows faster because of severe current flowing across its melting element.

2.2. Islanding Detection

The detection of islanding in CIGRE network is performed by using a technique which is based on the phase angle difference between grid voltage and the DG side voltage. This islanding technique is described with the help of the block diagram shown in **Figure 6**.

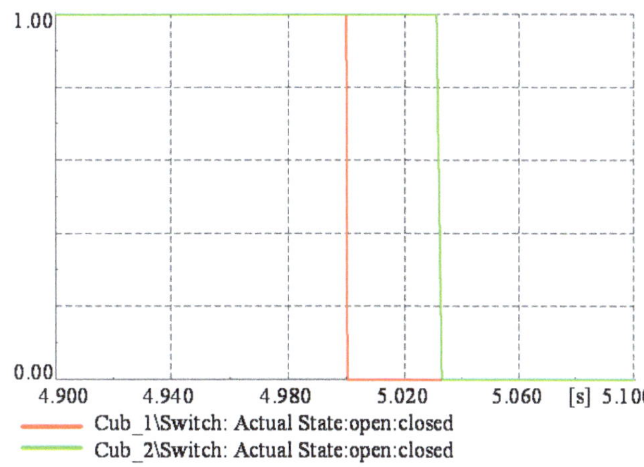

Figure 5. The time of operation of fuse and circuit breaker.

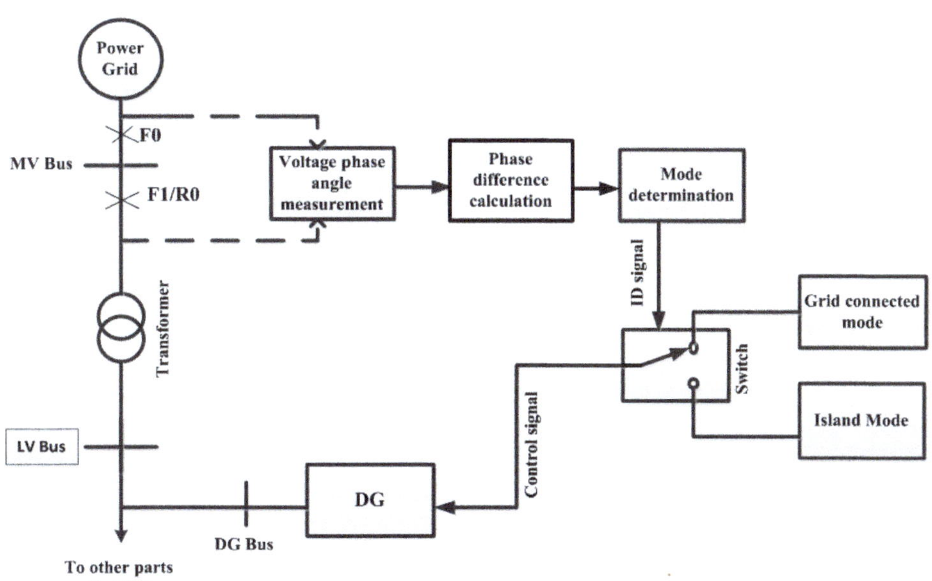

Figure 6. The block diagram of the island detection technique and control mode determination.

The simple block of power network is shown in **Figure 6**. It can be seen in **Figure 6** that the phase angles of the voltages at the grid side and DG side are measured and difference of the phases is calculated. The network changes its mode of operation from the grid connected to islanding according to the difference of the voltage phase angles. The island is detected in the network if there is a phase angle difference between the two voltages as shown in **Figure 7**.

It can be seen in **Figure 7** that the phase angle difference of the voltages at the mentioned points in **Figure 6** is zero in the normal operating conditions. This difference of the voltage phase angles changes when fuse F0 is blown out.

The Island Detection (ID) signal is shown in **Figure 8**. This figure describes that the system is connected to the grid when there is no difference between voltage phase angles. The network switches to island mode when the difference in the voltage phase angles is appeared. Hence, LOG is detected at 5.0001 s in the CIGRE low voltage network.

This method of LOG detection performs well during sudden switching/variation of the loads and in the transient conditions somewhere in other parts of the CIGRE network. To see the effects on the transients and check whether this method may not cause false detection in case of the fault, the study is performed by applying a 3-phase fault on bus R11 at the time equal to t = 5 s. The voltage phase angles in this case can be read in **Figure 9** and the ID signal is shown in **Figure 10**.

It can be seen in **Figure 9** that voltage phase angles at the time of short circuit fault *i.e.*, t = 5 s changes but their difference is zero. Hence, system remains connected to the grid and there is no false detection by using this method as shown in **Figure 10**. This is the main advantage of this method over other methods of LOG detection.

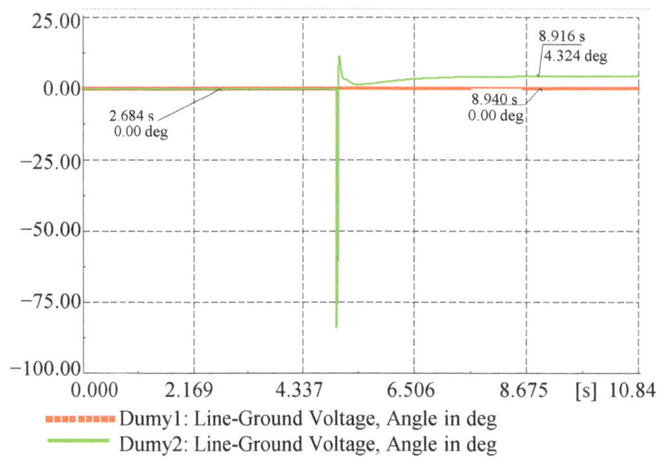

Figure 7. Phase angle difference between the voltages.

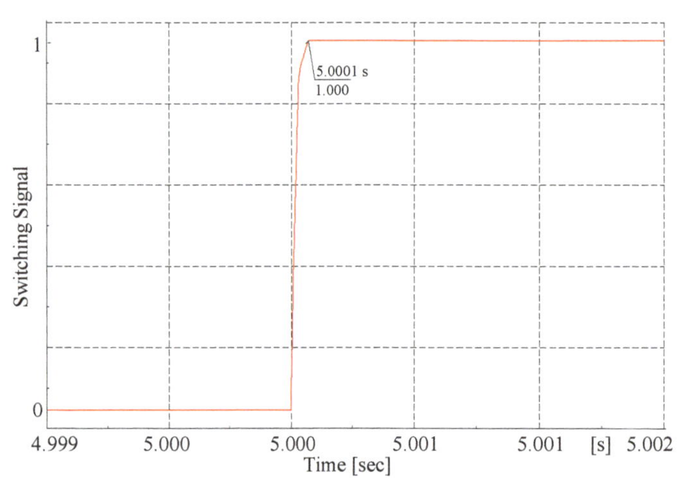

Figure 8. Island Detection (ID) signal.

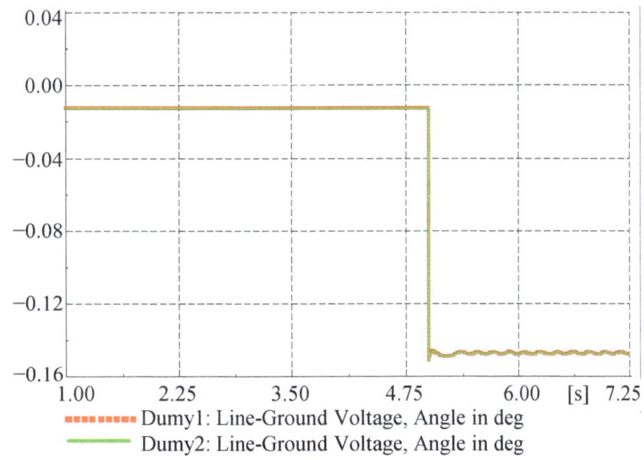

Figure 9. Phase angle difference between the voltages.

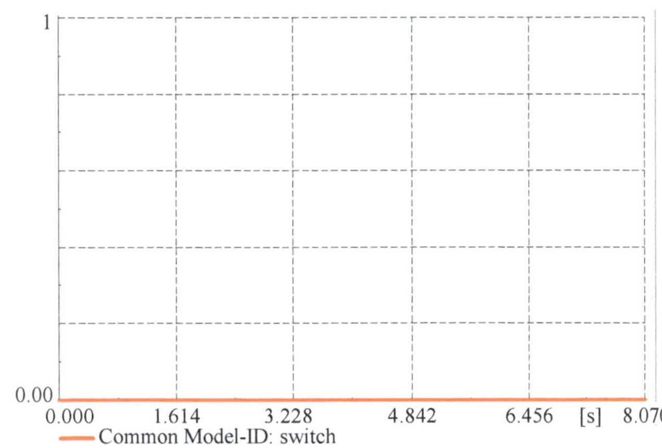

Figure 10. ID signal.

3. Conclusion

The protection against the bi-directional fault current in low voltage CIGRE network has been described in this paper by using current based and voltage based protection devices. The overview of the available island detection techniques is presented and the detection of LOG is proposed by the technique which is based on voltage phase angle differences. The working of this method has been validated by various simulations plots and its main advantages over the other methods are also described with the help of the plots. The problems of voltage and frequency in island Micro Grid (MG) will be tackled in future.

Acknowledgements

We are thankful to Kai Strunz for providing the CIGRE test network for LV distribution network. I, Ghullam Mustafa am thankful to Aalborg University and Quaid-E-Awam University of Engineering Sciences and Technology, Nawabshah, Sindh, Pakistan for the support and funding.

References

[1] Jenkings, N., Allan, R., Crossley, P., Kirschen, D. and Strbac, G. (2000) *Embedded Generation* (*IEE Power and Energy Series* 31), The Institution of Engineering and Technology, London.

[2] Ropp, M., Aaker, K., Haigh, J. and Sabhah, N. (2000) Using Power Line Carrier Communications to Prevent Islanding. *Proceedings of* 28*th IEEE Photovoltaic Specialists Conference*, Anchorage, 15-22 September 2000, 1675-1678. http://dx.doi.org/10.1109/PVSC.2000.916224

[3] Xu, W., Zhang, G., Li, C., Wang, W., Wang, G. and Kliber, J. (2007) A Power Line Signaling Based Technique for

Anti-Islanding Protection of Distributed Generators—Part I: Scheme and Analysis. *IEEE Transactions on Power Delivery*, **22**, 1758-1766. http://dx.doi.org/10.1109/TPWRD.2007.899618

[4] Wang, G., Kliber, J., Zhang, G., Xu, W., Howell, B. and Palladino, T. (2007) A Power Line Signaling Based Technique for Anti-Islanding Protection of Distributed Generators—Part II: Field Test Results. *IEEE Transactions on Power Delivery*, **22**, 1767-1772. http://dx.doi.org/10.1109/TPWRD.2007.899620

[5] O'Kane, P. and Fox, B. (1997) Loss of Mains Detection for Embedded Generation by System Impedance Monitoring. *6th International Conference on Developments in Power System Protection*, Nottingham, 25-27 March 1997, 95-98. http://dx.doi.org/10.1049/cp:19970037

[6] Usta, O., Redfern, M.A. and Barrett, J.I. (1994) Protection of Dispersed Storage and Generation Units against Islanding. *Proceedings of 7th Electrotechnical Conference on Mediterranean*, Antalya, 12-14 April 1994, 976-979. http://dx.doi.org/10.1109/MELCON.1994.380938

[7] Affonso, C.M., Freitas, W., Xu, W. and da Silva, L.C.P. (2005) Performance of ROCOF Relays for Embedded Generation Applications. *IEE Proceedings of Generation, Transmission and Distribution*, 10 January 2005, 109-114.

[8] Ding, X., Crossley, P.A. and Morrow, D.J. (2007) Islanding Detection for Distributed Generation. *Journal of Electrical Engineering & Technology*, **2**, 19-28. http://dx.doi.org/10.5370/JEET.2007.2.1.019

[9] Salman, S.K., King, D.J. and Weller, G. (2001) New Loss of Mains Detection Algorithm for Embedded Generation Using Rate of Change of Voltage and Changes in Power Factors. *Proceedings of 7th International Conference on Developments in Power System Protection*, Amsterdam, 9-12 April 2001, 82-85.

[10] Ding, X., Crossley, P.A. and Morrow, D.J. (2007) Islanding Detection for Distributed Generation. *Journal of Electrical Engineering & Technology*, **2**, 19-28. http://dx.doi.org/10.5370/JEET.2007.2.1.019

[11] Cooper, C.B. (1989) Standby Generation Problems and Prospective Gains from Parallel Running. *Proceedings of International Power System Protection Conference*, Singapore, 1-6.

[12] CIGRE Task Force C6.04.02 (2011) Benchmark Systems for Network Integration of Renewable Energy Resources, Version 7.

[13] Mustafa, G., Bak-Jensen, B., Bak, C.L. and Pillai, J.R. (2013) Protection of Low Voltage CIGRE Distribution Network. *Smart Grids and Renewable Energy*, **4**, 489-500.

[14] Medium Voltage Products, VD4 Medium Voltage Vacuum Circuit Breakers (12 - 36 kV). www.abb.com

15

The Necessity of the Development of Standards for Renewable Energy Technologies in Nigeria

Vincent Nnaemeka Emodi, Samson D. Yusuf, Kyun-Jin Boo

Department of Technology Management, Economics and Policy Program, College of Engineering, Seoul National University, Seoul, South Korea
Email: emeka50@snu.ac.kr

Abstract

Clean energy is vital for the sustainability of any economy in the world. Many industrialized nations have increased their production capacity of renewable energy while other countries lacking the technical expertise and resource have resorted to import these technologies. The imported technologies mostly have standards that are followed by the manufactures while others are manufactured cheaply and exported to developing countries that do not have adequate standards and certification bodies. Nigeria which is a fast growing country has no existing standards to check the influx of renewable energy technologies into the country. This study stresses the need for the development of standards for renewable energy technologies in order to prevent the importation of substandard renewable energy technologies in Nigeria. The study reviews the renewable energy potentials in Nigeria, introduces the concept of standardization and discusses the development of standards for renewable energy technologies.

Keywords

Standards, Standardization, Renewable Energy Technologies, Substandard

1. Introduction

Energy is an important factor for socioeconomic development and economic growth of any nation. Energy is usually stored in energy system which provides energy services. Energy services are desired and useful products, processes or services that result from the utilization of energy like lighting, powering of home-based appliances such as air-conditioner, refrigerators, and cookers for cooking. The energy chain to deliver these cited services begins with the collection or extraction of primary energy, which is converted to energy carriers suitable for the final consumers. These energy carries are utilized in energy end-user technologies to provide the desired energy services [1].

Energy sources are divided into conventional energy which is crude oil, natural gas and coal while non-conventional energy or renewable energy is solar, wind, hydro and biomass energy [2]. About 90% of the total energy supply in the world comes from conventional energy. Conventional energy contains mostly carbon which is the leading cause of global warming when released from crude oil products, gas and coal to the atmosphere [3]. The increase in global warming has indulged both the developed and developing countries to make a shift to renewable energy resources.

Most energy projection shows that the current and future world energy demand patterns are definitely not sustainable. However, the demand for energy will increase dramatically with developing countries consuming more and more energy. The only solution is to decouple the economic activities associated with conventional energy consumption [4]. This can be achieved by increasing the share of renewable energy in the energy consumption balance to enhance sustainability and improve the security of energy supply [5].

Renewable energy development in developing country aims at providing sustainable energy which will ensure the proper mitigation of energy scarcity and global warming. Renewable energy development has led to the increase in the manufacturing of renewable energy technologies such as solar panels, wind turbines and small hydropower (SHP). These technologies transform the energy stored in the sun, wind and hydro to energy which can be harnessed for human consumption.

Renewable energy that is being manufactured by different industries in various countries of the world has their specification and standards. When these technologies are imported into another country, standards are created by standards bodies in the importing country to ensure that the renewable energy technologies are reliable, compactable and safe for its citizens. Countries that don't have the technical expertise to manufacture renewable energy technologies have to import from other countries like China, South Korea, Japan, and the US etc. The countries that lack technical expertise are mostly in the African continent and they rely on imported technologies. These technologies are sometimes substandard and unreliable while others are even dangerous to the consumers and can cause fire outbreak which can destroy lives and property.

Nigeria, which is located in the West African region has been importing renewable energy technologies but has failed to ensure the establishment of standards for renewable energy by the Standards Organization of Nigeria (SON). This has led to the import of substandard renewable energy technologies into the country and has posed some dangers to her citizens who pay with their money hard earning but get low value for their money [6].

The question that arises now is: what should be done to ensure that this menace of the influx of substandard renewable energy technologies into the country is checked and how can a renewable energy standard be developed in Nigeria. This study seeks to answer the research questions by first reviewing the renewable energy resources in Nigeria, identifying the barriers to its development, provides strategies to enhance its development and then discusses on the development of renewable energy standards. The remainder of this paper is organized as follows. Section 2 reviews the renewable energy potentials in Nigeria including the policies, barriers and strategies to enhance its development. The concept of standardization is given in Section 3 which also includes the roles of standards and the standards regulatory agency in Nigeria. Section 4 discusses the development of renewable energy standards in Nigeria while conclusions are provided in Section 5.

2. Renewable Energy Resources in Nigeria

The global mindfulness of the phenomenon of climate change along with the anticipation of conventional energy scarcity have prompted many countries to develop a more sustainable energy system to cater for economic development and growth. Clean and environmental friendly energy can only be achieved through the proper utilization of renewable energy technology [7]. The renewable energy resources in Nigeria are as enormous as they are diverse [8]. However, the problem lies with the level of utilization which is very low. A long term commitment from the Government of Nigeria (GON) is crucial in implementing any kind of policies which will lead to renewable energy development, this can be seen in countries like Germany, Denmark, Japan and more recently, South Korea [9].

Nigeria's opportunity to improve the standard of living for its citizens, ensuring socioeconomic and political growth depends on the nation's ability to increase energy supply and proper utilization of its energy resources starting from the grassroots level [10]. Renewable energy technology has great potentials in alleviating the staggering energy situation currently being experienced in Nigeria. The potentials of renewable energy resources in Nigeria is about 1.5 times that of conventional energy resources in energy terms [11]. **Table 1** [12] shows the renewable energy resources in Nigeria and they are discussed in the following subsections. It should be noted that biomass energy was not discussed in this study because biomass energy technology includes biomass plants and therefore this study only focused on solar, wind and hydropower.

2.1. Solar Energy

Among all the renewable energy resources available, solar is the most promising of them all due to its apparent limitless potentials. The energy that is radiated from the sun is about 3.8 × 1023 KW per second and most of it is transmitted radially as electromagnetic radiation which gives about 1.5 KW/m^2 at the boundary of the atmosphere, a square meter of the earth's surface can receive solar power of about 1.5 KW which averages about 0.5 daily [13].

Nigeria is located within a high sunshine belt and solar radiation is well distributed within the country. The intensity of solar radiation exhibits remarkable variation from the northern region to the southern region but is higher in the northern region as shown in **Figure 1** [14]. With an average domestic load demand of 2324 Wh/m^2 per day [15], Nigeria has a solar radiation potential of about 6500 Wh/m^2 in the far north and 4000 Wh/m^2 in the southern part of the country. Given an average solar radiation level of about 5500 Wh/m^2, and prevailing efficiencies of commercial solar electric generators, then if solar collectors were employed to cover about 1% of Nigeria's land area of 923,773 KM, about 1850 × 103 GWh of solar electricity could be generated per year [16] [17].

The Energy Commission of Nigeria (ECN) has made some effort to harness the solar energy within Nigeria through the direct coordination of research and development activities undertaken by the Sokoto Energy Research Center (SERC) and the National Center for Energy Research and Development (NCERD) [18]. However, this effort has seen little or no significant impact in improving the energy situation in Nigeria. There is need for a strong industrial infrastructure which will be effective in the complete utilization of solar energy in Nigeria.

2.2. Wind Energy

Wind is a natural phenomenon related to the movement of air masses caused primarily by the movement of air masses caused primarily by the different solar heating of the earth's surface. Seasoned variations in the energy received from the sun affects the strength and direction of the wind. The ease with which aero turbines transform energy in the moving air to rotary mechanical energy suggest that the use of electrical devices to convert wind

Table 1. Renewable energy resource in Nigeria (source: [12]).

Resource type	Reserves		Production	Domestic utilization (natural units)
	Natural units	Energy units (Btoe)		
Small hydropower	3500 MW	0.34 (over 40 years)	30 MW	30 MW
Large hydropower	11,250 MW	0.8 (over 40 years)	1938 MW	1938 MW
Wind	2 - 4 m/s at 10 m height (mainland)	0.0003 (4 m/s@12% probability, 70 m height, 20 m rotor, 0.1% land area,40 years)	-	-
Solar radiation	3.5 - 7.0 kWh/m^2/day (4.2 million MW h/day using 0.1% land area)	5.2 (40 years and 0.1% land area)	6 MW h/day	6 MW h/day
Biomass fuel wood	11 million hectares of forest and wood land excess of 1.2 m ton/day	-	0.120 million ton/day	0.120 million t/day
Animal waste	211 million assorted animals waste	-	0.781 million ton of waste/day	None
Energy crops and agricultural residue	28.2 million hectares of arable land (=30% of total land)	-	0.256 million ton of assorted crops/day	None

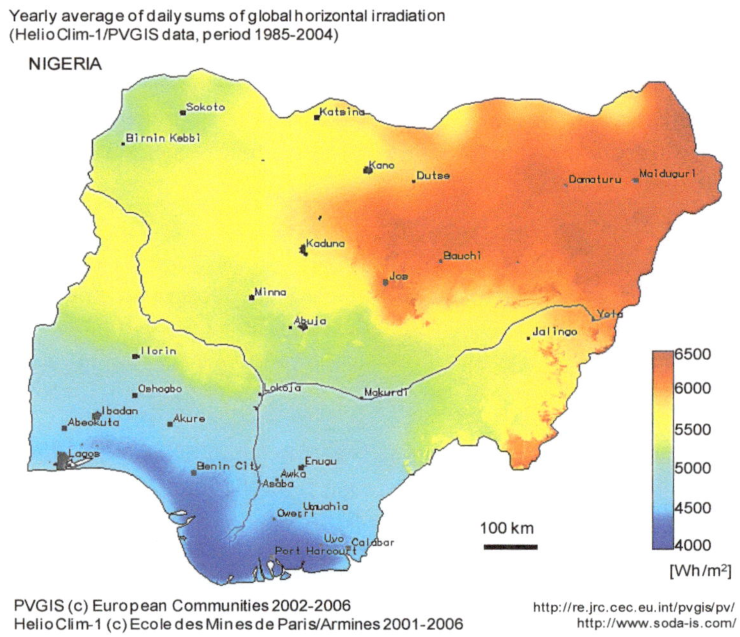

Figure 1. Yearly average of daily sun in Nigeria (source: [14]).

energy to electricity. For decades, wind has been used for water pumping and milling of grains [1].

Wind speed in Nigeria ranges from 4.0 to 6.0 m/s in the northern part of the country and 2.5 to 4.0 m/s in the southern part of the country as shown in **Figure 2** [19]. With this amount of wind energy potential, small scale wind turbine could be installed to boost electricity supply and also be integrated into the national grid [8].

Despite Nigeria's exploitable wind energy resources, the present share of wind energy in the national energy consumption has been low with no available commercial wind farms but only standalone wind power plants for pumping water which were installed in 5 northern states during the 1960s. Also, a 10 MW wind turbine is currently installed in Katsina state to tap into the wind energy potentials in northern Nigeria. More effort should be put in to adequately harness the wind energy potential in Nigeria.

2.3. Hydropower

Hydropower generation is an important option to meet the growing demand for energy worldwide. In 2004, the world hydropower capacity was 2810 TWh and is projected to increase to 4903 TWh by the year 2030, with a growth rate of 1.8% per year, but its share still remains at 2% of the world energy supply [19]. Hydropower resources in Nigeria was first harnessed in 1962 by the Niger Dams Authority. Hydropower generation in Nigeria has substantial potential like the small, mini and micro water capacity for electricity generation but its total power contribution has declined in recent years due to some technical reasons [20]. Moreover, the use of oil and gas for power generation has had a negative impact on hydropower development in Nigeria [21].

Notwithstanding, small hydropower (SHP) is fast gaining rapid consideration due to its inherent advantages like in-excessive topography problems, low environmental impact, minimal civil works and high possibility for power generation along with irrigation and flood prevention. Research has been conducted by [22] and shows that about 738 MW of SHP could be harnessed from 278 sites based on the 1980 survey. However, the SHP potential in Nigeria is estimated at 3500 MW which represent about 23% of the total hydropower potential in the county.

Hydro capacity depends mainly on annual rainfall levels, with its distribution as well as the river systems which is subject to seasonal drought. In the northern part of Nigeria, the total rainfall is about 500 mm depth with a total precipitation which last for 3 months in a year while the southern part of the country has 3400 mm with a precipitation which may be less than 8 months a year [23]. **Figure 3** shows the various water ways in Nigeria [24]. The government should do more to develop SHP to its full extent in order to improve power generation and reduce fiscal loads.

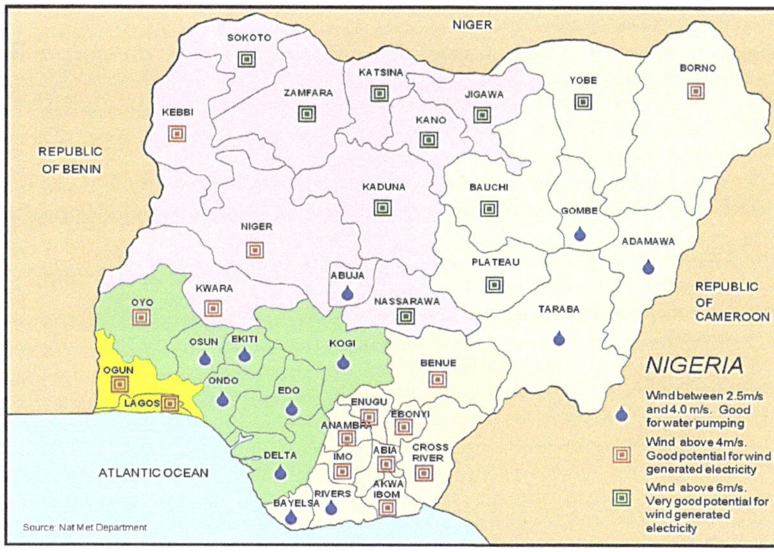

Figure 2. Wind energy locations in Nigeria (source: [19])

Figure 3. Nigerian water ways (source: [24]).

2.4. Renewable Energy Policy

The Energy Commission of Nigeria (ECN) has a mandate as an agency to develop and promote renewable energy technologies in Nigeria with strategic energy planning, policy coordination and performance monitoring for the entire energy sector. The ECN also has the responsibility of providing guidelines for the utilization of the various energy types for the different purpose and develop recommendations on new exploitable sources of energy. This includes renewable energy [25].

The Federal Government of Nigeria (FGN) in 2003 approved the National Energy Policy (NEP) which was facilitated by the ECN to articulate the sustainable exploitation and utilization of all energy sources. The key elements of the NEP on the development and utilization of renewable energy and its technologies are as follows:

- To develop, promote and harness the renewable energy resources of Nigeria and incorporate all viable ones

in the national energy mix;
- To promote decentralized energy supply, especially in rural areas, based on renewable energy resources;
- To deemphasize and discourage the use of wood as a fuel;
- To promote efficient methods in the use of biomass energy resources;
- To keep abreast of the international development in the renewable energy technologies and applications [26].

In 2006, the ECN also formulated the Renewable Energy Master Plan (REMP) as part of the strategy to reduce Greenhouse Gas (GHG) emission in Africa and address the challenges hindering the development of clean, reliable, secured and competitive energy supply [27] [28]. The REMP main objectives are given below:
- To develop and implement strategies that will achieve a clean reliable energy supply mechanism to develop the sector based on international best practices, to show case viability for private sector participation;
- To provide a comprehensive framework for developing renewable energy that will ensure:
- ➢ Expanding access to energy services to Nigerians;
- ➢ National agenda on emission reduction;
- ➢ Raising the standard of living, especially in the rural areas;
- ➢ Stimulating economic growth, employment and empowerment;
- ➢ Increasing the scope and quality of rural services, including schools, information, health services, entertainment, and water supply and reducing rural urban migration [29].

The REMP aims at a 10% renewable energy contribution to the national energy mix by 2020 through the adoption of a renewable portfolio standard (RPS). A RPS is a requirement for electric utilities to supply a specific amount of electricity to customers. This can be achieved through the purchase of renewable energy certificates from suppliers with a larger share of renewables in their energy mix. Other measures considered are the creation of innovative fiscal and market incentives to promote renewable energy industries, as well as preferential customs duty exemptions for imported renewable energy technology components. However, the lack of implementation of the master plan has meant that the 10% target of renewable energy mix in the energy supply cannot be achieved. The REMP is presently being subjected to a review, likely resulting in the setting of new targets. It will be essential that any future targets set for the attainment of a RE energy mix should be backed by legislation to ensure compliance, which is presently lacking.

Similarly, the National Policy and Guidelines on Renewable Electricity (NPGRE) was produced in 2006 with the main aim to expand the market for renewable electricity by 5% of the total electricity generation by 2016. The strategy for achieving this target included: encouraging local manufacture and assembly of renewable energy components, provision of subsidies, and establishment of technical standards for RE components and introduction of feed-in-tariffs. The strategy is yet to be fully adopted as the reforms in the energy sector are still ongoing and decisions on tariffs and subsidies for renewable energy and other incentives have not yet been taken [8].

2.5. Barriers to Renewable Energy Development

Renewable energy in Nigeria has been inconsistent with minimal or no significant impact in the lives of Nigerians. Many barriers hinder the development and diffusion of renewable energy in Nigeria, including the following [30]:
- The lack of proper institutional framework which is due to the weak coordination between government ministries and agencies. These institutions are responsible for the promotion and development of renewable energy technologies in both the rural and urban areas;
- Low human capacity in the field of renewable energy development. Capacity building in the areas of training to install, operate, maintain and manufacturing of renewable energy is lacking in Nigeria;
- Non-existing Power Purchasing Agreements (PPA) plan for renewable energy generation to the utility grid. The PPA set the terms in which electricity is marketed and determines the delivery location, power characteristics, pricing, quality, schedule and terms of agreements including the punishment for breach of contract;
- Affordability of renewable energy technologies involves high cost of capital as compared to conventional energy. This has posed a lot of problems to investors who face the problem of high transaction costs and restricted to funds or capital. Consumers on the other side also face difficulty in purchasing these technologies because of its expensive cost of purchase;
- The lack of public awareness has negatively affected the development of renewable energy in Nigeria. Most

Nigerians view renewable energy technologies as new technologies that are only for the wealthy in the society. This has made the most of the population who are low income earners to depend on conventional energy which is cheaper;
- Uncertified/poor quality renewable energy technologies now flood the markets in Nigeria. This has led to the withdrawal of most potential consumers because most renewable technologies easily damaged due to their low quality and the replacement parts or services usually cost a lot to obtain. Nigeria presently lack the proper standards in renewable energy technologies.

2.6. Strategies to Enhance Renewable Energy Development

As Section 2.5 highlighted the barriers to renewable energy development, this section will offer some strategies which will promote renewable energy development in Nigeria:
- Placing subsides on renewable energy technologies and also improving access to micro finance loans by potential consumers;
- The government should do more in the area of public awareness campaign to promote renewable energy technologies;
- Research and development in the area of renewable energy should be encouraged. This will enable the growth of human capacity and ensure that the technologies which are mostly imported can be manufactured in the country;
- A legally binding long term Power Purchase Agreement plan should be established in order to encourage developers of renewable energy. Also, other renewable energy development mechanisms like Feed-in-Tariffs (FiTs), Renewable Obligation (RO), Renewable Energy (green) Certificate (REC) and Renewable Portfolio Standard (RPS) should be initiated to facilitate the adoption of renewable energy technologies in Nigeria;
- Development of Renewable Energy Standards by the Standards Organization of Nigeria to check the influx of substandard renewable energy technologies into the country.

3. Standardization

The process of establishing and applying standards, standardization is defined by the International Organization for Standardization as "the process of formulating and applying rules for an orderly approach to a specific activity for the benefit and with the cooperation of all concerned, and in particular for the promotion of optimum overall economy, taking due account of functional conditions and safety requirements". Standardization can be applied to specific products, as well as to, for example, norms, requirements, methods, terms, and designations commonly used in international trade and in science, engineering, industry, agriculture, construction, transportation, culture, public health, and other spheres of the national economy [31]. Standardization has a significant influence on the rate of development and level of production. Standardization provides the following advantages:
- Better product quality, reliable and durable life service;
- Components or parts mass production at a low cost;
- Availability of parts for replacement and maintenance;
- Less time and effort (productivity is high) for manufacture;
- Sizes and grades variations correction.

The concept of standardization uses numbers to limit unnecessary variations in sizes and grades of products, the general requirements indicates that such requirements are satisfied when it follows a geometric series. In international business, standards are applied to penetrate the markets in such a way that the products are made to fit into the requirements of any country. To do so, the need to follow the internationally recognized guidelines arises. Examples of products standard includes fuel economy and airbag specifications.

3.1. Definition of Standards

A "standard" can be defined as a specification (or a set of specifications) that relates to a product's attributes [32]. The International Organization for Standardization (ISO) defines standards as "a document, established by consensus and approved by a recognized body that provides, for common and repeated use, rules, guidelines or characteristics for activities or their results, aimed at the achievement of the optimum degree of order in a given context" [33]. In the WTO, definition for standards differentiates voluntary standards and technical regulation, where standards are voluntary and technical regulations mandatory with administrative provisions [34].

The IRENA defines standards as "a repeatable, harmonized, agreed and documented way of doing something. Standards contain technical specifications or other precise criteria designed to be used consistently as a rule, guideline, or definition. They help to make life simpler and increase the reliability and the effectiveness of many of the goods and services we use" [35]. Some economists have given various definition of standards, among them are; Alfred Marshall who defined standards as "customs" that recorded technical progress, firstly in an informal manner and then in a formal manner starting in industrial era [36]. David *et al.* (1990) defined standards as "a set of technical specifications adhered to be a producer, either tacitly or as a result of formal agreement" [37]. Richard *et al.* (1995) defined standards as "agreed external points of reference to which the physical and performance characteristics of technologies can be compared" [38].

From the above literatures we can understand that standards are published documents that establish specifications and procedures designed to maximize the reliability of the materials, products, methods, and/or services people use every day. Standards address a range of issues, including but not limited to various protocols to help maximize product functionality and compatibility, facilitate interoperability and support consumer safety and public health.

Standards form the fundamental building blocks for product development by establishing consistent protocols that can be universally understood and adopted. This helps fuel compatibility and interoperability and simplifies product development, and speeds time-to-market. Standards also make it easier to understand and compare competing products. As standards are globally adopted and applied in many markets, they also fuel international trade.

It is only through the use of standards that the requirements of interconnectivity and interoperability can be assured. It is only through the application of standards that the credibility of new products and new markets can be verified. In summary standards fuel the development and implementation of technologies that influence and transform the way we live, work and communicate.

3.2. Process of Standards Development

The process of developing a standard is typically facilitated by a Standards Development Organization (SDO), which adheres to fair and equitable processes that ensure the highest quality outputs and reinforce the market relevance of standards. SDOs, such as IEEE, IEC, ISO, and others, offer time-tested platforms, rules, governance, methodologies and even facilitation services that objectively address the standards development lifecycle, and help facilitate the development, distribution and maintenance of standards. While the goals of each SDO are essentially the same, each SDO applies its own rules, processes, and terminology to the standards development process. Typically, each SDO is comprised of boards, committees and staff who establish and maintain the policies, procedures and guidelines that help ensure the integrity of the standards development process, and the standards that are generated as an outcome of this process.

The development of a new standard is typically triggered by a formal request, submitted to an SDO by a Sponsoring Body (individual or entity, such as an industry society) for review and evaluation. The SDO mandates, oversees, and helps facilitate the process for standards development. The Sponsor for the standards project assumes responsibility for the respective area of standards development, including the organization of the standards development team and its activities. Once the SDO approves the request to develop a new standards development project, the Sponsor follows the SDOs rules and processes to recruit and assemble a collaborative team or "Working Group" to engage in active standards development (note: the term "Working Group" is an IEEE term. Working Groups may be called different names by the various SDOs, and may follow slightly different processes). Working Groups are comprised of individuals and/or entities (people, companies, organizations, non-profits, government agencies) who volunteer to support the development of standards [39].

Collectively, these volunteer participants carry a specific interest in a specific area of development as producers, sellers, buyers, users and/or regulators of a particular material, product, process or service. When a Working Group is formed the Working Group officers may either be elected by the Working Group members or appointed by the Sponsor. Consult the Sponsor's policies and procedures and the Sponsor's Working Group policies and procedures for details. Working Group officers oversee the standards development project in adherence to SDO rules and process, and remain accountable to the project Sponsor and the governance structure of the SDO itself.

Based on the rules and criteria established by the SDO, participants may contribute at varying levels to the standards development process. For example, the IEEE Standards Association (IEEE-SA) has established rules related to membership and participation, and employs a separate "Entity Standards Development Process" for standards that are sponsored by entities (such as corporations, governments, non-profits, associations or other organizations, etc.). Such rules help ensure that highly dedicated individuals lead participation and no one interest dominates the standards development process. Working Groups leverage these rules and guidelines and establish their own individual, organizational, communications and meeting structures, and govern work process, activities, consensus building, decision making, balloting and even financial reporting in accordance with SDO rules. To build consensus through democratic means, participants engage in meetings, draft and review position pieces, create and review presentations, examine data and engage in active discussion and debate to resolve outstanding issues.

These activities fuel the gradual definition of each standard, which is compiled into a draft standard that may undergo multiple revisions. Once a draft standard has been finalized, reviewed, and approved by the Working Group, it is submitted to the Sponsor for Sponsor balloting. Upon successful completion of the Sponsor ballot, the draft is submitted to the Review Committee (RevCom). The balloted draft is reviewed by RevCom and then submitted to the Standards Board for approval. After submission, review and acceptance, the approved standard is published and made available for distribution and purchasing within in a number of outlets, including through the SDO itself [39] [40].

It is important to remember that standards are "living documents", which may initially be published and iteratively modified, corrected, adjusted and/or updated based on market conditions and other factors. At any given point in time, therefore, a standard may be referred to as having a number of different "status" classifications.

These include (see **Figure 4**):

- Approved Project—An initial project request is approved, in stages of group formation.
- Active Project—An active standards development project.
- Withdrawn Project—A cancelled standards development project.
- Approved Standard—The standard is approved and published for public use.
- Withdrawn Standard—The standard is no longer market relevant or active.
- Superseded Standard—The standard has been replaced by a new standard.

In keeping with the standards development lifecycle, Working Groups may also go through periodic stages of activity or dormancy. Depending on where a standard is in its lifecycle a standard may be accompanied by supplemental documents that are produced by its respective Working Group. These may include errata (which address errors in the standard), amendments (which modify sections of the standard), corrigenda (which only correct errors or ambiguities in a standard), handbooks, tutorials and other related materials. Supplemental documents help interested parties better understand and apply the standard [39].

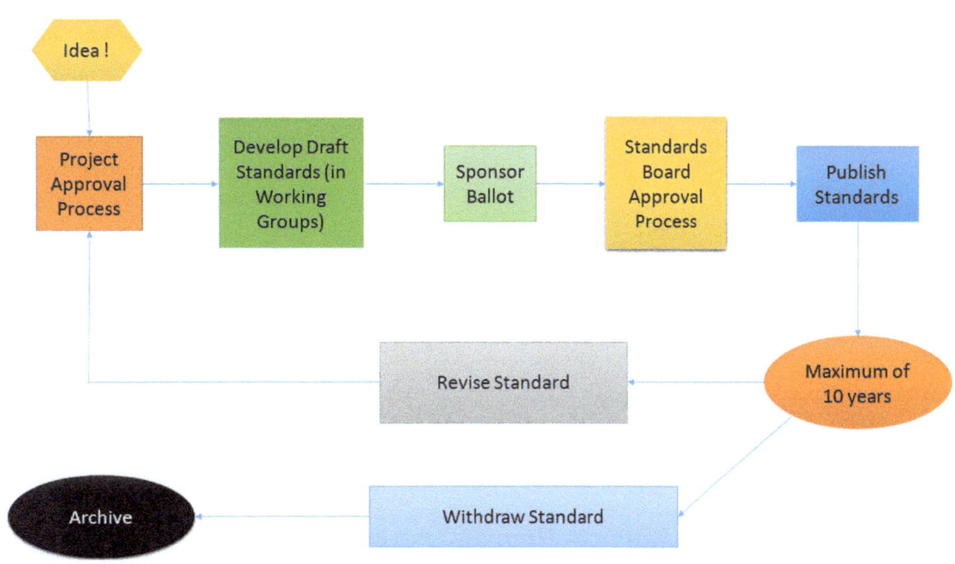

Figure 4. Process of standards development.

3.3. Roles and Functions of Standards

Standards are established through government policies or consumer protection organizations to protect the consumers of various goods and services as most consumers have little or no choice in what they are offered. In rural communities in developing countries, consumers do not generally have the luxury of comparing features and selecting their suppliers or products from the Internet. Hence, the role of standards is to ensure that whatever product or service is provided to the consumers is fit for purpose, safe and has value. An important aspect of this protection is to ensure the product or service delivers as claimed, performs as specified, and is reliable, durable and safe. In many cases renewable energy standards and conformity assessment can be the catalyst in providing alternatives to systems which are sometimes unsafe, operate with low efficiency, and use fossil fuel energy with possible detrimental effects on health. This is achieved by providing confidence and trust in renewable energy products and those who provide energy-related services. Standards also have the ability to allow those not typically trained in these energy sources to reach a level of understanding that allows them to provide, install or operate systems for themselves or under reduced supervision. This is achieved by providing guidance and best practice in designing, specifying, installing and maintaining the systems. A good example is the set of international standards for rural electrification. Standards also provide an effective framework for harmonizing information flow, understanding technical product design, manufacturing and service requirements, as well as establishing common rules and requirements. Standards should enable all these functions to take place while ensuring there is flexibility for the product, service, system provider and user [35].

3.4. The "De Facto" and "De Jure" Standards

A "de facto" standard is one created almost by default because of one dominant player in that industry. A "de facto" standard can also be described as a standard that achieved its status as a result of market competition. For example, Windows for personal computers or VHS for video cartridge recorders are typical de facto standards. A process that a product becomes a standard after it is brought to the market and exposed to competition is called the "classical model". For this instance, such case as Microsoft which maintains monopoly in the market is quite unusual. It is more common that several companies form a group, and those several groups make competition among each other and the winner is determined. A process that a winner is determined after products are brought to the market imposes a heavy burden on companies as they already have completed their capital investment. Consumers also get lost, as even if they purchased products, those styles may no longer be seen in the market some time later. A good example is that consumers felt inconvenience if they had bought video tapes of beta system for some time as they only saw those of VHS system some time later and those tapes previously bought could not be used anymore. Other de facto standards include QWERTY keyboard, Microsoft Word format for documents and Adobe Postscript languages for laser printers [41] [42].

The opposite of an informal "de facto" standard is the "de jure" standard, which means "by force of law". The de jure standards are standards that are developed after going through a certain procedure or official organization. Those standards developed by ISO or IEC are developed through a procedure that experts gather from all over the world and discuss on standardization proposals, which are finally approved through voting by national standards institutes. Japan's standards JIS are also developed going through a procedure that JISC first deliberate a proposal and competent ministers finally make decision to incorporate the proposed standards in the JIS. Korea's Standards KS are developed by Korean Standards Association which is a public organization under the Korean Ministry of Trade, Industry & Energy (MOTIE) to carry out industrial standardization, certification, and quality promotions thereby contributing to national economy of Korea. However, de jure standards are developed through a transparent and democratic procedure. As companies bring their products to the market based on de jure standards, they do not impose excessive burden on corporations or consumers. However, development of de jure standards takes a long time. They are not suitable for products with short cycle such as IT-related products. Examples of de jure standards include Internet TCP/IP protocol, Wireless 801.11n and Unicode international character encoding [41] [42].

As a result, a new type of standards "forum standards" has been introduced recently, which adopts the positive parts of both de fact standards and de jure standards. Influential companies get together to form a standard before bringing products to the market, so that neither companies nor consumers need to take risks [41].

3.5. Standards Regulatory Agency in Nigeria

The only standards regulatory agency in Nigeria is the Standards Organization of Nigeria (SON) which was established by Act. No. 56 of 1971 which vested it with the authority for: standards elaboration, specification, quality assurance system commodities, manufactured industrial and imported products and services. The Act No. 20 of 1976 which amended the previous one conferred on the Honorable Minister of Industry the power to declare Mandatory Industrial Standards in respect of products or processes recommended by the Nigerian Standards Council.

The Act No. 32 of 1984 changed the name of the Organization to Standards Organization of Nigeria (SON) from Nigeria Standards Organization (NSO) to eliminate conflicting identity with the then Nigerian Security Organization. Finally, the Act No. 18 of 1990 conferred on SON partial autonomy from the Ministry of Industry. This amendment gave far-reaching transformation to the organization succession and a common seal, and may sue or be sued in its corporate name. The statutory functions of Standards Organization of Nigeria by Section 3, subsections (1) of 1971 Act No. 56 are as follows:

- To organize test and do everything necessary to ensure compliance with standards designated and approved by the council;
- To undertake investigations as necessary into the quality of facilities, materials and products in Nigeria, and establish a quality assurance system including certification of factories, products and laboratories;
- To ensure reference standards for calibration and verification of measures and measuring instrument;
- To compile an inventory of products requiring standardization;
- To compile Nigeria Industrial Standards;
- To foster interest in the recommendation and maintenance of acceptable standards by industry and the general public;
- To develop method for testing of materials, supplies and equipment including items purchased for use of departments Government of the Federation or State and Private establishment;
- Register and regulate standard marks and specification;
- To undertake preparation and distribution of standard samples;
- To establish and maintain such number of laboratories or other institutions as may be necessary for the performance of its functions under this Act;
- To compile and publish general scientific or other data:
- Resulting from performance of its function under this Act; or
- From other sources when such data are of importance to scientific or manufacturing interest or to the general public and are not available;
- To advise departments of government of the Federation or state on specific problems relating to standards.

Nigeria represented by SON is a member of following International Standards Bodies:

- International Organization for Standardization (ISO);
- International Electrochemical Commission (IEC);
- African Organization for Standardization (ARSO).

Presently, Nigeria (SON) is a member of the International Organization for Standardization (ISO) with Participating (P-Member) and Observer Member (O-Member) in various Technical Committees (TC) and Project Development Committees (PDCs) [43]. Among all the standards available in the SON, there is none that is responsible for the standards of renewable energy technologies as at present. However, the next section describes a developmental process in which a renewable energy standard could be formulated and adapted by the SON.

4. Development of Renewable Energy Standards

Currently, the SON maintains 33 Standing Technical Committees (TC) which also have sub technical committees that deals with specific issues. The various TC usually creates a forum wherein the various experts are harnessed in the development of standards. A draft standard is usually prepared by the SON Technical Officers in the Organization depending upon the priority or work programme of the directorate and on the need or demand from the stakeholders. The officers, thereafter collect the relevant scientific data related to the draft standard under development from technical literatures including journals, laboratory tests results and factory inspections among others. The draft is then sent out to the TC members for review. The comments are received and collate the and the meeting of the relevant TC is held to review the comments and evaluated all quality parameters being prescribed are accepted by consensus to be adequate for approval by the Nigeria Standards Council or in its absence the Honorable Minister of Supervision Ministry of the Standards Organization of Nigeria, Approval of

National Standards. The responsibility of authorization of National Standards as per the Act establishing SON is vested in the Organization's Governing Council also referred to as Nigeria Standards Council. In absence of the Council, the Honorable Minister, Federal Ministry of Trade and Investments has the responsibility to approve a standard for use. Apart from development of new standards, existing standards are revised or reviewed so that the outdated matches with the current quality parameters. The procedure for review of standards is the same with that of drafting new ones. Standards are reviewed after three years, if there is technological development in the sector concerned or where inadequacies are observed. When standards reviewed and are no longer relevant for the purposes they were developed are withdrawn [43].

As earlier discussed in this study, the SON does not have any form of standards for renewable energy technologies present. However, development of renewable energy standards would be a welcomed development. The testing and proper certification of renewable energy equipment for off-grid or small scale application will still face some challenges. These challenges could come from the political class or importers who import substandard products into the country. Most renewable energy products are imported from China into the Nigerian market without proper product certification. Some of the imported technologies include solar panels, inverters, energy storage systems, small hydropower (SHP) units and wind turbines.

The deployment of renewable energy depends on some issues which are external to standards like Feed-in-Tariffs (FiTs), planning, environmental factors, building regulation and economic development. Standards will be used along with various policies, certification and regulation to open up renewable energy market.

Developing renewable energy standards involves high cost and is time consuming, this has hindered actors from participating in the process of renewable energy standards development. Development of renewable energy standards can be achieved if the SON collaborate with the ECN and other relevant bodies in engaging more external stakeholders who will make use of reports, case studies, research articles and various workshops to ensure that proper standards are developed for renewable energy technologies. These activities will improve the mix of experts that will in turn improve the quality of the standards under development. The SON will need to increase its TC to accommodate experts in the field of renewable energy. The SON should also engage specific projects related to renewable energy standards development as well as promote knowledge dissemination in renewable energy technologies.

4.1. Cost of Standards

Some challenges are faced by standard bodies in the developing countries is have the right standards in the first place and to have access to purchase the standards and also to keep them up to date. The cost implications for the standard organizations in the developing countries and emerging markets can make the difference between adopting the already existing internationally developed standards, producing local requirements or the complete rejection of the standards. The cost of standards may be seen as a barrier to the adoption of standards, but in the case of Nigeria, it should be seen as means to ensure that her citizens are protected from poor quality renewable energy technologies manufactured with the best practice of standardization. However, some recommendations for small renewable energy and hybrid systems are provided in **Table 2** for rural electrification (see [35]).

4.2. Recommended Standards for Renewable Energy Technologies

Below are some recommended standards currently available through the International Organization for Standardization (ISO). These recommended standards should be used as an immediate measure to address the issue of substandard importation of renewable energy technology.

- **ISO/TC 180 Solar Energy**

Standardization in the field of solar energy utilization in space and water heating, cooling, industrial process heating and air conditioning.

- **IEC/TC 82 Solar Photovoltaic Energy System**

Includes the entire field from light input to a photovoltaic cell to and including the interface with the electrical system(s) to which energy is supplied.

- **IEC/TC 88 Wind Turbines**

These standards address design requirements, engineering integrity, measurement techniques and test procedures. Their purpose is to provide a basis for design, quality assurance and certification. The standards are concerned with all subsystems of wind turbines, such as mechanical and internal electrical systems, support struc-

Table 2. Recommendations for small renewable energy and hybrid systems for rural electrification [35].

Standards No.	Standard Name	Cost of Standards (USD), [Cost in NGN]*
IEC/TS 62257-9-1	Recommendations for small renewable energy and hybrid systems for rural electrification—Part 9-1: Micro-power systems	252.31 [41441.58]
IEC 60364 (all parts)	Low-voltage electrical installations Note: All parts would equate to 32 standards documents	4284.00 [703641.27]
IEC 60364-5-53:2001 included in all parts above	Electrical installations of buildings—Part 5-53: Selection and erection of electrical equipment—Isolation, switching and control	Not Available**
IEC 60529	Degrees of protection provided by enclosures (IP Code)	262.50 [43115.27]
IEC/TS 62257-2:2004	Recommendations for small renewable energy and hybrid systems for rural electrification—Part 2: From requirements to a range of electrification systems	283.22 [46518.51]
IEC/TS 62257-4:2005	Recommendations for small renewable energy and hybrid systems for rural electrification—Part 4: System selection and design	188.81 [31011.79]
IEC/TS 62257-5:2005	Recommendations for small renewable energy and hybrid systems for rural electrification—Part 5: Protection against electrical hazards	188.81 [31011.79]
IEC/TS 62257-6:2005	Recommendations for small renewable energy and hybrid systems for rural electrification—Part 6: Acceptance, operation, maintenance and replacement	94.41 [15506.72]
IEC/TS 62257-7-1:2006	Recommendations for small renewable energy and hybrid systems for rural electrification—Part 7-1: Generators—Photovoltaic arrays	293.71 [48241.47]
IEC/TS 62257-7-3:2008	Recommendations for small renewable energy and hybrid systems for rural electrification—Part 7-3: Generator set—Selection of generator sets for rural electrification systems	209.79 [34457.73]
IEC/TS 62257-9-2:2006	Recommendations for small renewable energy and hybrid systems for rural electrification—Part 9-2: Micro-grids	251.75 [41349.6]
IEC/TS 62257-9-4:2006	Recommendations for small renewable energy and hybrid systems for rural electrification—Part 9-4: Integrated system—User installation	25.87 [4249.11]
	Total	**6435.18 [1056969.7]**

Note: *NGN means Nigerian naira. The currency conversation was made on 4th October, 2014; **The cost information was not available at the time of publication.

tures and control and protection systems. They are intended to be used together with appropriate IEC/ISO standards.

- **IEC/TC 4**

  Hydraulic rotating machinery and associated equipment allied with hydro-power development.
- **ISO/TC 238 Solid Biofuels**

  Standardization of terminology, specifications and classes, quality assurance, sampling and sample preparation and test methods in the field of raw and processed materials originating from arboriculture, agriculture, aquaculture, horticulture and forestry to be used as a source for solid biofuels. Excluded: areas covered by ISO/TC 28/SC 7 Liquid biofuels, ISO/TC 193 Natural gas.
- **ISO/PC 248 Sustainability Criteria for Bioenergy**

  Standardization in the field of sustainability criteria for production, supply chain and application of bioenergy. This includes terminology and aspects related to the sustainability (e.g. environmental, social and economic) of bioenergy.
- **ISO/TC 255 Biogas**

  Standardization in the field of biogas.

4.3. Recommendations to Standards Organization of Nigeria

The Standards Organization of Nigeria which is the standard body in Nigeria will need to make some adjustment to its policy in order to protect the citizens of Nigeria from the influx of poor quality renewable energy products into the country. Some recommendations are given below:

- Develop standard compliance cooperation with foreign countries where most of the renewable energy products are manufactured;
- Development of renewable energy standards with the cooperation of international standard bodies in the field of renewable energy;
- Promote the training of experts through development programmes, constructing standard databases, designating a permanent agency or commission for the promotion and implementation of renewable energy product standard;
- Implement laws for the promotion of renewable energy standardization plans, organization, budget and human resource development;
- Revise a number of laws and articles to ensure that renewable energy standardization is embedded in the political framework;
- Ensure that each renewable energy product imported into the country by an importer are given After Sales Service (ASS) for up-to 5 years from the time of purchase to the consumer.

5. Conclusions

For sustainable energy to be achieved in any country, clean energy development has to be a national priority. The issues of climate change have alerted most governments around the world to mitigate global warming by moving towards renewable energy adoption and incorporation into their energy mix. The manufacturing of renewable energy technologies like solar panels, solar thermal, inverters, energy saving systems, various kinds of wind turbines and hydropower systems has increased in recent times. Most countries like South Korea, Japan, China and the US manufacture these renewable energy technologies according to their own specification and standards and export it to other countries that do not have the technical expertise or resources to produce theirs. While this may sound good, some countries do manufacture substandard renewable energy technologies and export it to developing countries that may not have proper standards bodies to check the imported products into their market.

Nigeria has abundant renewable energy resources but do import the technology which is able to harness the renewable energy potential. These technologies are usually substandard and are dangerous to her citizens. This study has discussed the necessity of the development of standards for renewable energy technology which is imported, stressed the need to develop local content in the field of renewable energy and given some recommended standards that could be applied. Nigeria has a standard body which is the Standards Organization of Nigeria (SON) but no standards have been made available for renewable energy technologies. However, the country depends on standards set by other countries that export the technologies.

However, some immediate measures should be put in place by the SON to protect Nigerian citizens from the various substandard renewable energy technologies which include solar panels, inverters, and energy-saving systems. As an immediate measure, the SON should ensure that the technologies which are imported into the country are checked for international standards which are usually granted by International Electrochemical Commission (IEC) and also apply the given recommendation in this study. This will stop the influx of substandard renewable technologies while the SON develops its own standards.

This study strongly believes that the development of standards for renewable energy technologies is of great importance to secure the citizens of Nigeria and also to fast track the development of renewable energy in Nigeria. The study focused on the necessity for the development of standards for renewable energy technologies and recommended some standards but did not propose any standards. Thus, more studies should be undertaken to come up with a proper standard that can be submitted to the SON.

References

[1] Sambo, A.S. (2005) Renewable Energy for Rural Development: The Nigerian Perspective. *ISESCO Science and Technology Vision*, **1**, 12-22.

[2] Oyedepo, S.O. (2012) Energy Efficiency and Conservation Measures: Tools for Sustainable Energy Development in Nigeria. *International Journal of Energy Engineering*, **2**, 86-98.

[3] Okafor, E.N.C. and Joe-Uzuegbu, C.K.A. (2010) Challenges to Development of Renewable Energy for Electric Power

Sector in Nigeria. *International Journal of Academic Research*, **2**, 211.

[4] Patlitzianas, K., Flamos, A., Doukas, H., Kagiannas, A. and Psarras, J. (2007) Promoting Renewable Energies and Energy Efficiency through the CDM Funding Options. 2004 *New and Renewable Energy Technology*, 1.

[5] European Commission (2003) Second ECCP Progress Report: Can We Meet Our Kyoto Targets? European Commission, Brussels.

[6] Why Solar Energy Is Underutilized in Nigeria—Suleiman (2014). www.leadership.ng/news/347736/solar-energy-underutilised-nigeria-suleiman

[7] Shaaban, M. and Petinrin, J.O. (2014) Renewable Energy Potentials in Nigeria: Meeting Rural Energy Needs. *Renewable and Sustainable Energy Reviews*, **29**, 72-84. http://dx.doi.org/10.1016/j.rser.2013.08.078

[8] Vincent, E.N. and Yusuf, S.D. (2014) Integrating Renewable Energy and Smart Grid Technology into the Nigerian Electricity Grid System. *Smart Grid and Renewable Energy*, **5**, 220-238. http://dx.doi.org/10.4236/sgre.2014.59021

[9] Sambo, A.S. (2009) Strategic Developments in Renewable Energy in Nigeria. International Association for Energy Economics, 16.

[10] Uduma, K. and Arciszewski, T. (2010) Sustainable Energy Development: The Key to a Stable Nigeria. *Sustainability*, **2**, 1558-1570. http://dx.doi.org/10.3390/su2061558

[11] Nadabo, S.L. (2010) Renewable Energy as a Solution to Nigerian Energy Crisis.

[12] Energy Commission of Nigeria (ECN) (2009) Renewable Energy Master Plan.

[13] Ilenikhena, P.A. and Ezemonye, L.I.N. (2010) Solar Energy Applications in Nigeria. WEC MONTREAL.

[14] Yearly Average of Daily Sun in Nigeria. http://www.soda-is.com/eng/nigeria_en.html

[15] Adeoti, O., Oyewole, B.A. and Adegboyega, T.D. (2001) Solar Photovoltaic-Based Home Electrification System for Rural Development in Nigeria: Domestic Load Assessment. *Renewable Energy*, **24**, 155-161. http://dx.doi.org/10.1016/S0960-1481(00)00188-9

[16] Bugaje, I.M. (2006) Renewable Energy for Sustainable Development in Africa: A Review. *Renewable and Sustainable Energy Reviews*, **10**, 603-612. http://dx.doi.org/10.1016/j.rser.2004.11.002

[17] Oyedepo, S.O. (2012) On Energy for Sustainable Development in Nigeria. *Renewable and Sustainable Energy Reviews*, **16**, 2583-2598. http://dx.doi.org/10.1016/j.rser.2012.02.010

[18] Energy Commission of Nigeria (2006) Energy Commission of Nigeria (ECN) Renewable Energy Master Plan. Energy Commission of Nigeria, Government of Nigeria, Abuja.

[19] Nigeria Wind Energy. http://www.neenigeria.com/Nigeria_wind_NEW.png

[20] Bada, H.A. (2011) Managing the Diffusion and Adoption of Renewable Energy Technologies in Nigeria. *World Renewable Energy Congress*, Linkoping, 8-13 May 2011, 2642-2649.

[21] Mohammed, Y.S., Mustafa, M.W., Bashir, N. and Mokhtar, A.S. (2013) Renewable Energy Resources for Distributed Power Generation in Nigeria: A Review of the Potential. *Renewable and Sustainable Energy Reviews*, **22**, 257-268. http://dx.doi.org/10.1016/j.rser.2013.01.020

[22] Sambo, A.S. (2009) The Place of Renewable Energy in Nigeria Energy Sector. *World Future Council Workshop on Renewable Energy Policies*, Addis Ababa, 10 October 2009, 5.

[23] Ohunakin, O.S. (2010) Energy Utilization and Renewable Energy Sources in Nigeria. *Journal of Engineering and Applied Sciences*, **5**, 171-177. http://dx.doi.org/10.3923/jeasci.2010.171.177

[24] The Nigerian Water Ways (2014). http://www.safty4sea.com/

[25] Iloeje, O. (2002) Renewable Energy Development in Nigeria: Status & Prospects. *Proceedings of a National Workshop on Energizing Rural Transformation in Nigeria: Scaling up Electricity Access and Renewable Energy*.

[26] Ikuponisi, F.S. (2004) Status of Renewable Energy in Nigeria. *One Sky Energetic Solutions Conference*, Nigeria, 21-27 November 2004, Background Paper.

[27] Abumere, S.I., Okafor, S.I. and Oluwasola, O. (2002) Rural Infrastructure and the Development Process in Rural Nigeria (No. 36). Development Policy Centre.

[28] Sambo, A.S. (2010) Renewable Energy Development in Nigeria: A Situation Report. *World Future Council/Strategy Workshop on Renewable Energy*, Accra, 21-24 June 2010.

[29] Spalding-Fecher, R., Winkler, H. and Mwakasonda, S. (2005) Energy and the World Summit on Sustainable Development: What Next? *Energy Policy*, **33**, 99-112. http://dx.doi.org/10.1016/S0301-4215(03)00203-9

[30] Efurumibe, E.L. (2013) Barriers to the Development of Renewable Energy in Nigeria. *Scholarly Journal of Biotechnology*, **2**, 11-13.

[31] Standardization (2014). http://www.encyclopedia2.thefreedictionary.com/

[32] Sykes, A.O. (1995) Product Standards for Internationally Integrated Goods Markets. Brookings Institution, Washington DC.

[33] International Standard Organization (ISO). http://www.iso.org/

[34] World Trade Organization (WTO). http://www.wto.org/

[35] REP, IRENA (2013) International Standardisation in the Field of Renewable Energy.

[36] Marshall, A. (1919) Industry and Trade: A Study of Industrial Technique and Business Organization and of Their Influences on the Condition of Various Classes and Nations. MacMillian and Co. Limited, London, New York.

[37] David, P.A. and Greenstein, S. (1990) The Economics of Compatibility Standards: An Introduction to Recent Research. *Economics of Innovation and New Technology*, **1**, 13-41.

[38] Hawkins, R., Mansel, R. and Skea, J., Eds. (1995) Standards, Innovation and Competitiveness: The Politics and Economics of Standards in Natural and Technical Environment Edward Elgar, Aldershot, Brookfield, p. 1.

[39] IEEE Standards Association (2014) How Are Standards Made? http://www.standards.ieee.org/develop/process.html

[40] Try Standards: Global Standards Education and Standards Search. http://www.trystandards.org/mod/page/view

[41] The Difference between De Facto Standards and De Jure Standards (2014). http://www.ncos.gr.jp/English.files/e_newpage51

[42] De facto and De Jure Standards (2014). http://www.teach-ict.com/as_a2_ict_new/ocr/A2_G063/333_networks_coms/standards/miniweb/pg3

[43] Standards Organization of Nigeria (2014). http://www.son.gov.ng/

16

Performance of a Refrigeration Absorption Cycle Driven by Different Power Sources

Sameh Alsaqoor[1], Khaled S. AlQdah[2*]

[1]Mechanical Engineering Department, Tafila Technical University, Tafila, Jordan
[2]Mechanical Engineering Department, Faculty of Engineering, Taibah University, Medina, KSA
Email: sameh@wp.pl, [*]dr.khaled_qdah@yahoo.com

Abstract

In this study, performance assessment of absorption refrigeration cycle has been carried out under variable power sources namely electric, conventional fuel and renewable energy sources. The ammonia-water absorption cycle was used in this work, the temperatures at each point in the cycle such as generator, absorber, evaporator and condenser have been measured and with using absorption device system. The coefficient of performance and efficiency of the plant were measured and then compared. The results showed that when the cycle driven by electricity, the coefficient of performance varied 0.694 to 1.032 along the test time and the generator temperature changes from 48.1˚C to 101.5˚C with the average efficiency of 57.1% and average coefficient of performance of 0.78. When methane used as a fuel to generate power the coefficient of performance varied between 0.686 and 0.94 under the generator temperature of 123.3˚C and 127.4˚C and average efficiency of 40.02% with coefficient of performance of 0.735. Solar energy used as the alternative source of power which is the clean and safe power source and when the plant driven by the solar thermal energy, the coefficient of performance reached to 0.801 under the generator temperature of 91˚C, but the system efficiency about 11.68% along the test time. Solar energy can be used efficiently and replaced the conventional power sources to drive the absorption refrigeration unit.

Keywords

Absorption, Solar Energy, Performance, Generator, COP

1. Introduction

Many people in the world are currently experiencing dramatic shifts in lifestyle as their economies make the transition from subsistence to an industrial or service base. The largest increases in energy demand will take place in developing countries where the proportion of global energy consumption is expected to increase from 46 to 58 percent between 2004 and 2030 [1]. Energy consumption in developing countries is projected to grow at an average annual rate of 3 percent from 2004 to 2020. Energy consumption in developing regions is projected to surpass that in industrialized regions and about half of the increase in global energy demand by 2030 will be for power generation and one fifth for transport needs—mostly in the form of petroleum-based fuels (EIA, 2007).

Conventional air-conditioning systems require high quality energy, electricity, generated from primary energy resources, however, in spite of CFC-free cooling liquids, and energy consumption and pollution level surprising. Therefore, from ecological point of view needs to be implemented as the green house gases effect remains threat to the fact, most of the buildings' cooling demands in summer are associated with high solar energy availability, which offers an opportunity to further exploit solar energy for cooling. Solar cooling technology provides an important contribution to both economical and ecological energy supply. Thermally driven refrigeration systems are providing golden energy by using heat as motive energy. This heat can be obtained from combined heat and power systems, waste-heat sources or solar en energy. Since Jordan in a none oil producing country and the energy issues and the seek for new alternative energy sources one of the most priority for us, therefore we will focus on the alternative energy sources to operate the absorption refrigeration cycle that may be used for cooling and refrigeration processes. Solar cooling and absorption cooling system has been witnessed a lot of development and studies by researcher in the world and the almost available studies are:

Hattem and Dato (1983) [2] described of an active solar cooling system, using a LiBr-H_2O absorption machine. Bong *et al.* (1987) [3] studied the performance a solar powered air conditioning system. Yeung *et al.* (1992) [4] investigated the performance of a solar powered air conditioning system in Hong Kong. Chinnappa *et al.* (1993) [5] investigated the solar-assisted vapor compression/absorption cascaded air-conditioning systems. Best and Ortega (1999) [6] studied the solar cooling and refrigeration. Chen and Hihara (1999) [7], a new absorption refrigeration cycle using solar energy has been carried out. Deng and Ma (1999) [8] conducted an experimental studies on the characteristics of an absorber using LiBr/H_2O solution as working fluid. Li, Z. and Sumathy (2000) [9] reviewed the technology development in the solar absorption air-conditioning systems. Rivera and Xicale (2001) [10] studied the heat transfer Coefficients in two-phase flow for the water/lithium bromide mixture used in solar absorption refrigeration systems. Grossman (2002) [11] conducted an experimental study of solar powered systems for cooling, dehumidification and air conditioning. Lokurlu and Müller (2005) [12] built the first solar cooling system based on trough collectors combined with double effect absorption chillers. Syed *et al.* (2005) [13] carried out an experimental investigation of a solar cooling system in Madrid. Balaras *et al.* (2007) [14] cited a total of 54 solar air-conditioning projects in Europe are inoperation, 33 of them are working with lithium bromide-water absorption chillers. Kim (2007) [15] conducted a theoretical study on solar absorption cooling in Delft University of Technology, Netherlands. Zambrano *et al.* (2007) [16] developed a new model and validation of a solar cooling plant. Khaled AlQdah (2011) [17] conducted an experimental study to evaluate the performance of aqua ammonia auto air conditioner system using exhaust waste energy.

In recent years, many researchers have been carried out on refrigeration systems driven by solar energy, which achieves conservation of conventional energy and a reduction of environmental pollution. Although many kinds of refrigeration systems have been developed and studied. It can be seen that very little work has been done to compare the performance of refrigeration or air conditioning system under different power sources which is the motivation of our work that will compare the performance of absorption refrigeration under different climatic conditions in Jordan. The structure of this manuscript starts by discussing the related research from the body of literature in Section one, while Section two presents a description of absorption cycle. Section three displayed experimental apparatus and setup. While the experimental methodology was presented in Section four. The results were presented and discussed in section five. Finally; section six summarized the study findings through the conclusion.

2. Absorption Cycle

The absorption cycle is a process by which refrigeration effect is produced through the use of two fluids and some quantity of heat input, rather than electrical input as in the more familiar vapor compression cycle [18]. Both vapor compression and absorption refrigeration cycles accomplish the removal of heat through the evaporation of a refrigerant at a low pressure and the rejection of heat through the condensation of the refrigerant at a higher pressure. In the absorption system, a secondary fluid or absorbent is used to circulate the refrigerant. As the name implies, absorption refrigeration systems involve the absorption of a refrigerant by a transport medium. The most widely used absorption refrigeration system is the ammonia-water system shown in **Figure 1**, where ammonia (NH_3) serves as the refrigerant and water (H_2O) as the transport medium. Absorption machines are commercially available today in two basic configurations. For applications above 32˚F (primarily air conditioning), the cycle uses lithium bromide as the absorbent and water as the refrigerant. For applications below 32˚F, an ammonia/water cycle is employed with ammonia as the refrigerant and water as the absorbent.

3. Experimental Setup and System Description

A series of experiments were carried out in Tafila Technical University mechanical engineering laboratories using absorption ammonia—water cycle device as depicted in **Figure 2**. The system is similar to vapor-compression system, except that the compressor has been replaced by a complex absorption mechanism consisting of an absorber, a generator, a control valve, and a rectifier. Once the pressure of NH_3 is raised by the previous components, it is cooled and condensed in the condenser by rejecting heat to the surroundings, is throttled to the evaporator pressure, and absorbs heat from the refrigerated space as it flows through the evaporator.

The heat on the generator (Boiler) transferred to the ammonia vapor and the result is driven it from the solution. This hot vapor rises into the rectifier and a portion of the water condenses and flows by gravity into the absorber. The hot ammonia vapor continues to rise into the condenser where it gives up its heat to the surrounding air and condenses into a liquid. The liquid ammonia enters by gravity into the evaporator, where it is mixed with hydrogen gas. Circulation of hydrogen gas causes a reduction in pressure within the evaporator. The low pressure causes the ammonia liquid to boil into a gas (evaporating) and absorbing heat in the process (refrigerating effect).

The mixture of hydrogen/ammonia vapor that's carrying the absorbed heat is now drawn by gravity into the absorber. Because the water from the separator has a greater affinity for ammonia, it separates from the hydrogen gas. The hydrogen gas being very light rises and returns to the generator to start the cycle again.

4. Methodology

The system was prepared to test by three different types of energy sources as follows and as shown in **Figure 3**.

1. Wired layouts:

I. In the case of electrical energy, the plug connected to power network and switched the electric heater which enfolds the generator.

II. When the system heated by LPG gas, the combustion take place in the generator, then the heat transferred to solution by surface of combustion chambers.

III. The apparatus driven by thermal solar energy; at the help of tube connected it with the output of solar flat plate collector. The all systems collectors (with an angle 45˚ tracked in the sun direction) and absorber refrigeration located in open space at Tafila Technical University.

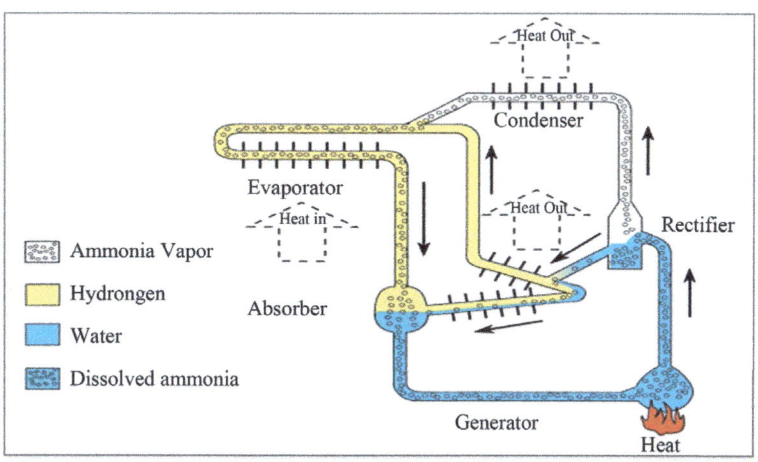

Figure 1. Absorption cycle schematic diagram.

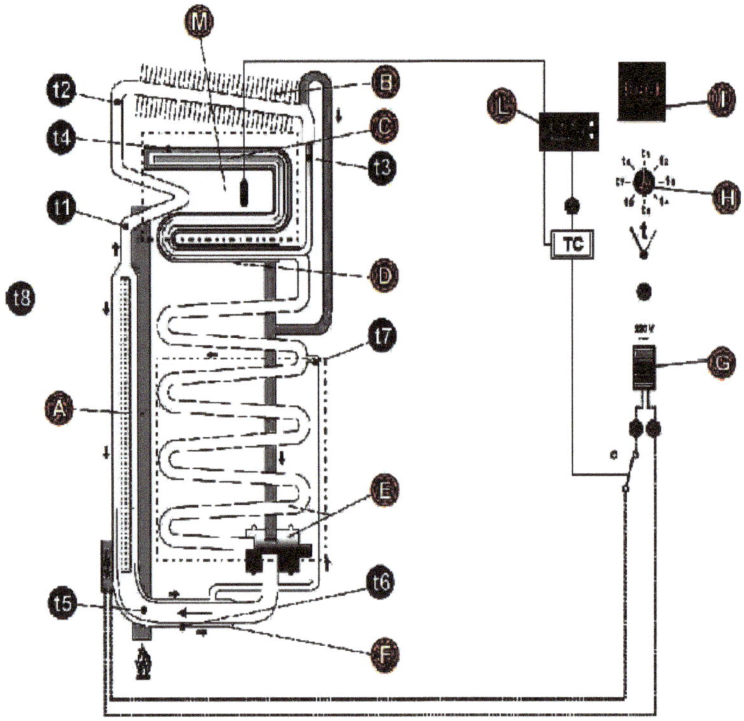

Figure 2. Absorption ammonia-water cycle device.

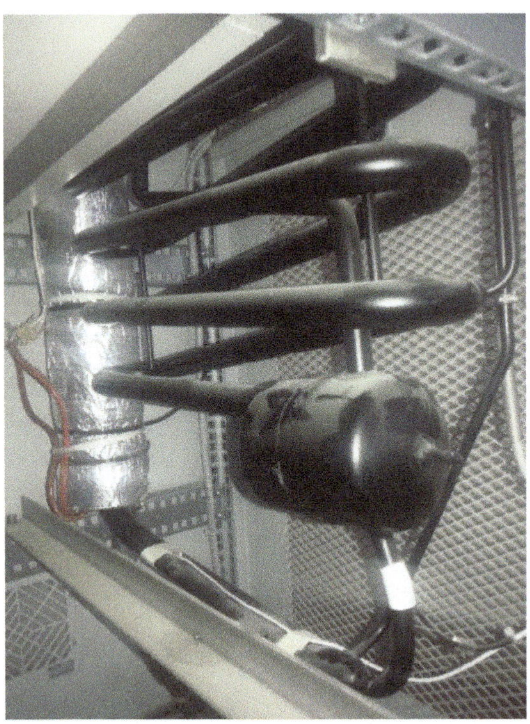

Figure 3. The absorption refrigeration cycle tested for different kinds of energy.

2. For each, and for all those cases, the measurement takes every 15 minutes, until the system obtaining the steady state conditions.

3. The results were compiled and presented in the form of figures.

The equations used for developed the measuring results covering plant efficiency, absolute coefficient of performance and real coefficient of performance can be represented as follows:

Absolute coefficient of performance (theoretical)

$$COP_{abs} = \frac{T_G - T_A}{T_G} \times \frac{T_E}{T_C - T_E} \quad [19]$$

where:

T_G: Temperature of the generator [°C]

T_A: Temperature of the absorber [°C]

T_E: Temperature of the evaporator [°C]

T_C: Temperature of the condenser [°C]

Real coefficient of performance (experimentally)

$$COP_R = \frac{\text{Heat removed}}{\text{Energy consumed}} = \frac{Q_c}{E} \quad [19]$$

where:

Q_C: Heat removed from cold room [kJ]

E: Energy consumed [kJ]

Heat removed from cold room

$$Q_c = V_{CR} \times \rho_{air} \times C_{P_{air}} \times (T_{amb.} - T_{CR})$$

where:

V_{CR}: Volume of the room [m^3]

ρ_{air}: Density of air = 1, 3 [kg/m^3]

$C_{P_{air}}$: Specific heat of air = 0.25 [Kcal/kg·°C].

T_{amb}: Ambient temperature [°C]

T_{CR}: Cold room temperature [°C]

The efficiency of the absorption refrigeration cycle can be calculated as:

$$\eta = \frac{COP_R}{COP_{abs}}$$

5. Results and Discussion

The absorption refrigeration system was used to test and compare the suitability of different sources of energy and their impact on working conditions at the absorption refrigeration process. So it has been used three types of energy sources: Electrical energy (coil), LPG gases and solar energy with the help of flat plate solar collector at Tafila Technical University in Jordan. The variation of temperature changes in the generator have been represented in **Figure 4** for all sources of energy. After 15 minutes of operation of the system, the temperature increases for a gas fuel to the value of 1230°C in the same time and the temperature ranged around 500°C for the other sources. After 30 minutes of operation the variation of temperature in the generator minimized and to the end of the measurements remained at the level of 1300°C for gas fuel and 1000°C for electrical energy, while in the case of solar energy the temperature is changed all the time and only at the end measurements equalized taken value about 800°C.

When it comes to the changes of temperature in an evaporator they have been represented in **Figure 5**. The temperature practically decreases linearly, when electrical energy is used to run the system, getting on the end of the experiment values −20°C. The temperature on the evaporator decreases to about −30°C when LPG fuel used to run the absorption refrigeration system, but in the case of the solar energy is used to run the system, the temperature has received at the level of 70°C.

Figure 6 shows the important value, it means Coefficient of Performance. It can be seen in the case of the COP_{abs} (Theoretical COP) which is calculated only on the basis of the measured temperature at various points in the system, as temperature on the generator, temperature on the absorber, temperature on the evaporator and temperature on the condenser almost all equal and varies between 0.69 to 0.8 for all kinds of sources of energy. While the situation is somewhat different in the case of value of real coefficient of performance (COPR), the differences is great, especially when used electrical energy (The COP reached about 0.46) to run the system and

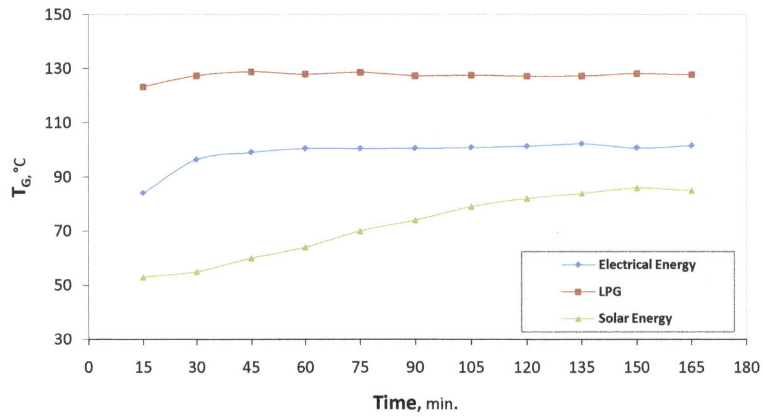

Figure 4. The changing of temperature at the generator of an absorption refrigeration cycle.

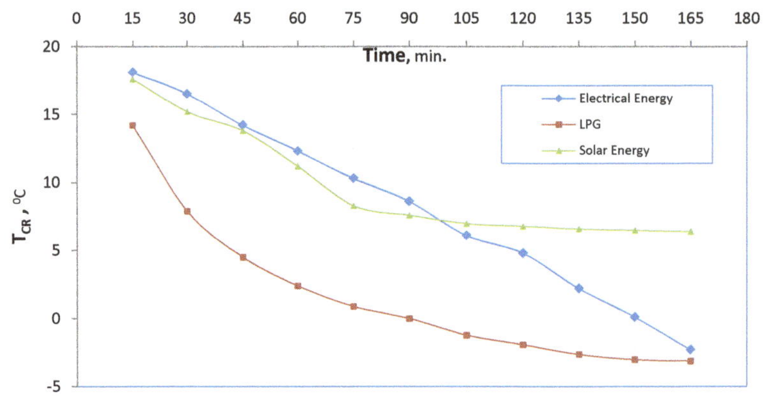

Figure 5. The changing of temperature at the evaporator (cooling room) of an absorption refrigeration cycle.

	Electrical Energy	LPG	Solar Energy
COPabs	0.733	0.686	0.801
COPR	0.463	0.244	0.109

Figure 6. The coefficient of performance for many types of energy runs the absorption refrigeration system.

solar energy ,the COPR about 0.10 whereas the COPR equal about 0.24.

From the **Figure 7**, it's evident that the maximum efficiency can be reached for an absorption refrigeration system runs by electrical energy achieved value of 63%, whereas receives on the level of 45% for the system runs by LPG gases, so the efficiency dramatically reduced for the absorption refrigeration system powered by flat plate solar collector.

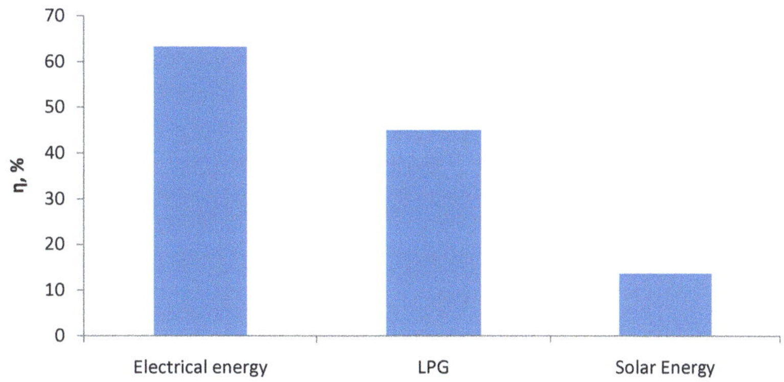

Figure 7. The efficiency of an absorption refrigeration cycle runs by many types of energy sources.

Even the fact that the solar energy has the lowest performance but has a lot of advantages, among others, it is free sources and does not affect the degradation of the environment, also in desert areas where there is no power grid, can be wins big role and application in refrigeration and air conditioning division.

Finally, it can be say that the COP_{abs}, depends on the temperature at various points in the system, for example, the flat plate solar collector cannot assure high temperatures in the generator which is one of the important components of the system and serious influence at the performance. Also consumed energy [E, kJ] losses for encounters various sources such as these losses are smaller in the case of electricity the coil is directly in the generator, in the case of solar energy the losses in the pipes connected two systems, it is the fact that the COP_{abs} is larger for solar energy source system.

6. Conclusion

The possibility of having a refrigeration system driven by the solar energy is a huge advantage in Jordan (as well as other parts of the world). There are many circumstances where people do not have access to electricity or gas to power a refrigeration system. An absorption refrigeration cycle employing an ammonia-water solution as the working fluid has been investigated in this work. The cycle has been powered by three different power sources to providing a high temperature heat source for absorption cooling unit. Based on the results of the investigations performed in this study, the concluding remarks are as follows: the renewable energy resources like solar energy can contribute effectively to the supply of heat to the refrigeration load. The efficiency of absorption cooling based on a type resource of heat reliant on the temperature to be formed inside the generator, this temperature using to dominate the efficiency and coefficient of performance COP. for absorption refrigeration system. For this we find the best of efficiency and COP was found when using the electricity as a power source for the absorption cycle, after that using of methane instead of electricity but with its environmental impact and pollution. The clean and safe power source when we using solar thermal energy in spite of lower COP and efficiency. A number of equipment requirements and limitations must be considered in the analysis and design of powered absorption systems.

The first limitation deals with the problem of reduced efficiency of the absorption cooler under start-up and transient conditions. Typical absorption coolers do not reach operating efficiency until after an hour or more of operation time, machine which is cycled on and off regularly will have a drastically reduced average coefficient of performance when compared to a machine in steady state performance, this problem has been overcome in at least one installation by the use of a cold storage unit. The second limitation is the high cost; which is attributable to the use of some additional equipment for the absorption cooling such as absorber and boiler. Research is under way to develop and cooling units at relatively low cost through a number of different factors may contribute to this subject.

References

[1] Ministry of Energy and Mineral Resources, Jordan (2011) Annual Report. Amman.

[2] Hattem, D.V. and Dato, P.A. (1981) Description of an Active Solar Cooling System, Using a LiBr-H_2O Absorption

Machine. *Energy and Buildings*, **3**, 169-196.

[3] Bong, T.Y., Ng, K.C. and Tay, A.O. (1987) Performance Study of a Solar Powered Air Conditioning System. *Solar Energy*, **39**, 173-182. http://dx.doi.org/10.1016/S0038-092X(87)80025-7

[4] Yeung, M.R., Yuen, P.K., Dunn, A. and Cornish, L.S. (1992) Performance of a Solar Powered Air Conditioning System Hong Kong. *Solar Energy*, **48**, 309-319. http://dx.doi.org/10.1016/0038-092X(92)90059-J

[5] Chinnappa, J.C., Crees, M.R., Murthy, S.S. and Srinivasan, K. (1993) Solar-Assisted Vapor Compression/Absorption Cascaded Air-Conditioning Systems. *Solar Energy*, **50**, 453-458. http://dx.doi.org/10.1016/0038-092X(93)90068-Y

[6] Best, R. and Ortega, N. (1999) Solar Refrigeration and Cooling. *Renewable Energy*, **16**, 685-690. http://dx.doi.org/10.1016/S0960-1481(98)00252-3

[7] Chen, G.M. and Hihara, E. (1999) Anew Absorption Refrigeration Cycle Using Solar Energy. *Solar Energy*, **66**, 479-482. http://dx.doi.org/10.1016/S0038-092X(99)00042-0

[8] Deng, S.M. and Ma, W.B. (1999) Experimental Studies on the Characteristics of an Absorber Using LiBr/H_2O Solution as Working Fluid. *International Journal of Refrigeration*, **22**, 293-301. http://dx.doi.org/10.1016/S0140-7007(98)00067-X

[9] Li, Z.F. and Sumathy, K. (2000) Technology Development in the Solar Absorption Air-Conditioning Systems. *Renewable Sustainable Energy Reviews*, **4**, 267-293. http://dx.doi.org/10.1016/S1364-0321(99)00016-7

[10] Rivera, W. and Xicale, A. (2001) Heat Transfer Coefficients in Two Phase Flow for the Water/Lithium Bromide Mixture Used in Solar Absorption Refrigeration Systems. *Solar Energy Materials and Solar Cells*, **70**, 309-320.

[11] Grossman, G. (2002) Solar Powered Systems for Cooling, Dehumidification, and Air Conditioning. *Solar Energy*, **72**, 53-62. http://dx.doi.org/10.1016/S0038-092X(01)00090-1

[12] Lokurlu, A. and Müller, G. (2005) Experiences with the World Wide First Solar Cooling System Based on Trough Collectors Combined with Double Effect Absorption Chillers. *Proceedings of the International Conference on Solar Air-Conditioning*, Bad Staffelstein.

[13] Syed, A., Izquierdo, M., Rodriguez, P., Maidment, G., Missenden, J., Lecuona, A. and Tozer, R. (2005) Anovel Experimental Investigation of a Solar Cooling System in Madrid. *International Journal of Refrigeration*, **28**, 859-871. http://dx.doi.org/10.1016/j.ijrefrig.2005.01.007

[14] Balaras, C.A., Grossman, G., Henning, H.M., Ferreira, C.A.I., Podesser, E., Wang, L. and Wiemken, E. (2007) Solar Air Conditioning in Europe—An Overview. *Renewable and Sustainable Energy Reviews*, **11**, 299-314. http://dx.doi.org/10.1016/j.rser.2005.02.003

[15] Kim, D.-S. (2007) Solar Absorption Cooling. Ph.D. Dissertation, Delft University of Technology, Delft.

[16] Zambrano, D., Bordons, C., Garcia-Gabin, W. and Camacho, E.F. (2008) Model Development and Validation of a Solar Cooling Plant. *International Journal of Refrigeration*, **31-32**, 315-327.

[17] AlQdah, K.S. (2011) Performance and Evaluation of Aqua Ammonia Auto Air Conditioner System Using Exhaust Waste Energy. *Energy Procedia*, **6**, 467-476. http://dx.doi.org/10.1016/j.egypro.2011.05.054

[18] Didacta Italia (2011) T108/6D, Educational Equipment on the Absorption Refrigeration Cycle User's Manual and Exercise Guide.

[19] Cengel, Y. and Boles, M. (2010) Thermodynamics: An Engineering Approach. 7th Edition, McGraw-Hill, New York.

Nomenclature

COP: Coefficient of performance
Cp_{air}: Specific heat for air
E: Energy consumed
${}_GW_A$: The work done between the generator and absorber
${}_CW_E$: The work done between the condenser and evaporator
Q_G: The heat removed from the generator
Q_E: The heat removed from the generator
P_E: The electric power of the generator [KW]
T_G: Temperature of the generator
T_A: Temperature of the absorber
T_E: Temperature of the evaporator
T_C: Temperature of the condense
T_{amb}: Room temperature [°C]
Δh = Change of enthalpy
ρ : Density
η : Efficiency

17

A User Friendly Approach for Design and Economic Analysis of Standalone SPV System

Sheeraz Kirmani

Department of Electrical Engineering, Jamia Millia Islamia, New Delhi, India
Email: sheerazkirmani@gmail.com

Abstract

This paper presents the use of solar energy for generation of electrical energy through solar photovoltaic (SPV) system to meet the load requirement of a domestic building. Complete design and economic analysis of SPV system for different costs of SPV module is done and compared with grid electricity for with and without storage conditions. The results of the study encourage the use of SPV system for a residential building and show that SPV system is an economically viable option to meet the exponentially growing electricity requirement for household applications in India.

Keywords

Household Electrification, Solar Photovoltaics, SPV System Design, Cost Analysis

1. Introduction

India faces a significant gap between electricity demand and supply. Demand is increasing at a very rapid rate compared to supply. According to the World Bank, roughly 40% of residences in India are without electricity. In addition, blackouts are a common occurrence throughout the country. The World Bank also reports that one-third of Indian businesses believe that unreliable electricity is one of their primary impediments of doing business. In addition, coal shortages are further straining power generation capabilities. In order to meet the situation, a number of options are considered. Power generation from freely available solar energy is one such option. Most parts of India receive bright sunshine around 3000 hours in a year except Kerala, northeastern states and Jammu & Kashmir where sunshine hours are appreciable low.

During monsoon, a significant decrease in sunshine occurs over the whole country except Jammu and Kashmir where the maximum duration of sunshine occurs in June and July, and minimum during January due to its location. The northeastern states and southeast peninsula also receive relatively less sunshine during October and November due to the northeast monsoons. As far as the availability of global solar radiation is concerned, more than 2000 kWh/m^2/year are received over Rajasthan and Gujarat, while east Bihar, northwest Bengal and the northeastern states receive less than 1700 kWh/m^2/year. The availability of diffuse solar radiation varies widely in the country. The annual pattern shows a minimum of 740 kWh/m^2/year solar radiations over Rajasthan increasing eastwards to 840 kWh/m^2/year in the north eastern state, and southwards to 920 kWh/m^2/year.

India is thus endowed with rich solar energy resource. The average intensity of solar radiation received in India is 200 MW/km^2. With a geographical area of 3.287 million km square, this amounts to 657.4 million MW. However, 87.5% of the land is used for agriculture, forests, etc., 6.7% for housing, industry, etc. and 5.8 % are barren, snowbound or generally inhabitable. Thus only 12.5% of the land area amounting to 0.413 million km square, in theory, can be used for solar energy installations. Even if 10% of this land can be used, the available solar energy would be 8 million MW, which is equivalent to 5909 million tons of oil equivalent.

Launching India's national action plan on climate change on June 30, 2008, the prime minister of India Dr. Manmohan Singh stated "Our vision is to make India's economic development energy efficient. Over a period of time, we must pioneer a graduated shift from economic activity based on fossil fuels and from reliance on non-renewable and depleting sources of energy to renewable sources of energy. In this strategy, the sun occupies centre-stage, as it should being literally the original source of all energy. We will pool our scientific, technical and managerial talents, with sufficient financial resources, to develop solar energy as a source of abundant energy to power our economy and to transform the lives of our people. Our success in this endeavor will change the face of India. It should also enable India to help change the destinies of people around the world."

The economic viability of a standalone PV system in comparison to the most likely conventional alternative system, *i.e.* a diesel-powered system, has been analyzed for energy demand through sensitivity analysis [1]. The analysis shows that PV-powered systems are the lowest cost option at a daily energy demand of up to 15 kWh, even under unfavorable economic conditions. When the economic parameters are more favorable, PV-powered systems are competitive up to 68 kWh/day. These comparisons are intended to give a first-order indication when a standalone PV system should be considered for application. As the cost of PV systems decreases and diesel costs increase, the break-even points occur at higher energy demand.

A methodology is presented [2] for determining the optimum size and location of installing the solar photovoltaic based DG system for supplying the active power at the node in a radial distribution system for loss reduction.

Techno-economic feasibility of three different energy-supplying alternatives, namely the solar photovoltaic (SPV) system, diesel generator system and extending the grid connection for energy supply to a remote village located around 15 km away from the place where grid supply is available, is suggested [3]. The study suggests that the SPV systems are cheaper in remote villages where grid extension is costlier.

Literature presents several studies on energy payback time and life cycle analysis of PV technologies. The analyses of the PV system with reference to a fuel oil-fired steam turbine and their GHG emissions and costs revealed that greenhouse gases (GHG) emission from electricity generation from the PV system is less than one-fourth that from an oil-fired steam turbine plant and one half that from a gas-fired combined cycle plant. From the life cycle energy use and GHG emission perspectives, the PV system is a good choice of power generation. However, it also indicates that large scale exploitation of PV could lead to other types of undesirable environmental impacts in terms of material availability and waste disposal [3].

Life cycle assessment of electricity generation by PV panels considering mass and energy flows over the whole production process starting from silica extraction to the final panel assembling, using the most advanced and consolidate technologies for polycrystalline silicon panel production is presented [4]. Briefly, the most important results of the analysis are the calculation of a gross energy requirement (GER) of 1494 MJ/panel (0.65 m^2 surface) and of a global warming potential (GWP) of 80 kg of equivalent CO_2 panel. The energy payback time (EPBT) has been estimated to be shorter than the panel operation life even in the worst geographic conditions. The results of the LCA support the idea that the photo-voltaic electric production is advantageous for the environment. The authors in [5] analyzed foreseeable technological advancements in current and emerging PV technologies over the next few decades are likely to lead to significantly lower per-kWh impact than the one that characterizes the current state of the art of the PV sector.

A techno-economic analysis of standalone solar photovoltaic system has been presented [6]. In this work, complete analytical methodology for optimum relationship between PV array and storage battery capacity to supply the required energy at a specified energy load fraction is carried out. To estimate the performance of solar photovoltaic system the solar radiation utilizability concept and the monthly average daily PV array efficiency have been used. The techno-economic optimization of a PV system has been done by using levelized energy cost computation based on the total number of battery replacements (brp's) through battery life-cycle model. It has been found that energy load fractions as well as the number of brp's have a significant impact on the selection of optimum sizing of a standalone PV system. From the techno-economic and environmental points of view, the feasible sites in Egypt to build a 10 MW PV-grid connected power plant and recommended few sites for large scale PV power generation [7].

A techno-economic comparison of rural electrification based on solar home systems and PV micro-grids to supply electricity to rural community for domestic purpose has been performed [8]. Based on study it is concluded that a micro grid might be a more attractive option financially for the user, energy service company and the society if the village has a large number of households, which is densely populated and lies in a geo-graphically flat terrain and more than 500 densely located households using 3 - 4 low power appliances (e.g. 9 W CFLs) for an average of 4 h daily. However, in rough terrains solar home systems might be a better option if the community is small and sparsely populated. The economic analysis has been performed on the grid connected SPV system connected to the Spanish grid [9]. Using net present value (NPV) and payback period (PP) parameters, the profitability of the system was studied. The system was evaluated for its economic as well as environmental benefits and the results clearly showed that the system was profitable enough to be invested in, but very long payback periods were dissuading the investors.

A strong case of standalone SPV systems has been built by conducting feasibility study in an island of West Bengal India by the name Sagar Deep based on socio-economic and environmental aspects. The generation costs of SPV systems and conventional power has been compared to show how conventional power systems suffer from diseconomy when power needs to be transmitted to extremely remote locations. The social viability of SPV was apparent from a conspicuous improvement in commerce, trade, education and increased participation of women in activities other than household chores in the island [10].

The paper uses the solar radiation data of a particular site for complete design and cost analysis of proposed Solar Photovoltaic (SPV) systems, to be installed in a building to supply electricity.

2. Domestic Electrical Load

The main electrical loads necessary for improving living conditions in the village are: household appliances (lighting, TV, refrigerator, washing machine, water pump, electric iron and fan, AC). The ratings of these appliances are given in **Table 1**.

3. Solar Radiation Analysis

Solar radiation data provide information on how much of the sun's energy strikes a surface at a location on earth during a particular time period. These data are needed for effective research into solar energy utilization.

The daily average solar energy incident over India varies from 4 to 7 kWh/m^2 with about 1500 - 2000 sunshine hours per year, depending upon location. Theoretically, a small fraction of the total incident solar energy

Table 1. Household electrical load.

Electrical Load	No. of Units	Operating Hours Per Day	Wattage Per Unit Used	Total Units Consumed Per Day
Lighting Lamp	5	4 lamps from 18 to 24 & 1 lamp from 0 to 6	60	1.8
Washing Machine	1	from 11 to 13	250	0.5
Refrigerator	1	from 0 to 24	180	2.4
Water Pump	1	For three hours daily	120	0.36
TV	1	from 17 to 24	80	0.56
Fan	3	from 0 to 24	60	1.44
Electric Iron	1	For one hour in two days	1500	0.75
AC 1.5 ton	2	6 hours each daily	2000	24
	Total		6530 Watts	31.81

(if captured effectively) can meet the entire country's power requirements. In view of this, measuring efficiently solar energy radiations in a big country like India and preparing data for various applications are major challenges. The network of solar energy measuring stations is rather scarce throughout the world. In India, only IMD (Indian Meteorological Department) Pune provides data for quite few stations, which is considered as the base data for research purposes. However, hourly measured data of solar irradiance is not available, even for those stations where measurement has already been done. Due to lack of hourly measured data, the estimation of solar energy at the earth's surface is an important study in the present scenario to meet the energy requirement from green energy sources. Monthly averaged data of global and diffuse solar irradiance at New Delhi is shown in **Figure 1**.

4. Design of Stand Alone SPV System

The schematic diagram of a PV power system with battery storage is presented in **Figure 2**.

In **Figure 2**, the function of the PV generator is to convert the sunlight directly into DC electrical power and that of the battery is to store the excess power through using the battery charger. The load and storage control is used to convert the DC electrical power into AC power to match the requirements of the common household AC appliances.

1) Sizing of the PV generator

The most appropriate SPV power system to cover such a load is illustrated in **Figure 2**.

The peak power of the PV generator (P_{PV}) is obtained as follows:

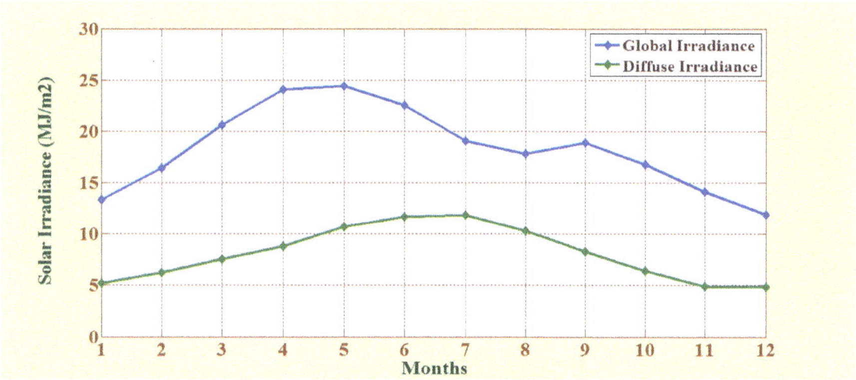

Figure 1. Monthly average daily global and diffuse solar irradiance on a horizontal surface at New Delhi.

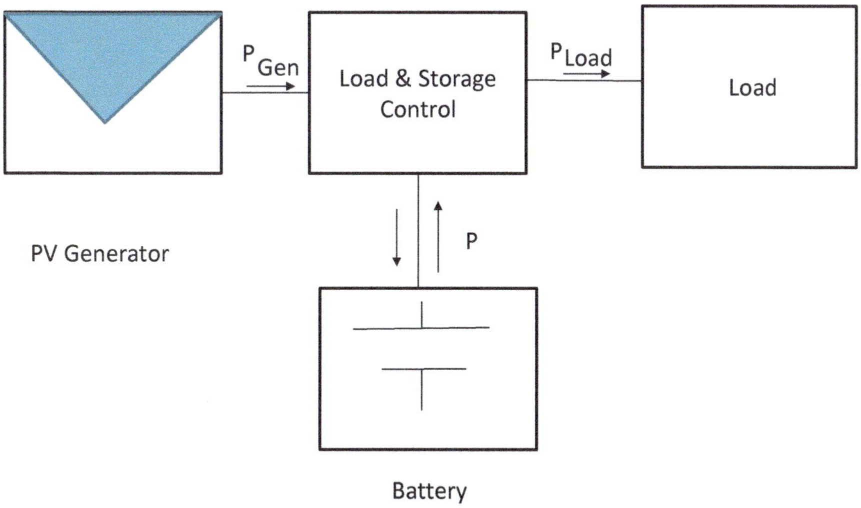

Figure 2. Schematic diagram of a PV power system with battery storage.

$$P_{PV} = \frac{E_L}{\eta_V \eta_R \text{PSH}} S_F$$

where E_L (daily energy consumption) = 31.81 kWh/day, (PSH) peak sun hours = 5. The efficiencies of the system components $(\eta_R = 0.92, \eta_v = 0.9)$ and the safety factor for compensation of resistive losses and PV-cell temperature losses $S_f = 1.15$.

Substituting these values in the above equation, we obtain the peak power of the PV generator:

$$P_{PV} = 8.836 \text{ kW}_p$$

To install this power, a mono-crystalline PV module type MBPV-CAAP (this module is of Moser Baer company) of a gross area of $A_{PV} = 1.65 \text{ m}^2$, peak power of $P_{mpp} = 200 \text{ Wp}$ is selected. The number of the necessary PV modules (N_{pv}) is obtained as

$$N_{PV} = \frac{P_{PV}}{P_{mpp}} = 44 \text{ modules are required}$$

V_{mp} for MBPV200 Watt Peak is 28.81 Volts.

No. of modules in series $= 220 \times 1.414/28.81 = 11$

Hence the number of modules in parallel = 4

For MBPV CAAP 200 W_p module

V_{oc} (open circuit voltage) = 35.99 Volts and

I_{sc} (short circuit current) = 7.65 Amperes

Now, we obtain an open circuit voltage and short circuit current for this SPV array as

$V_{oc} = 35.99 \times 11 = 395.89$ V

$I_{sc} = 7.65 \times 4 = 30.6$ A.

2) Sizing of the Battery Block

The storage capacity of battery block for such systems is considerably large. Therefore, special lead-acid battery cells (block type) of long life time, high cycling stability-rate and capability of standing very deep discharge should be selected. Such battery types are available but at much higher price than regulars batteries. The ampere hour capacity (C_{Ah}) and watt hour capacity (C_{wh}) of the battery block, which are necessary to cover the load demands for a period of 2.0 days without sun, is obtained as follows

$$C_{Ah} = \frac{1.5 \times E_L}{V_B \times \text{DOD} \times \eta_B \times \eta_V}$$

$$C_{wh} = C_{Ah} V_B$$

where V_B and η_B are voltage and efficiency of battery block, while DOD is the permissible depth of discharge rate of a cell. Assuming realistic values of $\eta_B = 0.85$, $\text{DOD} = 0.75$ and $V_B = 220$ V, we obtain:

$$C_{Ah} = 500 \text{ Ah}$$

$$C_{wh} = 110 \text{ kWh}$$

3) Design of the Battery Charge Controller

The battery charge controller is required to safely charge the batteries and to maintain longer lifetime for them. It has to be capable of carrying the short circuit current of the PV array. Thus, in this case, it can be chosen to handle 31 A.

4) Design of the Inverter

The used inverter must be able to handle the maximum expected power of AC loads. Therefore, it can be selected as 20% higher than the rated power of the total AC loads. Thus the rated power of the inverter becomes 7836 W. The specifications of the required inverter will be 7836 W, 220 VAC, and 50 Hz.

The cost of the equipments required in the proposed SPV system is presented in **Table 2**.

5. Life Cycle Cost Analysis

In this section the life cycle cost (LCC) estimation of the designed standalone PV system is discussed. The LCC

Table 2. Cost of equipments used in the proposed SPV system.

Item	PV	Battery Cell for 510 Ah	Charger	Inverter	Installation	M&O/Year
Cost	Rs. 60/W_p	Rs. 5500/cell	Rs. 258/A	Rs. 36/W	10% of PV cost	2% of PV cost

of an item consists of the total costs of owning and operating an item over its lifetime, expressed in today's money.

The costs of a standalone SPV system include acquisition costs, operating costs, maintenance costs, and replacement costs. All these costs have the following specifications:

- The initial cost of the system (the capital cost) is high.
- There are no fuel costs.
- Maintenance costs are low.
- Replacement costs are low (mainly for batteries).

The LCC of the PV system includes the sum of all the present worths (PWs) of the costs of the PV modules, storage batteries, battery charger, and inverter, the cost of the installation, and the maintenance and operation cost (M&O) of the system. The details of the used cost data for all items are shown in **Table 2**.

The lifetime N of all the items is considered to be 25 years, except that of the battery which is considered to be 5 years. Thus, an extra 4 groups of batteries (each of 6 batteries) have to be purchased, after 5 years, 10 years, 15 years and 20 years assuming inflation rate i of 3% and a discount or interest rate d of 10%. Therefore, the PWs of all the items can be calculated as follows [6].

PV array cost $C_{\text{wh}} = C_{\text{Ah}} V_B C_{\text{PV}} = 60 \times 44 \times 200 = \text{Rs. } 528000$

Initial cost of batteries cells $C_B = 110 \times 5500 = \text{Rs. } 605000$

The PW of the 1st extra group of batteries (purchased after $N = 5$ years) C_{B1PW} can be calculated, to be Rs. 435479, from

$$C_{\text{B1PW}} = C_B \left(\frac{1+i}{1+d} \right)^N$$

The PW of the 2nd extra group of batteries (purchased after $N = 10$ years) C_{B2PW}, that of the 3rd extra group (purchased after $N = 15$ years) C_{B3PW}, 4th extra group of batteries purchased after $N = 20$ years C_{B4PW} are calculated to be Rs. 313450.5, Rs. 225665, Rs. 162563.5, respectively.

Charger cost $C_C = 258 \times 31 = \text{Rs. } 8000$

Inverter cost $C_{\text{Inv}} = 36 \times 7836 = \text{Rs. } 282096$

Installation cost $C_{\text{Inst}} = 0.1 \times 968000 = \text{Rs. } 52800$

Operation and maintenance cost per year = Rs. 10560

The PW of the maintenance cost C_{MPW} can be calculated to be Rs. 125355.5, using the maintenance cost per year (M/yr) and the lifetime of the system ($N = 25$ years), from

$$C_{\text{MPW}} = (\text{M/yr}) \times \left(\frac{1+i}{1+d} \right) \times \left[\frac{1 - \left(\frac{1+i}{1+d} \right)^N}{1 - \left(\frac{1+i}{1+d} \right)} \right]$$

Therefore, the LCC of the system can be calculated, to be Rs. 2613054, from:

$$\text{LCC} = C_{\text{PV}} + C_B + C_{\text{B1PW}} + C_{\text{B2PW}} + C_{\text{B3PW}} + C_C + C_{\text{Inv}} + C_{\text{Inst}} + C_{\text{MPW}}$$

It is sometimes useful to calculate the LCC of a system on an annual basis. The annualized LCC (ALCC) of the PV system in terms of the present day Rupees can be calculated, to be Rs. 206116.8/yr, from,

$$\text{ALCC} = \text{LCC} \left[\frac{1 - \left(\frac{1+i}{1+d} \right)}{1 - \left(\frac{1+i}{1+d} \right)^N} \right]$$

Once the ALCC is known, the unit electrical cost (cost of 1 kWh) can be calculated, to be Rs. 17.5/kWh, from:

$$\text{Unit electrical cost} = \frac{\text{ALCC}}{365 E_L}$$

6. Result and Discussion

Now, using the same procedure as given above, the cost of the PV system without batteries can be calculated by excluding the terms $C_B + C_{\text{B1PW}} + C_{\text{B2PW}} + C_{\text{B3PW}} + C_{\text{B4PW}}$ from the above calculations.

The unit cost of electricity for with and without batteries for different rates of PV modules is calculated using the same procedure as above and tabulated in **Table 3**.

The cost of electricity generated from solar photovoltaic system with batteries is calculated to be Rs. 22/kWh, whereas the cost of utility electricity supplied to consumers is approximately Rs. 6.8/kWh for domestic purpose. However, the cost of electricity generated from solar photovoltaic is decreasing rapidly as the initial cost of PV modules is coming down with time as shown in **Table 3**. As observed from **Table 3** and **Figure 3** that the cost of power generation from SPV systems is decreasing at a very fast rate. The cost of one unit of electricity generated from SPV system was Rs. 28 and Rs. 17 for with and without batteries respectively in 2007, when the cost of the PV modules was Rs. 150/W_p. The cost of PV modules has decreased to Rs. 100/W_p in 2012, consequently the cost of one unit electricity generated from SPV has also come down to Rs. 22 and Rs. 11 for with and without batteries respectively as presented in this chapter.

As shown in **Table 3**, if the cost of PV modules decreases to Rs. 60/W_p by 2017 and to Rs. 40/W_p by 2022, the cost of unit electricity generated from SPV systems would decrease drastically to Rs. 18 and Rs. 7 by 2017 and to Rs. 15 and Rs. 4 by 2022 for with and without batteries respectively. However, the price of electricity generated from SPV systems discussed above are without the 30% subsidy as provided by MNRE, government of India, if this 30% subsidy is also considered then the unit of electricity generated from SPV will further decrease as shown in **Table 3** above. At the same time, the cost of electricity generated from grid supply is in-

Table 3. Cost trends for spv systems.

Year	PV Panel Cost/W_p	Unit Cost of Electricity (INR) With Battery		Unit Cost of Electricity (INR) Without Battery	
		Without Subsidy	With 30% Subsidy	Without Subsidy	With 30% Subsidy
2007	150	28	19.6	17	11.9
2012	100	22	15.4	11	7.7
2017	60	18	12.6	7	4.9
2022	40	15	10.5	4	2.8

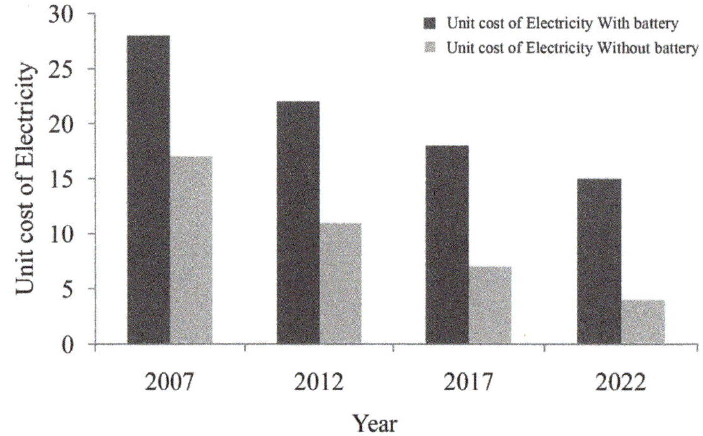

Figure 3. Cost trends for SPV systems.

creasing day by day, presently it is around Rs. 6.8/unit for domestic use. But in the near future, the cost of electricity generated from conventional fuels in India is expected to increase to around four times than its present value due to the rapid increase in the conventional fuel prices but the cost of unit electricity generated from SPV is decreasing as shown above. Hence SPV systems are expected to be a cheaper and good option for its use in residential purpose in the near future. Therefore the present results reflect the first order indication to energy planners in the near future. Hence the SPV power generation will be economically comparable to the grid connected power supply in the near future in India.

7. Conclusions

Proper utilization of renewable energy sources is very important worldwide especially in the developing countries like India. This study presents the complete design and the life cycle cost analysis of the PV system for single household application. The results of the study indicate that electrifying a single area household using PV systems is beneficial and suitable for long-term investments, because the initial prices of the PV systems are decreasing as the cost of PV panels is decreasing.

In this paper an electrification study for a single residential household is carried out using a standalone SPV system. The results of study reveal that although at present, the cost of electricity generated from SPV system is relatively high as compared to grid electricity but in the near future the cost of electricity from SPV systems will decrease rapidly and cost of electricity from grid is likely to increase. Also, the efficiency of the solar cells is increasing day by day which will make it further economical. As we know that solar power is one of the most promising and more predictable sources than other renewable sources and less vulnerable to change in seasonal weather. Whereas, generation of power from other renewable sources is limited to sites where these resources exist in sufficient quantities and can be harnessed, solar energy can produce power at the point of demand in both rural and urban areas. Hence, photovoltaic systems are considered as the most promising energy sources for these sites due to their high reliability and safety. Therefore, SPV system would be the better option for household electrification in India in the near future.

References

[1] Kolhe, M., Kolhea S. and Joshi, J.C. (2002) Economic Viability of Stand-Alone Solar Photovoltaic System in Comparison with Diesel-Powered System for India. *Energy Economics*, **24**, 155-165. http://dx.doi.org/10.1016/S0140-9883(01)00095-0

[2] Kirmani, S., Jamil, M. and Rizwan, M. (2011) Optimal Placement of SPV Based DG System for Loss Reduction in Radial Distribution Network Using Heuristic Search Strategies. *Proceedings of IEEE 2nd International Power and Energy Conference*, Bhubhaneshwar, 28-30 December, 1691-1695.

[3] Jamil, M., Kirmani, S. and Chatterjee, H. (2013) Techno-Economic Viability of Three Different Energy- Supplying Options for Remote Area Electrification in India. *International Journal of Sustainable Energy*, **33**, 470-482.

[4] Stoppato, A. (2008) Life Cycle Assessment of Photovoltaic Electricity Generation. *Energy*, **33**, 224-232. http://dx.doi.org/10.1016/j.energy.2007.11.012

[5] Raugei, M. and Frankl, P. (2009) Life Cycle Impacts and Costs of Photovoltaic Systems: Current State of the Art and Future Outlooks. *Energy*, **34**, 392-399. http://dx.doi.org/10.1016/j.energy.2009.01.001

[6] Kolhe, M. (2009) Techno-Economic Optimum Sizing of a Stand-Alone Solar Photovoltaic System. *IEEE Transaction on Energy Conversion*, **24**, 511-519. http://dx.doi.org/10.1109/TEC.2008.2001455

[7] EL-Shimy, M. (2009) Viability Analysis of PV Power Plants in Egypt. *Renewable Energy*, **34**, 2187-2196. http://dx.doi.org/10.1016/j.renene.2009.01.010

[8] Chaurey, A.A. and Kandpal, T.C. (2010) A Techno-Economic Comparison of Rural Electrification Based on Solar Home Systems and PV Microgrids. *Energy Policy*, **38**, 3118-3129. http://dx.doi.org/10.1016/j.enpol.2010.01.052

[9] Bernal-Agustin, J.L. and Dufo-López, R. (2006) Economical and Environmental Analysis of Grid Connected Photovoltaic Systems in Spain. *Renewable Energy*, **31**, 1107-1128. http://dx.doi.org/10.1016/j.renene.2005.06.004

[10] Chakrabarti, S. and Chakrabarti, S. (2002) Rural Electrification Programme with Solar Energy in Remote Region—A Case Study in an Island. *Energy Policy*, **30**, 33-42. http://dx.doi.org/10.1016/S0301-4215(01)00057-X

18

Design and Analysis of Different Types of Rotors for Pico-Turbine

Samia Tabassum, Mashudur Rahaman, Muhammad Shahriar Bashar, Saidul Islam, Afrina Sharmin, Abdullah Yousuf Imam, Azizul Hoque, Nahid Mahbub, Sayeda Khatun, Mahfuza Khanam

Institute of Fuel Research and Development, Bangladesh Council of Scientific and Industrial Research, Dhaka, Bangladesh

Email: shawon14@gmail.com

Abstract

Small-scale vertical axis wind turbine (VAWT) rotor is developed for use in areas lacking adequate energy infrastructure. The materials and methods of construction are selected to minimize cost as much as possible. The paper describes the design of different kinds of vertical axis wind turbine rotors having different number of blades and twist angle. The aim of the work is to study the influence of the different designs on rotational speed and power of rotor in different wind speed.

Keywords

Vertical Axis Wind Turbine, Savonius Rotor, Wind Energy, Power Co-Efficient

1. Introduction

The rising cost of fossil fuel and effects of climate change forces to enhance the demand of renewable energy sources in different sector. Like as other renewable energy sources, wind power has a key role in reducing greenhouse gas emissions [1]. Wind energy is a clean renewable energy source cheaper to maintain, saves fuel and can give decentralized energy [2]. A wind turbine is a device that taps the renewable kinetic energy of blowing wind, which converts it to useable mechanical, electrical or thermal energy. The main idea of wind power is to produce mechanical or electrical power economically without air pollution by using exhaustible natural resources. The function of a wind electric generator is to convert the kinetic energy of the wind into most useful form of energy *i.e.* electricity. The basic design requires the conversion done in the most efficient way and at lowest cost. This is done in few steps such as kinetic energy of the wind, converted through blades to mechanical energy of a rotating shaft and then transmitted through a gearbox to a generator, which converts the mechanical energy into electrical energy [3]. Today, the most commonly used wind turbine is the Horizontal Axis Wind Turbine (HAWT), where the axis of rotation is parallel to the ground. However, Vertical Axis Wind Turbine (VAWT), where, the axis of rotation is longitudinal to the ground is also an imperative issue in power generation sector. These devices can operate in flows coming from any direction, and take up much less space than a traditional HAWT [4], and VAWTs are definitely a credible source of energy for the future. VAWTs have a number of advantages over HAWTs, such as simple construction, low cost and can accept wind from any direction. In Bangladesh, wind speed is not so high and direction is highly variable [5]. That's the reason why vertical axis wind blade can be more effective for this country [6]. In this context, different types of vertical axis wind blade have been developed and analyzed [7]. A brief design description, details of the experiments conducted and the conclusions drawn are the prominent features of the present paper.

1.1. Basic Principle of Savonius Wind Turbine

Savonious wind turbine is vertical axis wind turbine (VAWT), which converts the force of the wind into torque on a rotating shaft. It is invented by the Finnish engineer Sigurd Johannes Savonius in 1922. The Savonius rotor consists of two (look like "S" shape in cross section) or three scoops. Because of the curvature, vertical axis wind turbines allow winds from any direction equally and force to rotate [8]. The savonius wind turbine works due to the difference in forces apply on each blade. The concave half part of the blade receives air and forces the blade to rotate around its central vertical shaft. At the same time, the convex half part of rotor hits the blade and causes the air to be deflected sideway around it, which is shown in **Figure 1**.

By placing two rotors on the same shaft, one designed to run clockwise and another one designed to run counter clockwise and exposing them to a flow, a significantly better performing rotor can be made. Both rotors should turn in the direction of the one with better starting characteristics. This type of vertical axis rotor is very robust and durable if built correctly, is relatively slow turning and can be easily built at home, without the hassles of aerofoil blade design and other problems associated with horizontal axis "propeller" type turbines.

1.2. Design and Theory

In this work, we have developed three different types of blade for investigating various properties. The first model (Blade-1), shown in **Figure 2**, is a twisted savonius with two helical blades arranged at 180° around the shaft. GI wires (15 swg) were placed on shaft (aluminum pipe with diameter of 0.0127 m) as frame of blade, which was covered by paper. Blade-2 is a drag-type savonius (**Figure 3**) made by plastic material. It has two sections. Each two-scoop would like an "S" shape in cross section. The two scoops in the top section of the rotor are rotated at 90° to the bottom ones. This ensures that there is always at least one scoop in a position to catch the wind and as a result, the turbine is self-starting. The third one (Blade-3) is also a drag type savonius with three blades (**Figure 4**). Here, the blades are arranged at 60° around the shaft. Each blade is triangular in shape and curved a bit with a support. GI sheets and aluminum pipe were used for the blades and shaft, respectively.

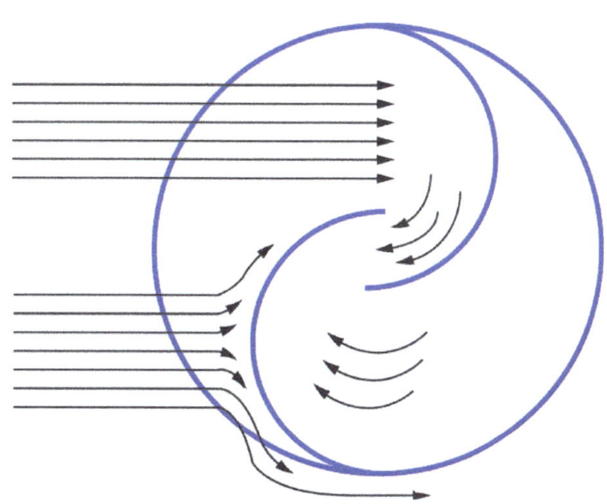

Figure 1. The effects of wind on savonius blade.

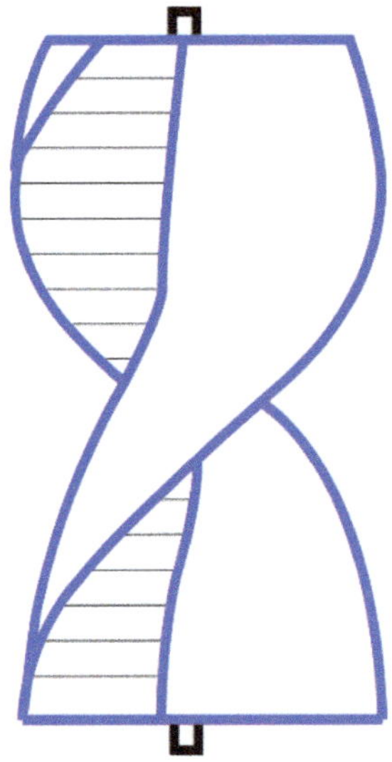

Figure 2. Twisted vertical axis blade (Blade-1).

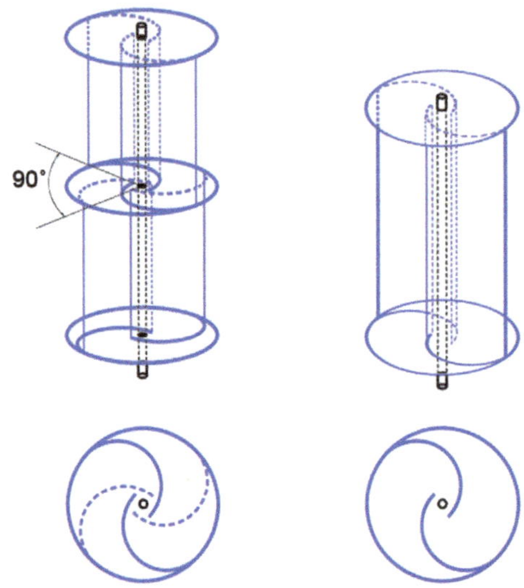

Figure 3. (a) Two steps savonius blade (Blade-2); (b) One step savonius blade.

Blade-1 Blade-2 Blade-3

Figure 4. Picture of different designed blades.

The amount of energy produced by a wind turbine primarily depends on the rotor area, also referred to as cross-sectional area, swept area, or intercept area. The swept area for savonius wind turbine can be calculated from the dimensions of the rotor

$$\text{Swept area} = A = H * D \tag{1}$$

where, H = rotor height; D = rotor diameter.

The tip speed ratio, λ is the ratio of the linear speed of rotor blade ω. R to the undisturbed wind speed; V. Here, ω is the angular velocity and R represents the radius revolving part of the turbine. High tip speed ratio improves the performance of wind turbine and this could be obtained by increasing the rotational rate of the rotor.

$$\lambda = \omega \cdot R/V. \tag{2}$$

The power coefficient (C_p) represents the amount of energy that can be harvested from the wind to convert it into mechanical energy. The power coefficient reaches its maximum for a unique tip speed ratio. It is the ratio of maximum power obtained from the wind to the total power available in the wind. This hypothesis shows the relationship between the power coefficient (C_p) and the wind speed (V), which expresses the basic theory of the Savonius wind turbine. Principally the power that the savonius rotor can extract from the wind (P_w) is less than the actual available from the wind power (P_a). The available power (P_a), which is also the kinetic energy (KE) of the wind, can be defined as:

$$P_a = 1/2\, m_a V^2$$

$$P_a = 1/2\, \rho A V^3$$

where, m_a (wind mass flow rate striking the swept area of the wind turbine (kg/sec)) $= \rho A V$.

ρ is density of air, A is the rotor area, V is the speed of wind.

The power coefficient (C_p) is given by:

$$C_p = \frac{\text{Extracted power from the wind}}{\text{Available power of the wind}} = \frac{P_w}{P_a}.$$

The extracted power from the wind of a wind turbine blade is determined by the following equation:

$$P_w = 0.5 \rho A V^3 C_p. \tag{3}$$

The fundamental laws of conservation of mass and energy allowed no more than 59% of the kinetic energy of the wind to be captured which is well-known as Betz' law. Practically, wind turbines cannot operate at this maximum limit. The C_p value is unique to each turbine type and is a function of wind speed that the turbine is operating in. In practice, values of obtainable power coefficients are in the range of 0.35 - 0.45 common even in the best designed wind turbines. This value below the theoretical limit is caused by the inefficiencies and losses attributed to different configurations, rotor blades and turbine designs. The maximum power coefficient, C_p for savonius rotor is 0.30. Hence, the C_p value used in this work is 0.30 [9]. However, density of air varies according to elevation, temperature and weather fronts. Density of air for 30°C is 1.16 kg/m^3 [10].

Wind speed was measured by cup anemometer (Lutron ABH-4224) and rotor speed in rotation per minute (RPM) was measured by using laser type tachometer (Lutron DT-1236L). The data was taken at room temperature. Swept area and obtained wind power is also important aspects for this purpose and it is calculated as well.

The wind tunnel was built to have interchangeable parts to test different combinations of design parameters. Experiments were conducted in this wind tunnel. The schematic diagram of wind tunnel made by GI sheet is shown in **Figure 5**. The wind tunnel is 172 cm long, which consists of contraction section, test section and diffuser section. The first part is contraction zone to produce a uniform wind velocity distribution to the test section. Rotors are placed in front of anemometer in the test section, where, wind speed data and rotational speed of rotor is measured. The exhaust fan is driven to flow the wind in tunnel in the diffuser section.

2. Results and Discussion

The aim of this experimental study is to obtain the aerodynamic performance of studied wind turbine rotor and also the detailed information about velocity field around the rotor. Different types of blade is developed and placed in wind tunnel, where, rotational speed is measured in variation of wind speeds. **Table 1** shows the summery of design parameters of blades.

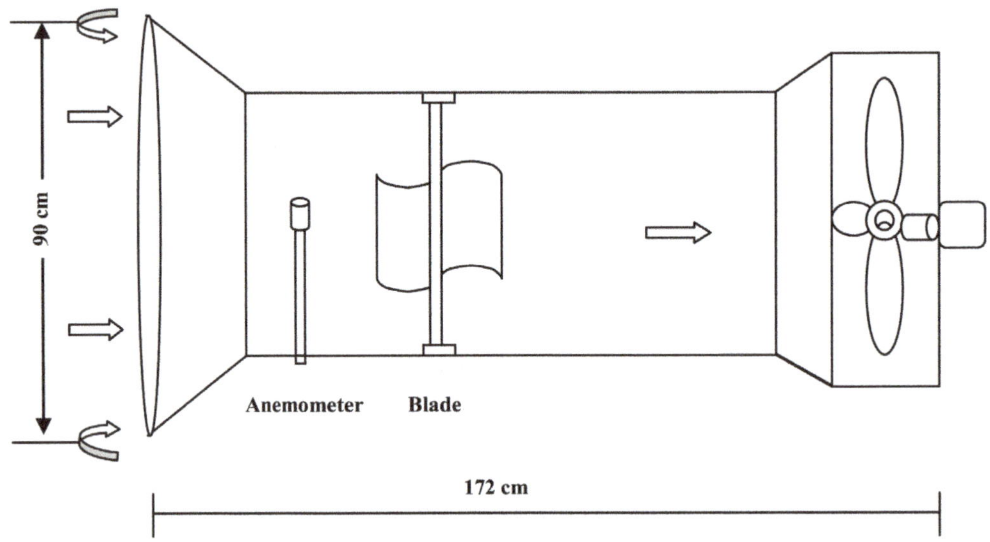

Figure 5. Schematic diagram of wind tunnel.

Table 1. Summary of design parameters of blades.

Blade	Height (m)	Diameter (m)	Aspect ratio (H/D)	Weight (kg)	Swept area A (m^2)	Tip speed ratio at 4 m/s
Blade 1	0.46	0.46	1.00	0.34	0.217	0.88
Blade 2	0.46	0.35	1.31	1.40	0.161	1.03
Blade 3	0.39	0.29	1.34	0.44	0.113	0.81

Figure 6 shows rotational speed of three blades in different wind speed. During the test, the wind turbine blades rotational speed varies from 72 rpm to 241 rpm and the upstream flow velocity varies from 2 m/s to 4.4 m/s. This figure state that the system has better performance for Blade-2 and 3. On the other hand, blade 1 showed low rotational speed. This may be due to the cause of slim rotor with small diameter which can get higher rotational speed but lower torque, and *vice versa* rotor with bigger rotational diameter produces a bigger torque but a lower rotational speed.

Figure 7 compares the theoretically calculated extracted power produced by the three different savonius rotors alone in different wind speed. Figure shows that the extracted power from the wind of the wind turbine blade mainly starts from 3 m/sec. Before that, power generation is very small. It largely increases when it turns from 3 to 4 m/sec. It also showed the variation of extracted power with available wind speed in air. This difference is due to the variation of swept area which has been already explained in Equation (3). Beside this, weight of turbine can play the role to change the extracted power with wind speed.

3. Conclusion

The average wind speed in coastal area of Bangladesh is not that much high (about 4 m/s). Hence, it can contribute effectively on our energy sectors depending on perfect designing and optimization of wind turbine. An advantage of VAWTs is that they can catch the wind from all directions and also can withstand much harsher environments. The paper presents the different designs of rotor and reveals the influence of design parameters on the mechanical performances of the rotor of wind turbines which can be used for small scale power generation. The experiments were carried out in wind tunnel using three different model turbines and performance was measured. The swept area and weight of the rotor can play an important role on the performance of rotor. The rotor with small diameter can get higher rotational speed but lower torque and lower power, and *vice versa* rotor with bigger rotational diameter produces a bigger torque and also extracts high power, but a lower rotational speed. Moreover, further accurate assembly and proper choosing of blade material will help to achieve better results. However, blades profile can be improved to have more aerodynamic rotation with an airfoil design and

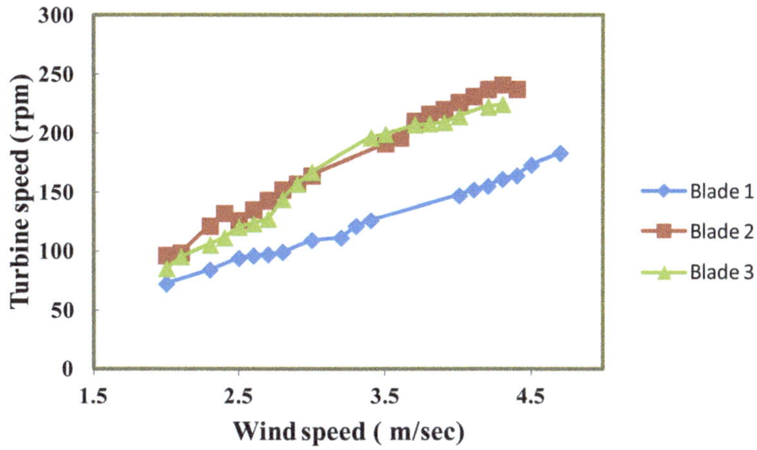

Figure 6. Data of rotational speed (rpm) for variation of wind speeds (v).

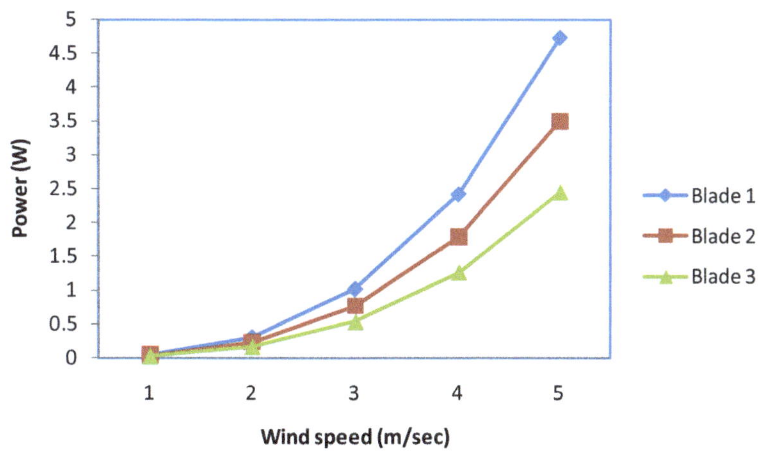

Figure 7. Power vs. wind speed curve.

providing a load can ensure a smoother result. Experimental results can be affected a bit while applying in coastal area with natural wind source and varying temperature as well.

References

[1] Ajedegba, J.O. (2008) Effects of Blade Configuration on Flow Distribution and Power Output a Zephyr Vertical Axis Wind Turbine. Thesis Submitted for a Degree of M.Sc. to University of Ontario Institute of Technology.

[2] Musgrove, P.J. (1987) Wind Energy Conversion: Recent Progress and Future Prospects. *Solar & Wind Technology*, **4**, 37-49. http://dx.doi.org/10.1016/0741-983X(87)90006-3

[3] Amirat, Y., Benbouzid, M.E.H., Bensaker, B. and Wamkeue, R. (2007) Condition Monitoring and Fault Diagnosis in Wind Energy Conversion Systems: A Review. *IEEE International Electric Machines and Drives Conference*, **2**, 1434-1439.

[4] Ali, M.H. (2013) Experimental Comparison Study for Savonius Wind Turbine of Two & Three Blades at Low Wind Speed. *International Journal of Modern Engineering Research* (*IJMER*), **3**, 2978-2986.

[5] Hyers, R.W., McGowan, J.G., Sullivan, K.L., Manwell, J.F. and Syrett, B.C. (2006) Condition Monitoring and Prognosis of Utility Scale Wind Turbines. *Energy Materials*, **1**, 187-203. http://dx.doi.org/10.1179/174892406X163397

[6] Kumar, A. and Grover, S. (1993) Performance Characteristics of a Savonius Rotor for Wind Power Generation—A Case Study, Alternate Sources of Energy. *Proceedings of Ninth National Convention of Mechanical Engineers*, IIT Kanpur.

[7] Ogawa, T., Yoshida, H. and Yokota, Y. (1989) Development of Rotational Speed Control Systems for a Savonius-Type Wind Turbine. *ASME Journal of Fluids Engineering*, **111**, 53-58. http://dx.doi.org/10.1115/1.3243598

[8] Shankar, P.N. (1978) Development of Vertical Axis Wind Turbines. Rural Technology, IISc, Bangalore, 145-162.

[9] Widodo, W.S., Chin, A.C., Sihombing, H. and Yuhazri, M.Y. (2012) Design and Analysis of 5 kW Savonius Rotor Blade. *Global Engineers & Technologists Review*, **2**.

[10] Bennouna, O., Heraud, N., Rodriguez, M. and Camblong, H. (2007) Data Reconciliation and Gross Error Detection Applied to Wind Power. *Proceedings of the Institution of Mechanical Engineers, Part I: Journal of Systems and Control Engineering*, **221**, 497-506. http://dx.doi.org/10.1243/09596518jsce266

19

Optimization of Building Layouts to Increase Wind Turbine Power Output in the Built Environment Assumed to Be Installed at Fukushima City and Tsu City in Japan

Akira Nishimura[1*], Masanobu Kakita[1], Junsuke Murata[1], Toshitake Ando[1], Yasunari Kamada[1], Masafumi Hirota[1], Mohan Lal Kolhe[2]

[1]Division of Mechanical Engineering, Graduate School of Engineering, Mie University, Tsu, Japan
[2]Faculty of Engineering and Science, University of Agder, Grimstad, Norway
Email: [*]nisimura@mach.mie-u.ac.jp

Abstract

It is very important to consider proper intelligent integration and locations of renewable energy sources into the built environment for developing smart cities. Wind speed distribution study in the built environment is very essential for analyzing the wind turbine performance located in the built environment. In this work, the building layout like nozzle is proposed and the objective is to optimize the building layout for increasing electrical energy output of wind turbine, assumed to be installed in actual cities of Japan. The wind speed distribution across buildings is numerically simulated by using CFD-ACE+. Wind turbine power output is estimated using the power curve of a real commercial wind turbine and wind speed distribution is simulated using CFD software. The meteorological data of Fukushima city and Tsu city of Japan are utilized for evaluating the wind speed distribution profile across the building and for finding the electrical energy output from wind turbine. The proposed building models, which have the angle between two buildings like nozzle of 90°, 135° and 180°, can provide the wind acceleration at the back of buildings for the wind blowing from the main wind direction and the angle of 135° is optimum building layout. In the case of installing the proposed building model in Fukushima city and Tsu city, the wind energy output in winter season is higher while that in summer season is lower irrespective of the buildings' angle. The interaction between the change in frequency distribution of wind speed and direction throughout the year and the location of open tip of building model decides the power generation characteristics of the proposed building model.

[*]Corresponding author.

Keywords

Built Environment Wind Turbine, Built Environment Wind Speed Distribution, Building Layouts for Built Environment Wind Turbine, Smart City

1. Introduction

For city planning, it is very important to propose the building model for integrating and locating environment friendly energy systems. It is effective to construct a smart city, consisting proposed buildings orientations for not only meeting the energy demand but also decreasing the amount of CO_2 emitted by power generation. In developing environment friendly smart cities, it is needed to analyze the intelligent integration with proper location of intermittent renewable energy sources in the built environment [1]-[3]. This study is proposing the built environment wind turbine locations as well as proper building layouts for utilizing the wind speed distribution effectively within the built environment. The city consists of many buildings and the buildings are thought to be an obstacle for natural wind flow. However, this wind distribution can be utilized for power generation of wind turbine if the wind blowing through building is managed through building orientations and layouts. In this study, horizontal axis wind turbine is considered for utilizing the effective wind speed distribution due to building layouts.

The wind speed profiles on the tall building rooftops are presented in the reference [4] for analyzing the usage of roof wind energy, but it has not considered the wind speed distribution due to the building topologies of the city. Numerical simulation on wind flow around building has been carried out by turbulent models such as standard k-ε, LES and DNS and they have been reported in [5]-[11]. In addition, the wind tunnel experiments on wind flow around building have been also reported in [12]-[15]. Although these reports have investigated the wind speed profile/distribution around various building models, there is no report considering building size dimension and layout in order to utilize the wind speed distribution around building for power generation of wind turbine [5]-[15]. To realize the wind energy utilization in the built environment, it is important to conduct a feasibility study on wind turbine power generation performance under the actual wind condition as well as considering impact of building layouts and wind turbine location.

It is observed through literature survey that no study has been reported yet for analyzing built environment wind turbine power output due to wind speed distribution considering building layouts [16]-[25]. In our previous studies [1] [2], the assessment of wind turbine power output in the built environment has been presented. It has also discussed the wind speed distribution in the built environment. The wind speed acceleration in the built environment has been confirmed when the building layout, like nozzle, is considered [2]. Using the wind speed data base of the Japan Meteorological Agency [26], the annual power generation characteristics of wind turbine have been investigated in the case of installing the building model in Tsu city, Japan. However, only one building layout has been investigated in it [2]. Therefore, the impact of building topologies/layouts has not been investigated on wind speed profiles and also on their utilization for wind turbine in the urban area/built environment [2]. It is important to optimize the building layout using actual meteorological data for increasing power output of built environment located wind turbine under the realistic condition.

In the present paper, the objective is to optimize building layout for increasing electrical energy output from wind turbine assumed to be installed in the actual city in Japan. In this work, the angle between two buildings, like nozzle (α), is changed by 90°, 135° and 180°. According to our previous study [2], the wind speed acceleration in the built environment is obtained at $\alpha = 90°$. In the present paper, the angle wider than 90° is investigated, since it is predicted that the larger amount of wind from many wind directions can inflow into the intervening space within the two buildings like nozzle. **Figure 1** shows the building model with $\alpha = 135°$. The wind speed distribution across the buildings is developed/simulated through a CFD (Computational Fluid Dynamics) software. The power output from wind turbine is estimated using the power curve of a real commercial wind turbine and the wind speed distribution around buildings. The present paper focuses on analysis of the wind speed distribution around the building at different building angles (*i.e.* nozzle angles) and to investigate the wind turbine power generation performance. Also, this work presents the change in wind speed profile and wind turbine electrical energy output with reference to the various building topologies in a smart city planning.

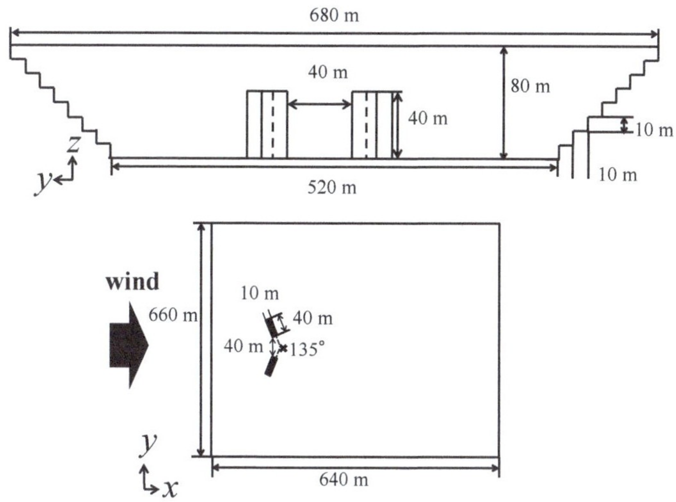

Figure 1. Simulation model for wind power generation.

The wind speed data base of the Japan Meteorological Agency [26] is used in this work. The meteorological data are utilized for evaluating the wind speed profile across the building and for finding the electrical energy output of wind turbine. The data for Fukushima city (located at east longitude of 140° and 28 arc-minute, north latitude of 37° and 45 arc-minute) and Tsu city (located at east longitude of 136° and 30 arc-minute, north latitude of 34° and 43 arc-minute) in Japan are used for the feasibility assessment of the proposed building model/layout. Fukushima city which is located in the northeast region in Japan is still suffering from the damage of the big earthquake and nuclear power plant accident occurred in 2011. This study believes that a smart city is suitable for the reconstructing of the Fukushima city. The Tsu city which is located in the center region in Japan is far from the Fukushima city. This study picks up these cities as representative middle size city in the Japan.

Two buildings are configured/arrayed as a nozzle (**Figure 1**) and considered building size is 10 m width, 40 m length and 40 m height. The representative length of this model L, which is a hydraulic diameter of horizontal cross area, is 16 m. The distance between two buildings is 40 m, which is optimized in our previous study [1]. In city planning, there will be several buildings, but this work considers only two buildings. In future work, multi buildings layouts will be considered for wind speed distribution in the downstream. Also it will consider the ground surface elevation/topologies etc. The results of this study may be useful for city planner for finding the proper locations of buildings for effective utilization of wind energy resources in the built environment.

2. Analysis of Procedure of Building Model

In this work, the wind speed distribution across the buildings is developed/simulated through the commercial CFD software CFD-ACE+ (ESI Japan, Ltd.). The previous works [1] [2] have also used this CFD-ACE+ for assessment of wind turbine power output in the built environment. The wind speed distribution model and simulation procedure, which have been used in this study, have been already presented in the Section 2 of our previous study [1]. In the present paper, the wind speed estimation procedure, which includes the several same steps as the previous study [2], is explained as follows:

The wind at the area at the back of buildings is thought to be available for power generation by wind turbine, since the wind speed would be accelerated by blowing through the buildings. Three points at the back of building which are apart from the building by 20, 30 and 40 m (x/L = 1.25, 1.875, 2.50) is assumed as the installation point of wind turbine. The wind speed for calculating the power generated by wind turbine is obtained on 1049 points located in the area where the rotor of wind turbine rotates, that is, the swept rotor area. The wind speed at each point on the swept rotor area is the averaged speed in the local area of 0.5 m × 0.5 m. By considering the wind speed distribution of this local wind speed, the wind energy can be calculated. Root mean cube wind speed to x axis direction is estimated by using the following equation:

$$U_{rmc} = \left(\frac{2Q_x}{N\rho A} \right)^{1/3} \quad (1)$$

where U_{rmc} is the root mean cube wind speed to x axis direction, Q_x is the summation of wind energy to x axis direction on each point for calculating wind speed distribution, N is the points for calculating wind speed distribution in the swept rotor area A (= 1049 points), ρ is the density of wind, A is the swept rotor area. Wind energy at each point on the swept rotor area is calculated by the following equation:

$$Q_x = \sum_{i=1}^{1049} Q_{x,i} = \sum_{i=1}^{1049}\left(\frac{1}{2}\rho A_i U_i^3\right) \quad (2)$$

where $Q_{x,i}$ is the wind energy to x axis direction at each point, A_i is the area of each point which is equal to 0.5 m × 0.5 m, U_i is the wind speed to x axis direction at each point for calculating wind energy. V_{rmc} which is the root mean cube wind speed to y axis direction is estimated by the same calculation way of U_{rmc}. The average root mean square wind speed to horizontal surface of the swept rotor area $U_{h,ave}$ is calculated by the following equation:

$$U_{h,ave} = \sqrt{U_{rmc}^2 + V_{rmc}^2} \quad (3)$$

Here, the wind speed and wind energy to z axis direction are ignored since the rotor of wind turbine can't move toward z axis direction and wind energy to z axis direction can't be utilized.

Building design procedure has been explained in reference [1]. Also the other parameters, e.g., wind turbine specifications, power curve for estimation of wind power output has been reported in reference [1] and they are also used in the present paper. **Table 1** lists the specification of wind turbine. In this study, the real commercial wind turbine is adopted for estimating the power generated by wind speed distribution. AEOLOS wind turbine of 50 kW class [27] is adopted in this study. The hub height and rotor diameter of this wind turbine is 30 m and 9 m, respectively, resulting that the height of building of 40 m is almost same as the axis height of wind turbine. **Table 2** lists the simulation condition in this study. Numerical simulation has been conducted under steady state

Table 1. Specification of wind turbine.

Characteristics	Figure
Rated power (kW)	50
Start wind speed (m/s)	3
Cut-in wind speed (m/s)	3
Cut-out wind speed (m/s)	25
Rotor diameter (m)	18
Rotor speed (rpm)	60
Hub height (m)	30

Table 2. Simulation condition of this study.

Condition	Configuration
Density of wind at inlet (kg/m^3)	1.166
Temperature of wind at inlet (K)	293
Pressure of wind at inlet (MPa)	0.10
Kinetic viscosity of wind (m^2/s)	1.56×10^{-5}
Wind speed at inlet (m/s)	$U = U_0 \times (z/30)^{0.25}$ (U_0 = 3.00 - 12.00)
Slip on side wall of building (m/s)	$V = (0.41 \times \lvert l \rvert)^{0.25} U$
Turbulent flow model	RNG k-ε model
Turbulent energy (m^2/s^2)	0.025
Dissipation rate (m^2/s^2)	$(1.58 \times 10^{-3})/z$
Calculation number (-)	10,000
Residue of each parameter (-)	$<1.0 \times 10^{-5}$
Calculation state	Steady state

by RNG k-ε model. Calculation number is set 10,000 times. This calculation number should be appropriate since the residue of each parameter under each numerical simulation condition keeps a stable small value after 500 times calculation. Wind speed at inlet of the model is set by the following equation:

$$U = U_0 \times \left(\frac{z}{30}\right)^{0.25} \tag{4}$$

where U is the wind speed in x direction, U_0 is the initial wind speed at $z = 30$ m, which is changed from 3.0 m/s to 12.0 m/s, z is height. $U_0 = 10.0$ m/s the rated wind speed of AEOLOS wind turbine of 50 kW class [27]. It is assumed that U equals to U_0 at $z = 30$ m which is the hub height of wind turbine when the wind reaches to the building.

In this work, wind speed data for Fukushima city and Tsu city are used from the Japan Meteorological Agency [26]. The wind speed data of five years (from 2010 to 2014) are used. Since the wind speed observation height by the Japan Meteorological Agency [26] is lower compared to the hub height of wind turbine, e.g., 2.6 m in the case of Fukushima city, the following correction formula [21] is used to derive the U_0:

$$\frac{U_o}{U_0} = \left(\frac{z_o}{30}\right)^{a} \tag{5}$$

where U_o is the wind speed at observation height for each city in the data base, z_o is the wind speed observation height for each city in the data base, a is the wind shear coefficient. In this study, $\alpha = 0.25$, corresponding to the country area with many trees, is adopted; since Fukushima city and Tsu city have large country area as well as urban area.

In this study, the buildings are located as nozzle, therefore the wind inflow direction is important for obtaining the wind blowing through the buildings. The layouts of the buildings are decided based on the wind speed direction. The wind speed directions and buildings with $\alpha = 135°$ are given in **Figure 2**. In this case, if the main wind direction is North (N), the wind from West-North-West (WNW), North-West (NW), North-North-West (NNW), North-North-East (NNE), North-East (NE) and East-North-East (ENE) including N can be utilized for blowing the wind among the buildings through nozzle. It is assumed that the wind blowing from the directions except for the above described wind directions can't be utilized for power generation of wind turbine. If the buildings with $\alpha = 90°$ are located, the wind from NW, NNW, NNE and NE including N can be utilized, which means that the winds from two wind directions can't be utilized compared to the buildings with $\alpha = 135°$. In this work, it is assumed that the winds from the same wind directions as the buildings with $\alpha = 135°$ can be utilized in the case of the buildings with $\alpha = 180°$. Since the buildings layout direction is parallel to the West (W) and East (E) in the case of the buildings with $\alpha = 180°$, it is considered that the winds from these wind directions can't be utilized.

Assuming the symmetry to the main wind direction, the wind speed distributions around the buildings for the inflow angles (β) of 22.5°, 45° and 67.5° are simulated to evaluate the effect of six angular inflows on the wind

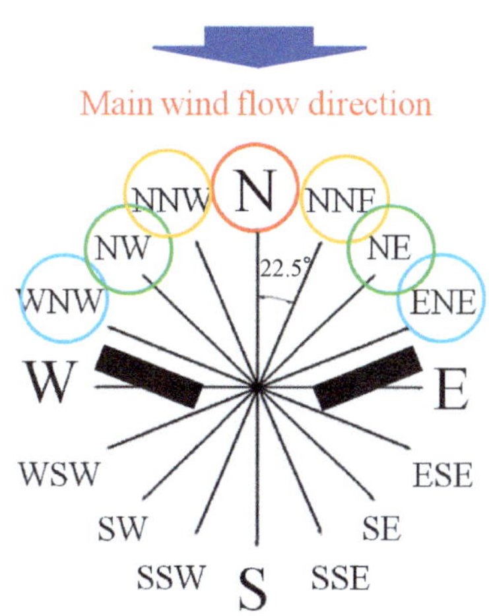

Figure 2. Wind direction for blowing in intervening space between buildings.

speed distribution in the case of buildings with $\alpha = 135^\circ$ and 180°. When α is 90°, the wind speed distributions around the buildings for $\beta = 22.5^\circ$ and 45° are simulated to evaluate the effect of four angular inflows on the wind speed distribution. Wind speed at inlet of the model is set by Equations (6) and (7), when the effect of inflow angle is considered.

$$U = \cos\beta \times U_0 \times \left(\frac{z}{30}\right)^{0.25} \tag{6}$$

$$V = \sin\beta \times U_0 \times \left(\frac{z}{30}\right)^{0.25} \tag{7}$$

where V is the wind speed in y direction, β is inflow angle.

3. Results and Discussion

3.1. Wind Speed Distribution around Buildings

The contour of wind speed distribution in x direction (U) around the buildings with $\alpha = 135^\circ$ for $U_0 = 10.0$ m/s at $z = 30$ m are given in **Figure 3** and they are on x-y cross section of the building. It shows the distribution of U in the case of $\beta = 0^\circ$, (*i.e.* the model faces the main wind direction). In this model, $x = 0$ m and $y = 0$ m is located at the center of distance between the nearest edge of adjacent two buildings. In this **Figure 3**, black lines mean the separation lines, which distinguish the different calculation domain in the model used for numerical simulation. It has been observed that the wind is accelerated within the intervening space between the buildings since some wind is over the U_0 of 10.0 m/s.

It is important to decide the location point of wind turbine in order to obtain the accelerated wind. In our previous studies [1] [2], the wind acceleration was confirmed and evaluated in the swept rotor area at $x/L = 1.25$, 1.875 and 2.50. Therefore, the present paper focuses on location points. Since the $U_{h,\,ave}$ is an important wind speed which is used for the power curve to calculate the power output of wind turbine, **Tables 3-5** list the

Table 3. Comparison of $U_{h,ave}$ and P among different U_0 at $x/L = 1.25$, 1.875, 2.50 for $\alpha = 90^\circ$ and $\beta = 0^\circ$.

U_0 (m/s)	3.0		4.0		5.0	
x/L (-)	$U_{h,ave}/U_0$ (-)	P (kW)	$U_{h,ave}/U_0$ (-)	P (kW)	$U_{h,ave}/U_0$ (-)	P (kW)
1.25	1.10	1.35	1.10	3.70	1.10	7.80
1.875	1.09	1.29	1.09	3.53	1.09	7.43
2.5	1.06	1.16	1.06	3.21	1.06	6.77
U_0 (m/s)	6.0		7.0		8.0	
x/L (-)	$U_{h,ave}/U_0$ (-)	P (kW)	$U_{h,ave}/U_0$ (-)	P (kW)	$U_{h,ave}/U_0$ (-)	P (kW)
1.25	1.10	14.10	1.10	7.80	1.10	14.10
1.875	1.09	13.43	1.09	7.43	1.09	13.43
2.5	1.06	12.26	1.06	6.77	1.06	12.26
U_0 (m/s)	9.0		10.0		11.0	
x/L (-)	$U_{h,ave}/U_0$ (-)	P (kW)	$U_{h,ave}/U_0$ (-)	P (kW)	$U_{h,ave}/U_0$ (-)	P (kW)
1.25	1.09	49.86	1.09	52.36	1.09	51.46
1.875	1.07	46.72	1.07	52.55	1.07	51.64
2.5	1.03	41.77	1.03	52.87	1.04	51.95
U_0 (m/s)	12.0					
x/L (-)	$U_{h,ave}/U_0$ (-)	P (kW)				
1.25	1.10	50.57				
1.875	1.08	50.71				
2.5	1.05	50.98				

Table 4. Comparison of $U_{h,ave}$ and P among different U_0 at x/L = 1.25, 1.875, 2.50 for α = 135° and β = 0°.

U_0 (m/s)	3.0		4.0		5.0	
x/L (-)	$U_{h,ave}/U_0$ (-)	P (kW)	$U_{h,ave}/U_0$ (-)	P (kW)	$U_{h,ave}/U_0$ (-)	P (kW)
1.25	1.09	1.29	1.09	3.56	1.09	7.52
1.875	1.08	1.24	1.08	3.45	1.08	7.30
2.5	1.05	1.12	1.06	3.24	1.06	6.86
U_0 (m/s)	6.0		7.0		8.0	
x/L (-)	$U_{h,ave}/U_0$ (-)	P (kW)	$U_{h,ave}/U_0$ (-)	P (kW)	$U_{h,ave}/U_0$ (-)	P (kW)
1.25	1.08	13.23	1.09	22.34	1.09	34.12
1.875	1.07	12.84	1.08	21.69	1.07	32.18
2.5	1.05	12.08	1.05	19.83	1.04	29.41
U_0 (m/s)	9.0		10.0		11.0	
x/L (-)	$U_{h,ave}/U_0$ (-)	P (kW)	$U_{h,ave}/U_0$ (-)	P (kW)	$U_{h,ave}/U_0$ (-)	P (kW)
1.25	1.08	48.01	1.09	52.39	1.08	48.01
1.875	1.06	45.27	1.08	52.48	1.06	45.27
2.5	1.03	41.37	1.05	52.68	1.03	41.37
U_0 (m/s)	12.0					
x/L (-)	$U_{h,ave}/U_0$ (-)	P (kW)				
1.25	1.09	52.39				
1.875	1.08	52.49				
2.5	1.05	52.68				

Table 5. Comparison of $U_{h,ave}$ and P among different U_0 at x/L = 1.25, 1.875, 2.50 for α = 180° and β = 0°.

U_0 (m/s)	3.0		4.0		5.0	
x/L (-)	$U_{h,ave}/U_0$ (-)	P (kW)	$U_{h,ave}/U_0$ (-)	P (kW)	$U_{h,ave}/U_0$ (-)	P (kW)
1.25	1.15	1.57	1.15	4.30	1.15	9.02
1.875	1.15	1.56	1.15	4.25	1.15	8.97
2.5	1.13	1.48	1.13	4.07	1.13	8.54
U_0 (m/s)	6.0		7.0		8.0	
x/L (-)	$U_{h,ave}/U_0$ (-)	P (kW)	$U_{h,ave}/U_0$ (-)	P (kW)	$U_{h,ave}/U_0$ (-)	P (kW)
1.25	1.15	16.33	1.15	26.74	1.15	40.50
1.875	1.15	16.15	1.15	26.53	1.15	40.28
2.5	1.13	15.39	1.13	25.25	1.13	38.44
U_0 (m/s)	9.0		10.0		11.0	
x/L (-)	$U_{h,ave}/U_0$ (-)	P (kW)	$U_{h,ave}/U_0$ (-)	P (kW)	$U_{h,ave}/U_0$ (-)	P (kW)
1.25	1.15	52.78	1.15	51.88	1.15	50.96
1.875	1.15	52.82	1.14	51.92	1.15	51.00
2.5	1.13	52.94	1.12	52.09	1.13	51.18
U_0 (m/s)	12.0					
x/L (-)	$U_{h,ave}/U_0$ (-)	P (kW)				
1.25	1.15	50.04				
1.875	1.15	50.08				
2.5	1.13	50.27				

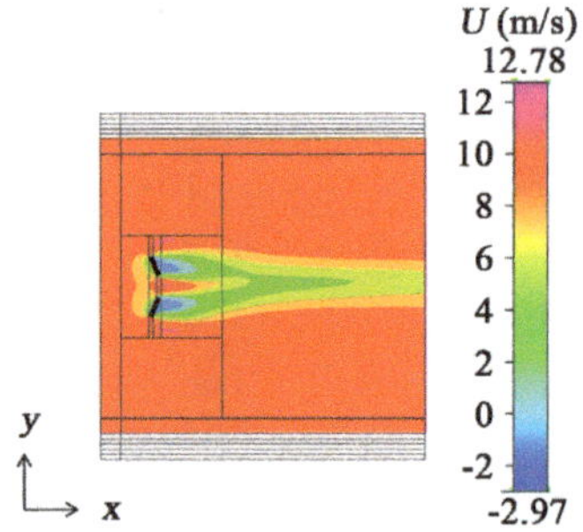

Figure 3. Contour of wind speed U distribution around buildings at z = 30 m on x-y cross section (α = 135°, β = 0°, U_0 = 10.0 m/s).

variation of $U_{h,\,ave}$ in the swept rotor area at x/L = 1.25, 1.875 and 2.50 under the different U_0 conditions for α = 90°, 135° and 180°, respectively. In these tables, $U_{h,\,ave}$ is estimated under the condition that β is 0°. The power output of wind turbine P corresponding to each $U_{h,\,ave}$ is also listed in these tables.

From these tables, it is seen that $U_{h,\,ave}$ is greater than U_0 for every condition. Hence, it is revealed that the proposed building model can provide the wind acceleration under the investigated conditions. Considering the location point of wind turbine, the highest $U_{h,\,ave}$ is obtained at x/L = 1.25 for every α and U_0 condition though P is relatively smaller compared to the other x/L under the condition that U_0 is over 10.0 m/s. Since the rated wind speed of supposed wind turbine is 10.0 m/s, the P decreases with the increase in wind speed gradually under the condition that U_0 is over 10.0 m/s according to the power curve [1]. This work intends to utilize the wind power by accelerating the low speed wind effectively, resulting that the location point of wind turbine is decided at x/L = 1.25. In the next section, the power generation characteristics of wind turbine in the built environment is evaluated at x/L = 1.25.

Since the wind blows from different directions actually, it is necessary to evaluate the effect of wind direction on wind speed distribution around buildings. In this study, the wind speed distributions around the buildings for β = 22.5°, 45° and 67.5° are simulated. As an example of the simulation results, the contour of wind speed U distribution around buildings for U_0 = 10.0 m/s at z = 30 m on x–y cross section for angular inflow direction of β = 45° are given in **Figure 4**. Since the wind speed at the inlet of the model is divided into two directions of x and y for the angular inflow case in this study, U at z = 30 m is 7.07 m/s when the wind reaches the building. Although the wind blows through the intervening space between the buildings, the wind acceleration is not high. Hence, the location of open tip of building model is important.

3.2. Power Generation Performance of Wind Turbine Located in Proposed Building Layouts in Actual Cities of Japan

To optimize the building layout for increasing electrical energy output of wind turbine assumed to be installed in actual cities of Japan, the wind speed data of five years (from 2010 to 2014) for Fukushima city and Tsu city are used from the Japan Meteorological Agency [26]. In estimation of the power generation characteristics of the building model, the location of wind turbine in proposed building layouts should be decided first. According to the wind speed data of the Japan Meteorological Agency [26], the main wind direction throughout the year for Fukushima city and Tsu city which is decided by the hourly measurement data from 2010 to 2014 is NW. In this study, the open tip of building model is located to be faced with the main wind direction (refer to **Figure 2**). For example, West-South-West (WSW), W, WNW, NNW, N and NNE in addition to NW are the available wind directions when α is 135°. Therefore, the proposed building layouts make the nozzle by utilizing the wind speed directions. In this work, it is assumed that the wind blowing from the directions except for the above described wind directions can't be utilized for power generation of wind turbine. The different inflow angle conditions are considered in the simulation for finding the suitable wind direction in the proposed building layouts.

The hourly data on wind speed and direction are obtained for Fukushima city and Tsu city from the Japan Meteorological Agency [26]. These data are used as inputs in the simulation for finding the wind turbine power output at a location of wind turbine in the proposed building layouts. The hourly power energy outputs of wind turbine located at location of the proposed building layout with α = 135° in the case of Fukushima city and Tsu city are given in **Figure 5** and **Figure 6**, respectively. In these figures, the hourly change in statistically-veraged

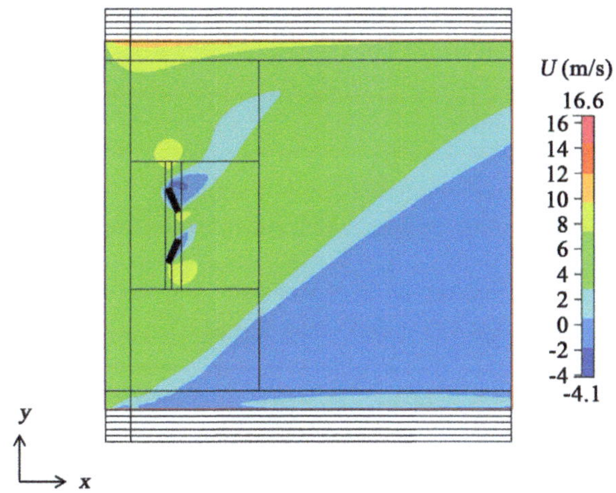

Figure 4. Contour of wind speed U distribution around buildings at z = 30 m on x-y cross section (α = 135°, β = 45°, U_0 = 10.0 m/s).

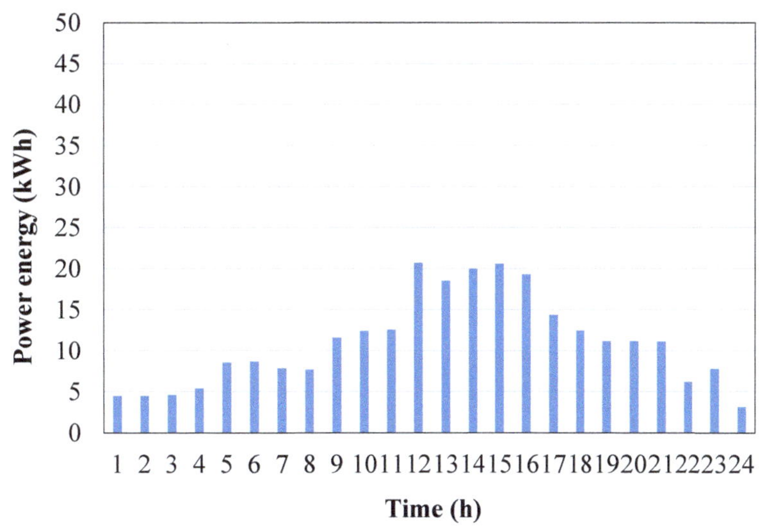

Figure 5. Time change in power energy of wind turbine in January in the case of installing building model in Fukushima city (α = 135°).

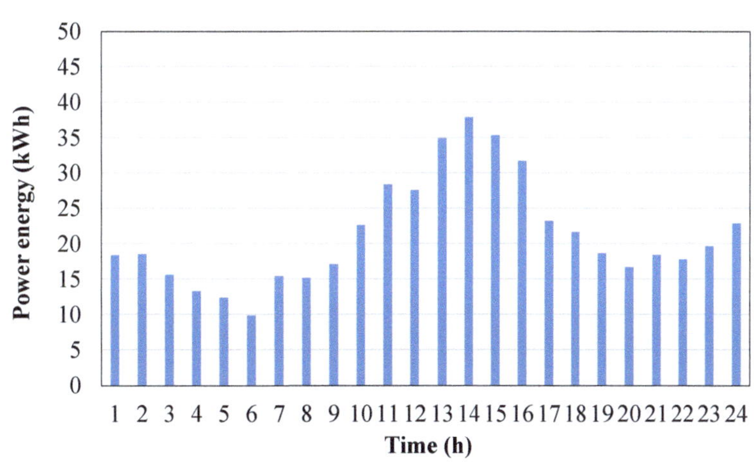

Figure 6. Time change in power energy of wind turbine in January in the case of installing building model in Tsu city (α = 135°).

day in January is shown as an example. From these figures, it is observed that the higher power energy of wind turbine is obtained in the daytime. This result is also confirmed from the other months. In these two cities, the wind energy is mainly obtained in the daytime. However, the hourly power energy output of wind turbine in the case of Tsu city has peaks in nighttime as well as daytime while there is only one peak in the case of Fukushima city. Since Tsu city is located facing Isewan bay, the wind blowing between land and bay is occurred resulting from the temperature difference between land and bay in the middle of daytime as well as that of nighttime.

Table 6 and **Table 7** list the monthly and annual wind power energy output of wind turbine in the building layouts with $\alpha = 90°$, $135°$ and $180°$ in the case of Fukushima city and Tsu city, respectively.

From these tables, it is observed that the wind power energy of wind turbine in the building layout with $\alpha = 135°$ is the highest among the investigated α conditions. Though $U_{h,ave}/U_0$ obtained under the different U_0 conditions for the building layout with $\alpha = 180°$ is relatively higher than that with $\alpha = 90°$ and $135°$ according to **Tables 3-5**, the difference among them is small. Therefore, it is believed that the optimum building layout is decided by the wind blowing from angular inflow directions.

Table 8 lists the comparison of P at $x/L = 1.25$ under different β conditions among $\alpha = 90°$, $135°$ and $180°$. In this table, P obtained under $U_0 = 10.0$ m/s condition are shown. It is found that P for $\beta = 45°$obtained under $\alpha = 135°$ condition is much higher compared to the other α conditions.

In addition, it is clarified from **Table 6** and **Table 7** that the wind power energy output in winter season is higher while that in summer season is lower irrespective of α as well as investigated city. The wind direction is changed throughout the year. **Figure 7** and **Figure 8** show the frequency distribution of wind speed direction in Fukushima city in January and August, respectively. From these figures, the direction of main wind speed in January and August are WNW and NE, respectively. On the other hand, the direction of main wind speed throughout the year is NW. In this study, it is assumed that the wind blowing from the restricted wind directions can be utilized for power generation of wind turbine while the amount of wind power energy production is estimated to be zero for the wind blowing from the other directions. WNW is the next direction to NW, resulting that the large wind energy can be obtained under every α condition. On the other hand, NE and ENE which are two directions conducing larger wind flow in August are neglected in the estimation of wind power energy, resulting in the lower wind power energy output in August.

Table 6. Monthly and annual wind power energy of wind turbine in the building layouts with $\alpha = 90°$, $135°$ and $180°$ for Fukushima city.

α(°)	January (kW·h)	February (kW·h)	March (kW·h)	April (kW·h)
90	8044	5545	6978	6430
135	8193	5734	7216	6685
180	5512	5940	9516	7608
α(°)	May (kW·h)	June (kW·h)	July (kW·h)	August (kW·h)
90	3417	1033	354	333
135	3826	1034	361	341
180	6100	855	299	1600
α(°)	September (kW·h)	October (kW·h)	November (kW·h)	December (kW·h)
90	1412	4993	4969	8086
135	1518	5367	5037	8342
180	1177	3294	3634	3604
α(°)	Year (kW·h)			
90	51,595			
135	53,655			
180	49,140			

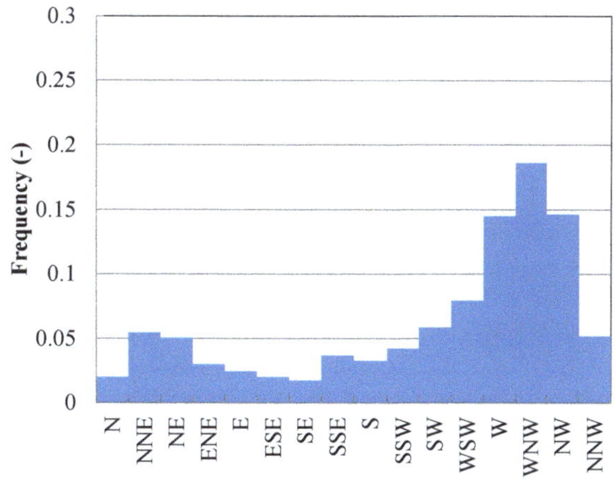

Figure 7. Frequency distribution of wind speed direction in January in Fukushima city.

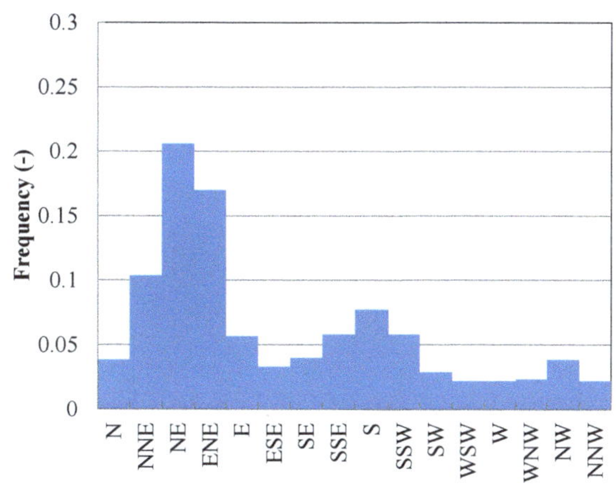

Figure 8. Frequency distribution of wind speed direction in August in Fukushima city.

Table 7. Monthly and annual wind power energy of wind turbine in the building layouts with $\alpha = 90^\circ$, 135° and 180° for Tsu city.

α(°)	January (kW·h)	February (kW·h)	March (kW·h)	April (kW·h)
90	14,900	13,220	12,696	11,212
135	15,898	14,608	13,367	12,164
180	14,408	12,249	12,022	10,376
α(°)	May (kW·h)	June (kW·h)	July (kW·h)	August (kW·h)
90	9004	936	4761	5223
135	9954	958	4833	5438
180	8563	843	4418	4824
α(°)	September (kW·h)	October (kW·h)	November (kW·h)	December (kW·h)
90	3953	7506	8756	13,866
135	4258	7905	9130	14,719
180	3792	6980	8158	13,020
α(°)	Year (kW·h)			
90	106,034			
135	113,233			
180	99,652			

Figure 9 and **Figure 10** show the frequency distribution of wind speed direction in Tsu city in January and August, respectively. Since the same phenomena as the Fukushima city can be observed in the case of Tsu city, the wind energy output in winter season is higher while that in summer season is lower.

Comparing the annual wind energy of output of turbine in the building layouts between Fukushima city and Tsu city, the annual wind energy for Tsu city is approximately twice as much as that for Fukushima city. According to the wind speed data of the Japan Meteorological Agency [26], the annual mean wind speed for Tsu city and Fukushima city are 3.4 m/s and 2.4 m/s, respectively. Since the wind power of wind turbine is derived from the wind consisting of the cube of wind speed, the gap of wind speed brings about the big difference of wind power energy between the two cities.

From the investigation in this study, it is revealed that the proposed building model performs the highest power generation production under $\alpha = 135^\circ$ condition. The power generation characteristics of wind turbine in the building layouts assumed to be located in actual cities have been clarified. The actual wind condition influences the power generation performance of proposed building model. This study will continue to verify the feasibility of the proposed model with more wind speed data in the near future.

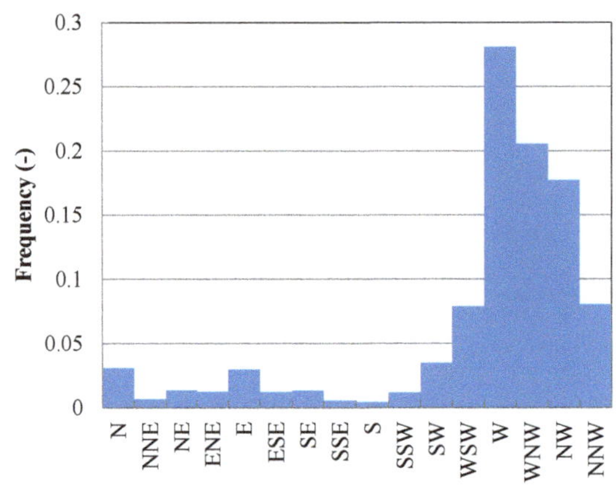

Figure 9. Frequency distribution of wind speed direction in January in Tsu city.

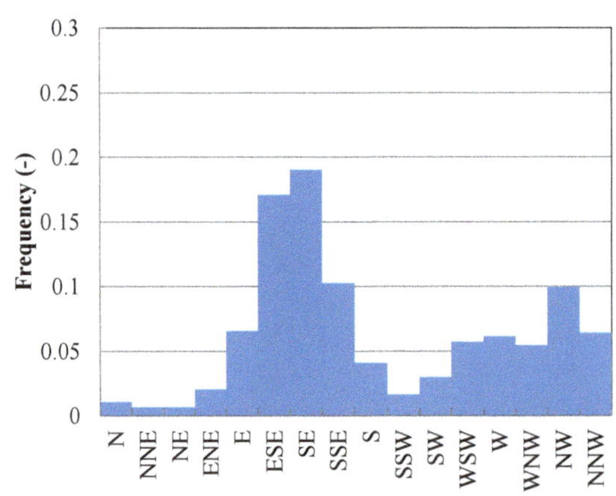

Figure 10. Frequency distribution of wind speed direction in August in Tsu city.

Table 8. Comparison of P under different β conditions among $\alpha = 90^\circ$, 135° and 180° ($U_0 = 10.0$ m/s, $x/L = 1.25$).

$\beta(^\circ)$	22.5	45	67.5
$\alpha(^\circ)$	P (kW)	P (kW)	P (kW)
90	5.29	0.62	-
135	5.29	1.26	0.21
180	5.34	0.21	0.37

4. Conclusions

In this study, building topologies/orientations in a smart city are investigated for finding out the wind speed profile in the built environment. The analysis of wind speed distribution in the built environment is very important for not only to find the mechanical wind stress but also to find the energy output from the built environment located wind turbines. This analysis is also useful for designing the building layouts in such a way to make a nozzle of the wind by using wind directions and then finding out the proper location of the wind turbine in the built environment. In this work, the objective is to optimize the building layout for increasing electrical energy output of wind turbine assumed to be installed in the actual cities of the Japan. The wind speed distribution across the buildings is simulated using CFD software. The output power of wind turbine is estimated by using the power curve of a real commercial wind turbine and the wind speed distribution around buildings using the wind speed data for Fukushima city and Tsu city. As a result, the following conclusions have been obtained:

1) The proposed building models with $\alpha = 90^\circ$, 135°, 180° can provide the $U_{h,\,ave}$ over U_0 at x/L = 1.25, 1.875 and 2.50 under the different U_0 conditions for the wind blowing from the main wind direction, *i.e.*, $\beta = 0^\circ$.

2) Though the wind blowing for angular inflow direction is not accelerated well by the proposed building models, the optimum α which is 135° is decided by the wind blowing from angular inflow direction.

3) In the case of installing the proposed building model in Fukushima city and Tsu city, the wind power energy output in winter season is higher while that in summer season is lower irrespective of α. The interaction between the change in frequency distribution of wind speed direction throughout the year and the location of open tip of building model decides the power generation characteristics of the proposed building model.

4) It is revealed that the annual wind power energy of wind turbine in the building layouts is influenced by the annual mean wind speed significantly.

Acknowledgements

This study was financially supported by MAYEKAWA HOUONKAI FOUNDATION. The authors thank for its cooperation.

References

[1] Nishimura, A., Ito, T., Murata, J., Ando, T., Kamada, Y., Hirota, M. and Kolhe, M. (2013) Wind Turbine Power Output Assessment in Built Environment. *Smart Grid and Renewable Energy*, **4**, 1-10. http://dx.doi.org/10.4236/sgre.2013.41001

[2] Nishimura, A., Ito, T., Kakita, M., Murata, J., Ando, T., Kamada, Y., Hirota, M. and Kolhe, M. (2014) Impact of Building Layouts on Wind Turbine Power Output in the Built Environment: A Case Study of Tsu City. *Journal of the Japan Institute of Energy*, **94**, 315-322. http://dx.doi.org/10.3775/jie.93.315

[3] Nishimura, A. and Kolhe, M. (2015) A Study on Assessment of Power Output by Integrating Wind Turbine and Photovoltaic Energy Sources with Futuristic Smart Buildings, Global Warming—Causes, Impacts and Remedies. In: Singh, B.R., Ed., *Global Warming—Causes, Impacts and Remedies*, InTech, Rijeka, Chapter 7, 139-167.

[4] Zhenglin, Z., Hongyan, W., Xiaolong, B. and Hongxiang, X. (2011) Study and Application of the Wind Power on the Tall Building Rooftops in Large Cities. *Proceedings of* 2011 *International Conference* (*IEEE*) *on Materials for Renewable Energy & Environment* (*ICMREE*), Shanghai, 566-569. http://dx.doi.org/10.1109/ICMREE.2011.5930876

[5] Yoshie, R., Tominaga, Y., Mochida, A., Kataoka, H. and Yoshikawa, H. (2005) CFD Simulation of Flow-Field around a High-Rise Building Located in Surrounding City Blocks Part (1) Outline of Experiment and Calculation, Influences of Various Calculation Conditions. *Wind Engineers*, **30**, 133-134.

[6] Murakami, S., Hibi, K. and Mochida, A. (1991) Three Dimensional Analysis of Turbulent Flow Field around Street Blocks by Means of Large Eddy Simulation (Part II). *Journal of Architecture, Planning and Environmental Engineering*, **425**, 11-19.

[7] Kataoka, H. and Mizuno, M. (1998) Numerical Simulation of Separating Flow around a Body Using Artificial Compressibility Method. *Journal of Architecture, Planning and Environmental Engineering*, **504**, 63-70.

[8] Nishimura, K., Yasuda, R. and Ito, S. (1999) An Experimental and Numerical Study of Concentration Prediction around a Building, Part II Numerical Simulation by k-ε Model. *Japan Society for Atmospheric Environment*, **34**, 103-122.

[9] Nozu, T. and Tamura, T. (2000) Application of Computational Fluid Technique with High Accuracy and Conservation Property to the Wind Resistant Problems of Buildings and Structures, Part 3 Analysis of the Flows and the Wind Pres-

sure around a High-Rise Building in the Turbulent Boundary Layer. *Journal of Structural and Construction Engineering*, **538**, 65-72.

[10] Nozu, T. and Tamura, T. (1998) Application of Computational Fluid Technique with High Accuracy and Conservation Property to the Wind Resistant Problems of Buildings and Structures, Part 2 Vortex Structures and Aerodynamic Characteristics around a Three-Dimensional Square Prism on a Ground Plane. *Journal of Structural and Construction Engineering*, **503**, 37-43.

[11] Kose, D.A., Fauconnier, D. and Dick, E. (2011) ILES of Flow over Low-Rise Buildings: Influence of Inflow Conditions on the Quality of the Mean Pressure Distribution Prediction. *Journal of Wind Engineering and Industrial Aerodynamics*, **99**, 1056-1068. http://dx.doi.org/10.1016/j.jweia.2011.07.008

[12] Meng, Y. and Hibi, K. (1997) Turbulence Characteristics and Organized Motions on the Flat Roof of a High-Rise Building. *Journal of Wind Engineering*, **72**, 21-34. http://dx.doi.org/10.5359/jawe.1997.72_21

[13] Meng, Y. and Oikawa, S. (1997) A Wind-Tunnel Study of the Flow and Diffusion within Model Urban Canopies. *Journal of Japan Society for Atmospheric Environment*, **32**, 136-147.

[14] Oikawa, S., Ishihara, T., Yasuda, R., Nishimura, K. and Hase, M. (1999) An Experimental and Numerical Study of Concentration Prediction around a Building, Part I Wind-Tunnel Study. *Journal of Japan Society for Atmospheric Environment*, **34**, 123-136.

[15] Zaki, S.A., Hagishima, A. and Tanimoto, J. (2012) Experimental Study of Wind-Induced Ventilation in Urban Building of Cue Arrays with Various Layouts. *Journal of Wind Engineering and Industrial Aerodynamics*, **103**, 31-40. http://dx.doi.org/10.1016/j.jweia.2012.02.008

[16] Ouammi, A., Dagdougui, H., Sacile, R. and Mimet, A. (2010) Monthly and Seasonal Assessment of Wind Energy Characteristics at Four Monitored Locations in Liguria Region (Italy). *Renewable and Sustainable Energy Reviews*, **14**, 1959-1968. http://dx.doi.org/10.1016/j.rser.2010.04.015

[17] Oyedepo, S.O., Adaramola, M.S. and Paul, S.S. (2012) Analysis of Wind Speed Data and Wind Energy Potential in Three Selected Locations in South-East Nigeria. *International Journal of Energy and Environmental Engineering*, **3**, Article No.: 7.

[18] Olaofe, Z.O. and Folly, K.A. (2013) Wind Energy Analysis Based on Turbine and Developed Site Power Curves: A Case-Study of Darling City. *Renewable Energy*, **53**, 306-318. http://dx.doi.org/10.1016/j.renene.2012.11.003

[19] Saito, S., Saito, K. and Sekizuka, S. (2006) A Discussion on Prediction of Wind Conditions and Power Generation with the Weibull Distribution. *JSME International Journal Series B*, **49**, 458-464. http://dx.doi.org/10.1299/jsmeb.49.458

[20] Khadem, S.K. and Hussain, M. (2006) A Pre-Feasibility Study of Wind Resources in Kutubdia Island, Bangladesh. *Renewable Energy*, **31**, 2329-2341. http://dx.doi.org/10.1016/j.renene.2006.02.011

[21] Firtin, E., Guler, O. and Akdag, S.A. (2011) Investigation of Wind Shear Coefficient and Their Effect on Electrical Energy Generation. *Applied Energy*, **88**, 4097-4105. http://dx.doi.org/10.1016/j.apenergy.2011.05.025

[22] Rocha, P.A.C., Sousa, R.C. and Andrade, C.F. (2012) Comparison of Seven Numerical Methods for Determining Weibull Parameters for Wind Energy Generation in the Northeast Region of Brazil. *Applied Energy*, **89**, 395-400. http://dx.doi.org/10.1016/j.apenergy.2011.08.003

[23] Usta, I. and Kantar, Y.M. (2012) Analysis of Some Flexible Families of Distributions for Estimation of Wind Speed Distributions. *Applied Energy*, **89**, 355-367. http://dx.doi.org/10.1016/j.apenergy.2011.07.045

[24] Mostafaeipour, A. (2010) Feasibility Study of Offshore Wind Turbine Installation in Iran Compared with the World. *Renewable and Sustainable Energy Reviews*, **14**, 1722-1743. http://dx.doi.org/10.1016/j.rser.2010.03.032

[25] Wang, L., Yeh, T.H., Lee, W.J. and Chen, Z. (2011) Analysis of a Commercial Wind Farm in Taiwan, Part I: Measurement Results and Simulations. *IEEE Transactions on Industry Applications*, **47**, 939-953. http://dx.doi.org/10.1109/TIA.2010.2101991

[26] Japan Meteorological Agency. http://www.jma.go.jp/jma/indexe.html

[27] AEOLOS Wind Turbine. http://www.windturbinestar.com/

20

Energy Efficiency in Smart Grid: A Prospective Study on Energy Management Systems

Hermes José Loschi, Julio Leon, Yuzo Iano, Ernesto Ruppert Filho, Fabrizzio Daibert Conte, Telmo Cardoso Lustosa, Priscila Oliveira Freitas

Department of Communications, Faculty of Electrical and Computer Engineering, University of Campinas (UNICAMP), Campinas/SP, Brazil
Email: eng.hermes.loschi@ieee.org

Abstract

The term Smart Grid has become a term to represent the benefits of a smart and sophisticated electrical grid, which can meet various social expectations related to sustainability and energy efficiency. The Smart Grid promises to enable a better power management for energy utilities and consumers, to provide the ability to integrate the power grid, to support the development of micro grids, and to involve citizens in energy management with higher levels of responsibility. However, this context comes with potential pitfalls, such as vulnerabilities to cyber-security and privacy risks. In this article, a prospective study about energy management, and exploring critical issues of modeling of energy management systems in a context Smart. Grid is presented along with background of energy management systems. An analysis of the demand response condition is also presented. Finally, the advantages and disadvantages of the implementation of energy management systems, and a comparison with the Brazilian electricity system are discussed.

Keywords

Smart Grid, Management, Cloud Computing, Energy, Efficiency

1. Introduction

Electric energy is essential to increase productivity and ensure a high quality of life; therefore, the relationship between electric power and economic growth is crucial. However, the consequence of the current worldwide economic growth and electricity demand is the depletion of energy resources. An essential and effective way to prevent the depletion of resources and promote economic growth at the same time is the application of the concept of energy efficiency through energy management systems, this is being the basic principle of the Smart Grid [1].

With the development of the Smart Grid, especially in the distribution grid, and with the possibility of load modeling, control over the peaks of energy demand becomes vital. The peaks of demand are serious problems and present themselves in the electrical system. The demand management in residential, commercial and industrial sectors can play an important role in reducing peak demand, reducing stress, overhead transmission and distribution lines. In many countries, there are various demand response programs, implemented for industrial and commercial loads [2].

There are few demand response (DR) programs in use for energy management in the residential sector. Direct load restriction is the most popular method used to reduce peak demand. However, when using direct control/restriction load, the consumer comfort can be compromised. In contrast, reduction of peak demand through load displacement can benefit consumers and energy utilities [2].

Peak demand of energy has caused adverse effects to the reliability and stability of the power system during recent decades. Reducing peak demand can reduce the risk of faults on transmission and distribution grid, consequently, the risk of interruptions. Demand response is one way to deal with peak events and avoid overload on the grid by providing the necessary flexibility through load displacement [2] [3].

Multiple energy management systems use the concepts of demand response; however, these DR systems are not broadly implemented due to complexity of the automation involved, particularly in industrial and commercial buildings. Although the residential sector is responsible for a significant amount of electrical energy demand, few DR programs are currently used in the residential field. Direct Load Control (DLC) is the method most used by energy utilities in Brazil to manage peak demand in the residential sector, where consumer loads are adjusted in time during peak network events [4] [5].

An energy management system that can automatically switch the operation of appliances during peak hours, can be used to management of peak demand without compromising the quality of supply to the consumer [6].

Up recognizing the very different perspectives and priorities of individuals and organizations involved in the electrical system, the authors propose a broad analysis through a prospective study to examine the different ways for the Smart Grid meeting the demands and developments of society. This study focuses on comparing the development of "Smart Grid" in different regions of the world and demonstrating the commitment of these countries to change the social and political contexts and expectations, which often are shaped by specific regions, goals and available resources.

2. Smart Grid Technologies

The term Smart Grid means more than a single technology or even a clear set of individual technologies. Is an "umbrella" term under which various technologies of electric power systems are considered, both in hardware and software. For some people, Smart Grid is characterized primarily as the addition of an information and communication technology (ICT), superimposed in a way on existing infrastructure. For others, Smart Grid represents the installation of new transmission lines, meters, and renewable energy generation [7].

However, in order for both conditions to comply, first it is necessary to understand the legacy electric systems worldwide. The current dominant infrastructure of electric power systems involves four basic elements:

- Generation: Electric power is generated in large-scale power plants;
- Transmission: High-voltage electrical energy is transported from the plant to substations closer to consumers;
- Distribution: Low voltage energy is distributed from substations to residences and commercial buildings;
- Consumer: Electricity used for consumer devices such as refrigerators, computers, lights, pumps and other devices used by residential, commercial, and industrial devices.

The main mechanism in power generation by current systems depends on the heat produced by burning fossil fuels, division of atoms in nuclear energy, or from the hydroelectric stations water movement. Except for solar cells, almost all other forms of power generation, including the burning of fossil fuels, nuclear, biomass, hydro, wind, concentrated solar, cogeneration, and need driving a turbine to produce electricity [7] [8].

The generation usually produces electricity with relatively low voltages ranging from 2 to 30 kilovolts (kV), depending on the size of the unit. Since electricity is generated, its tension is amplified before transmission. A critical step between the generation of electricity and long-distance transmission involves a transformer to increase the voltage. Often, the generation of electric energy occurs far from the places where the electricity is needed, making the long distances of high voltage transmission lines, a crucial part of the electrical system. The long-distance transmission voltage varies from 115 to 120 kilovolts (KV), so the transformer plays a crucial role in increasing the voltage for transmission [7] [8].

The high-voltage transmission lines carry electricity from generating plants to local substations, where the energy is "left over" for a lower voltage and then sent through electric energy distribution networks for local users, including the industrial, commercial, and residential consumers. From the substation, the electrical energy is distributed locally within a community to individual buildings and houses. The voltage is usually reduced at the point of use, to the standard voltage of that region, which varies in different countries (with most consumers receiving 110 - 120 V in the United States and 220 - 240 V in Europe) and with the requirements of electric power use [7] [8].

Main Technologies

The term Smart Grid represents the integration of digital technologies, sensors, and ICTs to empower and make the management on the use of electricity more reliable and efficient. Smart Grid includes technologies for the consumer (with which consumers interact) and the grid (transmission and distribution that are less visible to consumers). The Smart Grid technologies also include hardware and software [7] [8].

One of the definitions of the term Smart Grid is the integration of various technologies, products and services, from the generation, transmission, and distribution; using advanced communication and control technologies. **Figure 1** illustrates this concept.

Table 1 presents the main technologies proposals for Smart Grid and their definitions.

In addition to the above-mentioned Smart Grid technology, a more holistic approach to energy management systems is presented in the next section. A major challenge with current systems is the limited mechanisms for coordination and communication between the management of the different parts of the system. The management of the current systems' transmission and distribution are carried out through separate activities that occur in different parts of the network [8].

3. Energy Management Systems in Smart Grid

The vast majority of the energy management systems just consider the monitoring and data statistics of energy consumption of consumer electronics. For these systems, manual actuation is necessary in each device to reduce energy consumption. However, the Smart Grid technologies require management systems to be smarter and able to respond to demands related to the charge control, energy management, and timing systems with micro grids [9].

Some architectural solutions for energy management systems are being integrated into a concept of Smart Grid; these are presented blow. Chang-Sci Choi develops an architecture using AMI solutions for an energy management system entitled EMM (Energy Monitoring and Management), interoperating this system with the Smart Grid. Among the

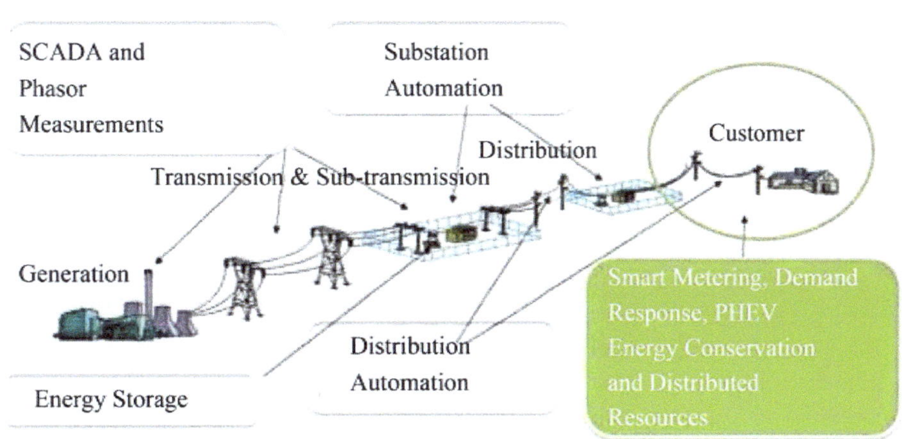

Figure 1. Concept for control and communication systems in Smart [9].

Table 1. Main technologies proposals for Smart Grid [7].

Phases	Technologies	Definition
Generation	Inverters "Grid Tie"	Use of "Grid Tie" inverter, connecting to mains, e.g. photovoltaic systems
	Synchrophasors	Measurement of Synchronized Phasor, *i.e.* sinusoidal measurements of AC magnitudes, and synchronized in time expressed in phasors. To determine useful information about grid performance
Transmission	FACTS (Flexible Alternate Current Transmission Systems) para HVAC (High Voltage Alternating Current) e HVDC (High Voltage Direct Current)	AC or DC (High Voltage ou Low Voltage) voltage, transmission from generation to distribution
	Static Deviation/Compensator VAR (Volt-amper reactive)	Used for static VAR compensation, have mutual inductance lines, consuming reactive power
Test and modeling of transmission	Software for Analysis of Transmitted Power	Full suite of tools to create, configure, customize and manage power transmission systems
Transmission in cases of distribution of HVDC and generation by renewable sources	Inverters and rectifiers	Used for conversion AC-DC and DC-AC
Transmission and distribution	Substation automation	Applications of automation: Control, timing, voltage transfer, reduction and detection of loads
Transmission, distribution and substation	Relays and circuit breakers	The relay is actuated to detect any failure in the power system, disarming the breakers of the circuits, avoiding damage to the equipment
Distribution (Distribution management system-DMS)	Fault location for distribution System.	Fault locators are devices and software, usually installed in substations to identify possible fault events, calculating the distance from the point of the failure to the source monitored identified in distribution system
Distribution	AMI (Advanced metering Infraestrutura)	AMI provides the communication with the smart meters and other power management devices
Distribution, information management.	Distribution automation	The system consists of equipment's, communication infrastructure and information technology, which are used as intelligence distribution system
	Management system of measurement data	Used for complex data collection processes measured for multiple data recording technologies
	RTU (Remote Terminal Unit)	The function of the RTU is the remote location of the system SCADA, for monitoring and control of the necessary equipment
Information management	SCADA (Supervisory Control and Data Acquisition)	Computational system that monitors and controls infrastructure and industrial plant processes based installation
	EMS (Energy Management System)	The EMS are tools used for operation of electrical grid, to monitor, control and optimize the performance of systems for generation, transmission and distribution
Information management and consumer economics	Smart Meters	The smart meter registers the range of the electric energy consumption

Continued

Information management: HAN (Home are network) and LCM (Load Control Modules).	PCT (Program Communication Thermostat)	Components of the control system, to detect the temperature of a system, controlling the same in the desired set point.
Residential Consumption	Load control receiver	Are devices used for load control, directly or indirectly, through voltage circuits, such as air conditioning thermostat.
Consumption of the generation, distribution and transmission	Short circuit current limiter	Current limitation is the practice in electrical or electronic circuits, the imposition of an upper limit to the current that can be delivered, avoiding damage to the equipment
Transformers generation, distribution and transmission	OLTC (On-Load Tap-Changer)	The OLTC is used to change the transformer voltage ratio, without interrupting the load service

assumptions adopted, the most important is the division of the stream of operation and settings of residences and buildings [9].

The sizing system EMM first considered complex apartments and residential environments popular in South Korea, with advanced infrastructure of home networks and several features, such as: electricity, gas, water, and cogeneration. **Figure 2** shows the architecture that is installed via the internet providing accessibility, mobility, and interoperation with other systems. Each EMM installed performs communication with smart watches "Smart Meters", and interface with networked home appliances. To report the current status of the house, the energy measurement system sends data in real time via the Internet [9].

For commercial buildings and industrial plants, the author proposes the BEMS (*Building Energy Management System*). However, these systems and its communication protocols depend on the market and corporate strategies at the time of construction, and its concepts of energy efficiency and energy management, which adopt generic and growing open communication standards and protocols, such as: BACnet, LonWorks, Modbus, KNX, WLAN, Zigbee, SNMP, IEC61850, DNP3, etc. To overcome these constraints, the author proposes the development of 2 CCL (*Common Communication Layer*) protocols. The first one is called BAS CCL, and the other one, EMM CCL. **Figure 3** illustrates the configuration of EMM for heterogeneous integration BEMS in each building [9].

In both systems, the server data is analyzed by software, providing information about weather and conditions of operation of the entire plant. This information can be broadly used and incorporated into a maintenance planning, among other applications [9] [10].

Levels of data analyzed by the software can be broadly classified as:

- Level 1, Management: Supervision of computers, servers and services data management servers;
- Level 2, Automation: Smart G/W (gateway);
- Level 3, Installation: Sensors, actuators and controllers.

The software operates considering a data group, composed of service server and data management that are installed on the building control center, as illustrated in **Figure 4**. The Smart Gateway is installed inside the building, collecting the data of the electric power consumption through intelligent devices such as meters, sensors and actuators, etc., that are engaged in the construction of an operating system, including electricity control systems, HVAC/HVDC, lighting systems, etc. [10].

This architecture allows the system to provide multiple functions for energy saving control, such as: control of maximum demand of energy, weather-based light and dimming control by means of sensors. The information generated by the system makes it possible to provide the function of energy efficiency through predictive analysis of the trend of energy consumption, comparing with the data in real time, through a comparative analysis with similar facilities. Another option is the provision of services, such as load shifting adapting to pricing/rates through Smart Grid function; connection with renewable energy sources; and energy exchange function. In addition, it allows a selection of ideal tariff systems appropriate to the construction standard and energy consumption of the installation [10].

4. Analysis of the Condition of Demand Response

To meet peak demand, high-cost generating stations are required. Adding more generation was the strategy used in

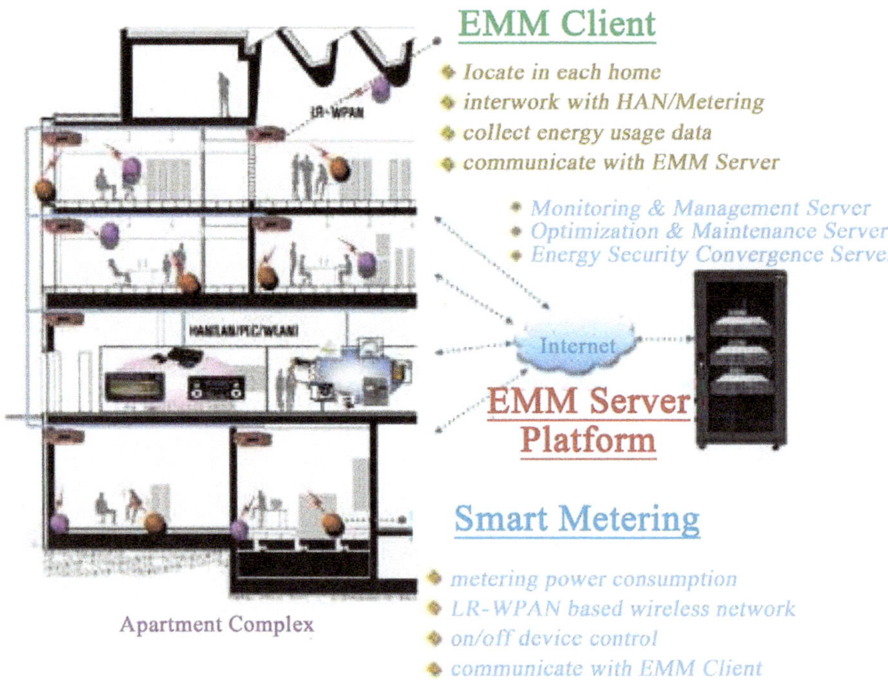

Figure 2. Systems configuration of EMM for residence [9].

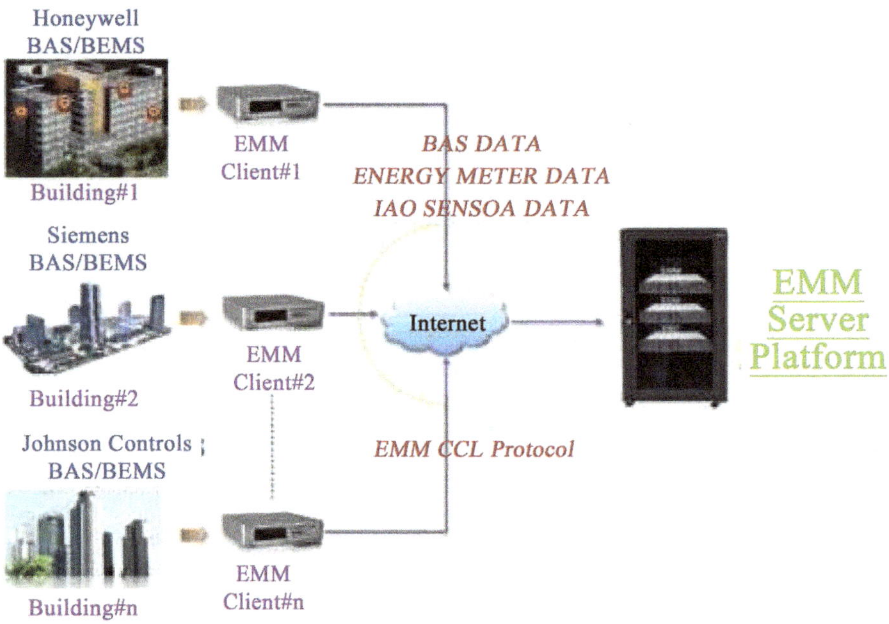

Figure 3. Systems configuration of EMM for integration with BEMS [9].

the past to meet the demand of electricity. Currently, the energy utilities have given more attention to demand management in order to reduce peak demand. Using the Smart Grid concept means more than a single technology or even a clear set of individual technologies for this [2].

Demand response (DR) is a key concept in energy demand management, which helps to reduce peak demand in critical situations. DR is defined as the changes in the use of electricity, for end consumers, of their normal consumption patterns in response to changes in the price of electrical energy over time, or the incentive payments intended to induce a better use of electricity at peak hours [11].

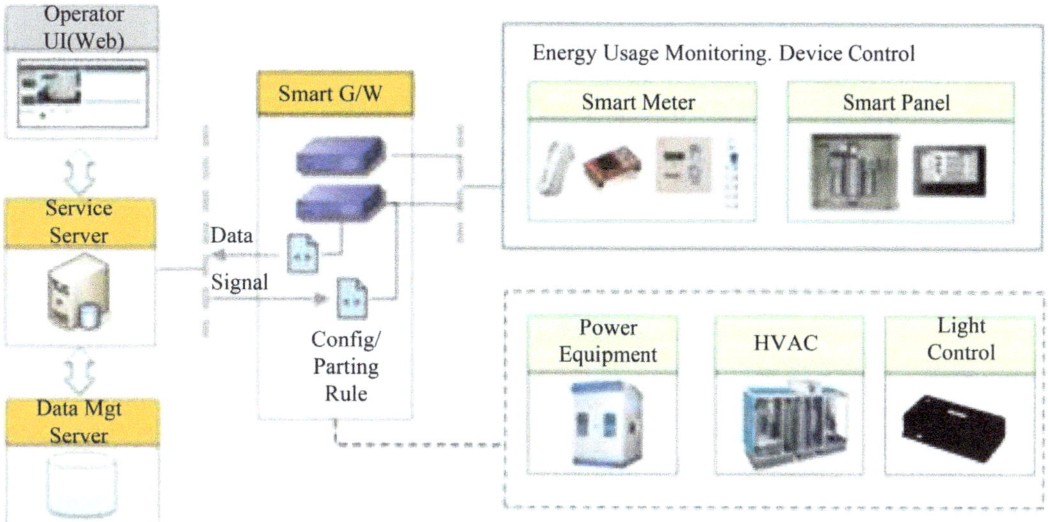

Figure 4. Performance software architecture [10].

Load management is defined as a set of objectives which aim to control directly or indirectly and/or modify patterns of electricity consumption of various consumers, aiming to reduce peak demand. This control and modification enable the supply system to meet the demand, making better use of its available generation and transmission capacity [12].

For leveling the peaks of demand, three common strategies for load management are used: "*Peak Clipping*", "*Load Shifting*" and "*Valley Filling*", as illustrated in **Figure 5**.

"Peak Clipping": Load reduction for short peaks and periods of use, usually performed by the direct load control. In this method, the energy utilities disconnect the consumer when there is a critical situation. This direct control can be used to reduce capacity requirements, operational costs and dependence on fossil fuel generation [12].

"Valley Filling": Creation of loads during the peak period. This helps to reduce the average price of electricity. One of the methods used in industrial production, which uses the loads generated by fossil fuels [12].

"Load Shifting": Moves the peak loads for other periods of time without necessarily changing the global consumption. This method combines the benefit of "Peak Clipping" and "Valley Filling" moving existing loads during off-peak hours [12].

In programs of DR, electricity consumers play an important role in the reduction of peak demand during peak hours. Consumers can move their loads and thus help the energy utilities to prevent failures and blackouts in the electrical system, reducing the probability of stress conditions of the system. Improve energy security through the DR increases productivity and customer satisfaction. The DR also eliminates the need for high-cost generators and eventually reduces the cost of electricity [2].

In order to inform consumers with real-time data, there must be a communication link between the energy utilities and consumers. Consumers must be able to measure their electrical energy demand, in real time, in order to act for demand response events. Advanced metering infrastructure implementations (AMI) and other technologies allows the user to measure the real-time energy demand and further enhance the use of resources of DR in daily operation [13].

Therefore, it is evident that there is a need for an automatic energy management system in DR programs, which will provide more flexibility consumers.

5. Cloud Computing Solutions in Smart Grid

The term cloud computing has many definitions; in scenarios as Smart Grid, it is defined as a model for enabling convenient, on-demand network access to a shared pool of configurable computing resources (e.g. networks, servers, storage, applications, and services) that can be rapidly provisioned and released with minimal management effort or service provider interaction. The characteristics of cloud computing include on-demand service,

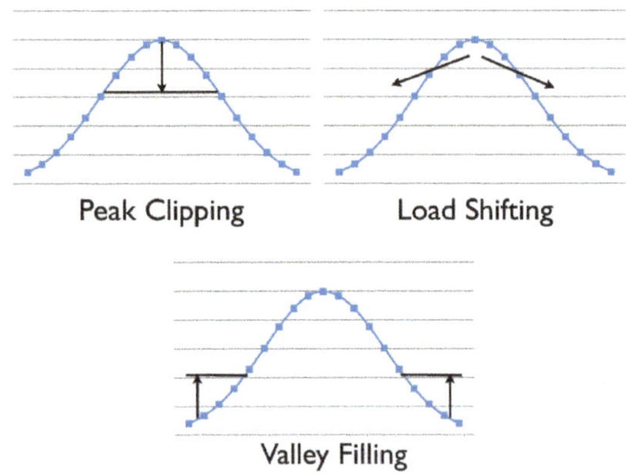

Figure 5. Strategies for load management [12].

ubiquitous network access, location independent resource pooling, rapid elasticity, and measured service.

Cloud computing is not a universal solution. It has strengths and weaknesses, and understanding them is the key to making a decision about whether it is right for a particular application [14]. The main advantages of cloud computing are: Economy of scale: sharing of computing resources between different customers; Pay per use: customers pay for the service instead of buying software licenses and hardware; On-demand usage/flexibility: cloud services can be used almost instantly and can easily be scaled up and down; External data storage: customers' data is stored externally at the location of the cloud computing provider; Highly reliable services: clouds manage themselves in case of failures or the performance degradation [15].

One of the most common applications for real-time data in manufacturing and process industries is SCADA, supervising remote processes over a network. With the growing popularity of cloud computing, many engineers and managers in the automation sector are looking at the possibility of using the cloud for SCADA. SCADA systems have evolved over time and have followed the progress of computing in general. As many view cloud computing as the next logical step in this evolution, enthusiastic visionaries foresee a fourth, "cloud" generation of SCADA, where an entire control system would be running in the cloud [16]. The cloud is creating a revolution in SCADA system architecture because it provides very high redundancy, virtually unlimited data storage, and worldwide data access all at a very low cost [17].

The cloud computing can support SCADA applications in two ways: The SCADA application is running on-site, directly connected to the control network and delivering information to the cloud where it can be stored and disseminated. The control functions of the SCADA application are entirely isolated to the control network. However, the SCADA application is connected to a service in the cloud that provides visualization, reporting, and access to remote users. These applications are commonly implemented using public cloud infrastructures (PaaS cloud service). The SCADA application is running entirely in the cloud and remotely connected to the control network. The controllers are connected via WAN links to the SCADA application running entirely in the cloud. These applications are commonly implemented using private or hybrid cloud architectures (IaaS cloud service) [18] [19].

Because of wide ranging variability of the entities in Smart Grids, there is a very high level of potential complexity in finding the optimal solution for each different Smart Grid. Smart Grid will eventually be deployed across all types of infrastructure using widespread Internet of Services, connecting all smart objects worldwide. It will become the major application domain of the Internet of Things, perhaps even referred to as the Internet of Energy [19].

6. Barriers of Brazilian Electrical Grid

Energy management systems in residential, commercial, and industrial sectors can play an important role in reducing peak demand of electric network. Eventually, it can help in reducing overhead and stress on transmission and distribution lines. In many countries, there are various demand response programs implemented for the industrial and commercial sector. With the installation of energy management systems, it becomes possible to perform load control, mainly through the models of RTP (Real Time Pricing) and TOU (Time-of-Use) [20].

Few demand response programs are used for management energy in the residential sector in Brazil. Direct restriction of loads is the most popular method used to reduce peak demand. But by direct control of the load, the consumer comfort can be compromised. In contrast, by the method of load displacement, the loads that have less impact to the consumer lifestyle are displaced outside of peak hours, ensuring a better stability of the network.

To analyze and consider the use of the concept of demand response is important to understand the load models "Hung" on the grid, for such function Smart Grid technologies play a fundamental role. This identification in conjunction with a proper communication between the consumer and the energy utilities and demand management of domestic charges are indispensable factors for high efficiency power management systems.

7. Final Considerations

Energy management systems, when developed in a context Smart Grid, have their functions enhanced, in regards of the technologies making up the system. Among the main features of an energy management system, the following stand out as the most enhanced:

- Real-time performance monitoring;
- Information to compose predictive maintenance planning;
- Energy management;
- The efficiency of the power system;
- Financial economics, avoiding any sanction from the energy utilities;
- To keep the tolerance of voltage and current to the extent allowed;
- Continuous monitoring for power quality;
- Load control and management with appropriate methodology;
- To keep consumption with the limit signed;
- Database of information that can assist future decision-making;
- Identify and correct the causes of energy disorder to avoid recurrences.

Among the main Smart Grid technologies, the most important and that directly impacts the design of an energy management system are the solutions on cloud computing. Cloud computing has established itself as an adequate means to provide resources to customers, primarily in energy management systems, with access to a large amount of information and computer storage. With cloud computing, customers do not have to manage and maintain their own information technology (IT), and are not bound to its local resources which often are limited. However, for customers and energy utilities, making sure that your cloud services are usable, an appropriate level of guarantees of Quality of Service (QoS) is needed.

In recent years, the creation of solutions in Data Center Networks (DCNS) came with rapid growth in scale and complexity, making possible hosting large applications, known as cloud hosting. This growth imposes enormous challenges to update the current datacenter infrastructure, especially considering a scenario of Smart Grid with cloud computing solutions, broadly used in energy management systems.

The proliferation of the adoption of cloud computing solutions in recent years is driven by the potential for obtaining benefits such as reduced costs, greater agility, and better use of resources. However, there are many challenges to ensure the success of these cloud-based services, and these need to be understood and managed before the major use in concepts such as Smart Grid.

However, the major current infrastructures are owned by a large number of Internet Service Providers (ISPs); and it is difficult to adopt new architectures without the agreement of all parties concerned. This includes the standardization of communication protocols and creating regulations for wide use of cloud computing solutions.

Acknowledgements

The authors would like to thank the *Coordenação de Aperfeiçoamento de Pessoal de Nível Superior* (CAPES), the *Concelho Naional de Desenvolvimento Científico e Tecnológico* (CNPq), the *Fundação de Amparo à Pesquisa do Estado de São Paulo* (FAPESP), the *Departamento de Comunicações* (DECOM), the *Faculdade de Engenharia Elétrica e de Computação* (FEEC), and the *Universidade Estadual de Campinas* (UNICAMP), for their support in the development of this research.

References

[1] Eissa, M. (2012) Energy Efficiency—The Innovative Ways for Smart Energy, the Future towards Modern Utilities.

[2] Ireshika, M.A.S.T. (2014) Home Energy Management System. Universitetet I Agder, Kristiansand & Grimstad.

[3] Khodaei, A., Shahidehpour, M. and Bahramirad, S. (2011) SCUC with Hourly Demand Response Considering Intertemporal Load Characteristics. *IEEE Transactions on Smart Grid*, **2**, 564-571. http://dx.doi.org/10.1109/TSG.2011.2157181

[4] Du, P. and Lu, N. (2011) Appliance Commitment for Household Load Scheduling. *IEEE Transactions on Smart Grid*, **2**, 411-419. http://dx.doi.org/10.1109/TSG.2011.2140344

[5] Gatsis, N. and Giannakis, G.B. (2012) Residential Load Control: Distributed Scheduling and Convergence with Lost AMI Messages. *IEEE Transactions on Smart Grid*, **3**, 770-786. http://dx.doi.org/10.1109/TSG.2011.2176518

[6] Li, J., Chung, J.Y., Xiao, J., Hong, J.W.-K. and Boutaba, R. (2011) On the Design and Implementation of a Home Energy Management System. *Proceedings of the* 6*th International Symposium on Wireless and Pervasive Computing*, Hong Kong, 23-25 February 2011, 1-6. http://dx.doi.org/10.1109/iswpc.2011.5751338

[7] Stephens, J., Wilson, E.J. and Peterson, T.R. (2015) Smart Grid (R) Evolution. Cambridge University Press, Cambridge.

[8] Jiang, T., Yu, L. and Cao, Y. (2015) Energy Management of Internet Data Centers in Smart Grid.

[9] Choi, C.-S., Ian, J.I., Park, W.-K., Jeong, Y.-K. and Lee, I.-W. (2011) Proactive Energy Management System Architecture Interworing with Smart Grid. *Proceedings of the IEEE* 15*th International Symposium on Consumer Electronics*, Singapore, 14-17 June 2011, 1-4.

[10] Park, K., Kim, Y., Kim, S., Kim, K., Lee, W. and Park, H. (2011) Building Energy Management System based on Smart Grid. *Proceedings of the IEEE* 33*rd International Telecommunications Energy Conference*, Amsterdam, 9-13 October 2011, 1-4.

[11] Balijepalli, V.S.K.M., Pradhan, V., Khaparde, S.A. and Shereef, R.M. (2011) Review of Demand Response under Smart Grid Paradigm. *Proceedings of the* 2011 *IEEE PES International Conference on Innovative Smart Grid Technologies-India*, Kollam, 1-3 December 2011, 236-243. http://dx.doi.org/10.1109/iset-india.2011.6145388

[12] Paracha, Z.J. and Doulai, P. (1998) Load Management: Techniques and Methods in Electric Power System. *Proceedings of the International Conference on Energy Management and Power Delivery*, Singapore, 3-5 March 1998, 213-217.

[13] Medina, J., Muller, N. and Roytelman, I. (2010) Demand Response and Distribution Grid Operations: Opportunities and Challenges. *IEEE Transactions on Smart Grid*, **1**, 193-198. http://dx.doi.org/10.1109/TSG.2010.2050156

[14] Bashir, A.K., Ohsita, Y. and Murata, M. (2015) Abstraction Layer Based Distributed Architecture for Virtualized Data Centers. *Proceedings of the Sixth International Conference on Cloud Computing, GRIDs, and Virtualization*, Nice, 22-27 March 2015, 62-67.

[15] Frey, S., Disch, S., Reich, C., Knahl, M. and Clarke, N. (2015) Cloud Storage Prediction with Neural Networks. *Proceedings of the Sixth International Conference on Cloud Computing, GRIDs, and Virtualization*, Nice, 22-27 March 2015, 68-72.

[16] Mcilvride, B. (2012) Will SCADA Envolve to the Cloud? http://real-timecloud.com/

[17] Combs, L. (2011) Cloud Computing for SCADA. http://www.controleng.com/single-ar

[18] Conway, G., Carcary, M. and Doherty, E. (2015) A Conceptual Framework to Implement and Manage a Cloud Computing Environment. *Proceedings of the Sixth International Conference on Cloud Computing, GRIDs, and Virtualization*, Nice, 22-27 March 2015, 138-142.

[19] Markovic, D.S., Zivkovic, D., Branovic, I., Popovic, R. and Cvetkovic, D. (2013) Smart Power Grid and Cloud Computing. *Renewable & Sustainable Energy Reviews*, **24**, 566-577. http://dx.doi.org/10.1016/j.rser.2013.03.068

[20] Albadi, M.H. and El-Saadany, E.F. (2007) Demand Response in Electricity Markets: An Overview. *Proceedings of the* 2007 *IEEE Power Engineering Society General Meeting*, Tampa, 24-28 June 2007, 1-5.

21

Implementation of Dynamic Line Rating in a Sub-Transmission System for Wind Power Integration

Saifal Talpur, Carl Johan Wallnerstrom, Patrik Hilber, Christer Flood*

School of Electrical Engineering, KTH Royal Institute of Technology, Fortum Distribution AB, Stockholm, Sweden
Email: *saifal@kth.se

Abstract

Based on conventional static line rating method, the actual current carrying capability of overhead conductors cannot be judged. Due to continuous increment in electricity demand and the difficulties associated with new line constructions, the overhead lines are therefore required to be rated based on a method that should establish their real-time capability in terms of electricity transmission. The method used to determine the real-time ampacity of overhead conductors not only can enhance their transmission capacity but can also help in allowing excessive renewable generation in the electricity network. In this research work, the issues related to analyzing an impact of wind power on periodical loading of overhead line as well as finding its static and dynamic ampacities with line current are investigated in detail. Moreover, the investigation related to finding a suitable location for the construction of a 60 MW wind farm is taken on board. Thereafter, the wind park is integrated with a regional grid, owned by Fortum Distribution AB. In addition to that, the electricity generated from the wind park is also calculated in this project. Later on, the work is devoted to finding the static and dynamic line ratings for "VL3" overhead conductor by using IEEE-738-2006 standard. Furthermore, the project also deals with finding the line current and making its comparison with maximum capacity of overhead conductor (VL3) for loading it in such a way that no any violation of safe ground clearance requirements is observed at all. Besides, the line current, knowing the conductor temperature when it transmits the required electricity in the presence of wind power generation is also an important factor to be taken into consideration. Therefore, based on real-time ambient conditions with actual line loading and with the help of IEEE-738-2006 standard, the conductor temperature is also calculated in this project. At the end, an economic analysis is performed to evaluate the financial advantages related to applying the dynamic line ratings approach in place of traditional static line ratings technique across an overhead conductor (VL3) and to know how much beneficial it is to temporarily postpone the rebuilding and/or construction of a new transmission line. Furthermore, an economic analysis related to

*Corresponding author.

wind power system is taken into consideration as well to get familiar with the costs related to building and connecting a 60 MW wind farm with the regional grid.

Keywords

Overhead Conductor, Static and Dynamic Ampacities, Real-Time Weather Conditions, Static and Dynamic Line Rating Techniques, Wind Power Integration, Regional Grid, Line Current. Conductor Temperature, Wind Power Economics, Dynamic Line Rating Economics

1. Introduction

Power system owners are facing major challenges with changed electricity consumption patterns and increase in distributed generation [1] [2]. Furthermore, quality regulations in many countries (e.g. UK, Sweden and Norway) have increased the demand for cost efficiency and at the same time the demand for higher reliability [3]. Smart structural refurbishments are thus needed to increase the overall efficiency in the future. Today, power system capacity is set with a static limit based on worst-case scenarios. One solution to significantly decrease and/or delay future investment costs and still to meet the same requirements (increase in capacity of overhead lines) is Dynamic Rating (DR) [4]-[7].

DR is therefore identified as an important topic in the quest for the development of the electrical grid [8]. The concept implies that the capacity of a component dynamically varies as a function of external parameters, like weather conditions and loading history [9]. Traditionally, the rating is statically set from simulations of worst-case scenarios referred to as static rating (SR) [10]. This leads to a margin between the actually available capacity and the rating itself, which is left unused [9]. The value of DR lies in utilizing existing equipment to a greater extent without passing loads which can lead to outages, broken components or extensive premature aging [9].

The DR concept can be used on different parts of power systems; in this paper, the specific case of applying DR on power lines is referred to as DLR. Wind power rarely reaches its maximum production. By implementing DLR and correlating the new ratings with wind power generation, the estimated time/risk of "overloading" can be significantly reduced [7]; for example, the probability is low that perfects wind conditions and extreme heat occur at the same time. This can remove or postpone some investment requirements [11] [12]. The methodology of this paper was developed around a real case study; adding 60 MW of wind power in a 130 kV sub-transmission system and its impact was investigated in terms of line loading. Besides this task, the SR and DLR of an overhead conductor were calculated and compared with the line current.

2. Theory behind Dynamic Line Rating

By using equations 2.1 - 2.15 (based on IEEE-738-2006 Standard), the DLR of overhead conductors (equation 2.34) can be calculated. SI units and the specifications of the parameters used in these equations are listed in **Table 1**. Based on these equations (2.1 - 2.15), it will be observed that conductor temperature actually varies with respect to heat input (gain rate) and heat output (loss) sources. The heat input (heat gained by the conductor) observed across the overhead conductor is due to ohmic losses and the solar heat gain (solar radiation or solar flux) [13]. After being heated up, the heat energy gained by the conductor is lost by means of two factors, *i.e.* convection and radiation [13], known as heat output sources.

According to the law of conservation of energy, there should always be a balance between heat gain and heat loss rates when not facing changes in conductor temperature, *i.e.* the heat balance equation 2.1 [14] must be followed at times of steady state.

$$\text{Heat}_{\text{GAIN}} = \text{Heat}_{\text{LOSS}} \tag{2.1}$$

The following heat balance equation [14] represents the balance amongst heat gain and heat loss rates, *i.e.*

$$P_{\text{loss}} + Q_{\text{solar}} = Q_{\text{convection}} + Q_{\text{radiation}} \tag{2.2}$$

The ohmic loss P_{loss} (W) causes heat gain in the conductor and is calculated based on Equations (2.3) and

Table 1. Parameters for the dynamic line rating.

PARAMETER (S)	SPECIFICATION (S)	SI Units
R_{T_C}	AC resistance of conductor i at temperature T_c	Ω/m
I_i	Current through conductor i	A
β_i	Solar absorptivity of conductor i	
φ	Total solar and sky radiated heat flux rate	W/m^2
θ	Effective angle of incidence of sun rays	degrees
$A_{p,i}$	Projected area of conductor i per unit length	m^2/m
D_i	Diameter of conductor i	Mm
T_A	Ambient temperature	°C
$T_{C,i}$	Temperature across conductor i	°C
T_{high}	Max. Conductor (i) temperature	°C
T_{low}	Min. Conductor (i) temperature	°C
σ	Density of air	kg/m^3
V_w	Speed of wind at conductor i	m/s
α	Thermal conductivity of air at temperature T_{film}	W/(m·°C)
ε	Dynamic viscosity of air	Pa·s
$\in_i$	Emissivity of conductor i	
K_{angle}	Wind direction factor	
ϕ	Angle between wind and conductor axis	Degrees
T_{film}	Film temperature	°C
h_e	Conductor elevation above sea	M

(2.4) [14], *i.e.*

$$P_{\text{loss}} = I_i^2 * R_{T_C} \tag{2.3}$$

$$R_{T_C} = \left[\frac{R_{T_{\text{high}}} - R_{T_{\text{low}}}}{T_{\text{high}} - T_{\text{low}}}\right] \times \left(T_{C,i} - T_{\text{low}}\right) + R_{T_{\text{low}}} \tag{2.4}$$

Similarly, the solar heat gain can be calculated with the help of equation 2.5 [14]. From this equation, it is observed that this heat gain $\left(Q_{\text{solar}}\right)$ depends upon four main factors, *i.e.* on the projected conductor area, ability of the conductor to absorb sun rays, the conductor latitude, the number of hours of the day (having co-relation with direct sun radiation).

$$Q_{\text{solar}} = \beta_i \varphi \sin(\theta) A_{p,i} \tag{2.5}$$

Furthermore, the heat loss rate is classified into two types, *i.e.* the heat loss due to convection and the heat loss due to radiation [13]. The convection heat loss rate is further classified into two types: natural convection and forced convection [9]. The natural convection heat loss rate Q_{NC} (W/m) is dependent upon conductor diameter, the temperature across the conductor, and the ambient temperature [14]. It can be calculated with the help of Equation (2.6) [14]:

$$Q_{\text{NC}} = 0.0205 \times \sigma^{0.5} \times D_i^{0.75} \times \left(T_{C,i} - T_A\right)^{1.25} \tag{2.6}$$

The forced convection heat loss rate mainly depends upon wind speed and its direction [13]. This heat loss rate is also classified into two categories, depending upon the magnitude of wind speed. At low wind speeds (lower than 4.47 m/sec [14]), the forced convection heat loss rate $Q_{\text{FC\_low}}$ (W/m) will be calculated based on Equation (2.7) [14]:

$$Q_{\text{FC\_low}} = \left[1.01 + 0.0372\left(\frac{D_i \times \sigma \times V_w}{\varepsilon}\right)^{0.52}\right] \times \alpha \times K_{\text{angle}} \times \left(T_{C,i} - T_A\right) \tag{2.7}$$

Similarly, at high wind speeds (higher than or equal to 4.47 m/sec [15]), the forced convection $Q_{\text{FC\_high}}$ (W/m) is calculated with the help of Equation (2.8) [10].

$$Q_{\text{FC\_high}} = \left[0.0119\left(\frac{D_i \times \sigma \times V_w}{\varepsilon}\right)^{0.6} \times \alpha \times K_{\text{angle}} \times \left(T_{C,i} - T_A\right)\right] \tag{2.8}$$

Moreover, in the case of low wind speeds, the larger of the two methods (natural or forced ventilation), should be used [14]. At zero wind speed, the forced convection heat loss rate will be zero. However, natural convection will help to reduce the conductor temperature [14]. To find the forced convection heat loss rate, the wind direction factor (K_{angle}) needs to be calculated, given in Equation (2.9) [14]:

$$K_{\text{angle}} = 1.194 - \cos(\phi) + 0.194 \times \cos(2\phi) + 0.368 \sin(2\phi), \tag{2.9}$$

where ϕ (angle between wind direction and conductor axis) is kept almost fixed, *i.e.* around 90°. Equation (2.10) provides the thermal conductivity of air (α) at temperature T_{film} [14].

$$\alpha = 2.424 \times 10^{-2} + 7.477 \times 10^{-5} \times T_{\text{film}} - 4.407 \times 10^{-9} \times T_{\text{film}}^2 \tag{2.10}$$

The average of ambient and conductor temperature (T_{film}) is given in Equation (2.11) [14]:

$$T_{\text{film}} = \frac{T_{c,i} + T_A}{2} \tag{2.11}$$

Equation (2.12) [14] represents the dynamic viscosity of surrounding air:

$$\varepsilon = \frac{1.458 \times 10^{-6} \left(T_{\text{film}} + 273\right)^{1.5}}{T_{\text{film}} + 383.4} \tag{2.12}$$

The air density can be calculated based on Equation (2.13) [15]:

$$\sigma = \frac{1.293 - 1.525 \times 10^{-4} \times h_e + 6.379 \times 10^{-9} \times h_e^2}{1 + 0.00367 \times T_{\text{film}}} \tag{2.13}$$

Equation (2.14) from IEEE-standard [14] gives the radiated heat loss rate $Q_{\text{radiation}}$ (W/m), *i.e.*

$$Q_{\text{radiation}} = 0.0178 \times D_i \times \epsilon_i \times \left[\left(\frac{T_{C,i} + 273}{100}\right)^4 - \left(\frac{T_A + 273}{100}\right)^4\right] \tag{2.14}$$

Like the forced convection heat loss rate, the radiated heat loss rate depends upon the difference in temperature between overhead conductors and ambient: the greater the difference, the higher the radiation heat loss rate. After establishing the heat gain and heat loss parameters, the ampacity of conductor i can be calculated based on Equation (2.15) [14]:

$$I_i = \sqrt{\frac{\left(Q_{\text{convection}} + Q_{\text{radiation}}\right) - Q_{\text{solar}}}{R_{T_C}}} \ (\text{A}) \tag{2.15}$$

Based on the ampacity of the conductor as shown in Equation (2.14), it is evident that the convection and radiation heat loss rates affect the capacity of the conductor by allowing higher currents. Similarly, its ampacity is increased when the solar radiation is lower. Moreover, the thermal AC resistance also plays its role in determining the ampacity of the overhead line.

3. Implementation of a Case Study

3.1. Objectives

An important goal of this paper is to evaluate the technical and economic aspects of dynamic line rating (DLR) across an overhead conductor (VL3) in a meshed 130 kV sub-transmission electricity network after integrating 60 MW wind power with the regional grid of Fortum.

The following specific objectives were formulated in this paper:

- Evaluation of wind speed and its direction (relative to the axis of line conductor) on wind power production.
- Development of a system model of wind power connection to a 130 kV system.
- Assessment of power flow study after wind power connection.
- Economic analysis of DLR implementation compared with traditional solutions.

3.2. Overall System Introduction

The area studied (area 160) belongs to the Värmland regional power system of Fortum Distribution AB and is located in western Sweden. The operational voltage is 135 kV and incorporates both industrial and residential loads. Initially, this power system was built by several large industries (paper and steel) around the mid-20th century and expanded during the exploitation of hydropower in the region. Now, in the north of the country there is large scale hydro power generation, whereas the south is dominated by consumption of electricity. Hence, this power system connects the electric power generation from the north with the loads in the south.

Besides connections with local generation, it is also connected with other network owners, such as Norway in the west and Vattenfall in the east. The network is also connected with the Swedish national grid via two 400-kVstations. The overhead line (VL3) connects an area with mainly power production in the north to a load-focused area in the east. Furthermore, the eastern area also includes a 400-kV connection.

High stress on VL3 is observed during periods of low industrial loads and high hydropower production. The total load pattern in Värmland is found to follow annual fluctuations, as described in **Figure 1**. For each month, the power system balance was chosen with a load reduction according to **Figure 1** and maximum power production. Thus, power system situations causing high stress on VL3 were achieved.

In **Figure 1**, the percentage indicates net power demand on monthly basis compared to the total connected load in the 130 kV regional network of Fortum, *i.e.* 910 MW in the whole year of 2012. From this figure, it is observed that during summer, power demand is comparatively lower than in the winter. The lowest power demand is observed in July with 70% load remained switched off. The real-time weather data related to ambient temperature, wind speed and its direction was first gathered before calculating the DLR. This data was based on Karlstad municipality in Sweden and was obtained from the Swedish Metrological and Hydrological Institute (SMHI) [16].

3.3. Wind Power Information, Data and Assumptions Used in the Case Study

3.3.1. Installation of Wind Turbines

Technological advancements in power electronics have increased the scope of wind power [17]. Due to the unstable nature of wind speed and its direction, keeping the system voltage and frequency to its constant level requires frequent use of power electronic equipment [17]. With the help of power electronic converters, constant electric power is sent to the active and/or passive loads at almost constant voltage [17]. Before installing wind turbines, a suitable location for the wind farm must be decided. In order to find such a location, many factors need to be considered; some important ones [18] are:

- A large and open flat area.
- Access to the road to carry out flexible installation and maintenance.
- Access to the national/regional grid.
- Low noise effect in the area.
- Compliance with air traffic rules.
- Consideration of the safety rules related to falling ice.
- Appropriate distance from protected areas.

Besides considering the technical benefits, it is important to take on board environmental and social perspectives as well, so that a widely accepted location of the wind farm is chosen. In this study, the wind farm is assumed to consist of 20 Enercon E–101 wind turbines [19], thus having an installed nominal power of 60 MW. The wind farm is assumed to be placed at Stöpsjön in close proximity to the regional grid. After considering the planned location for the wind farm, an optimal direction of the wind turbines had to be found. This decision was based on the results from wind roses. A wind rose is actually an important tool for determining the dominant

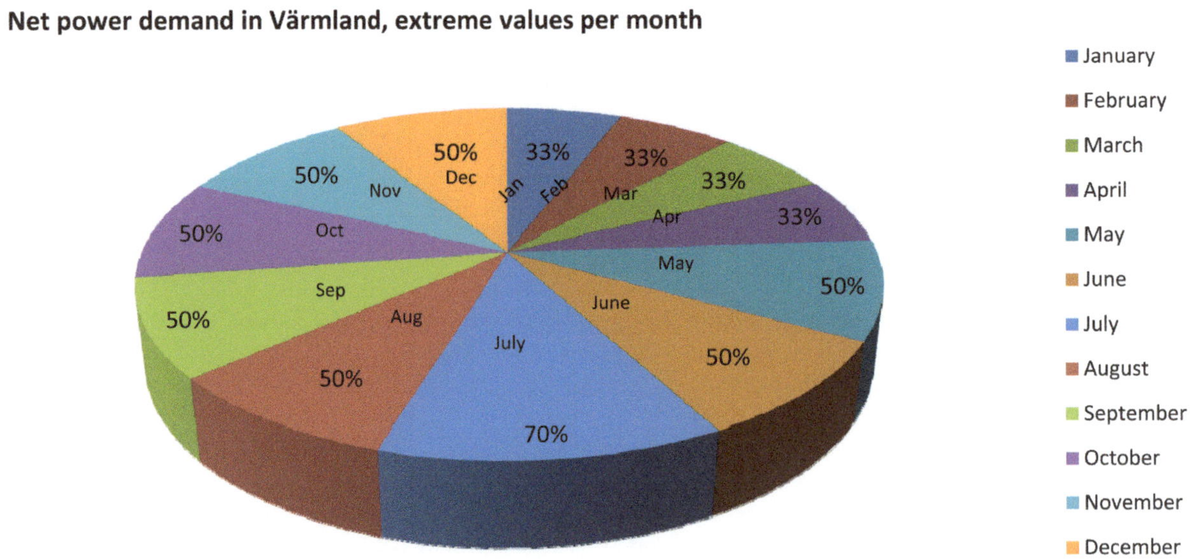

Figure 1. Net monthly power demand in Värmland regional network.

direction of wind speeds during a single year.

In this project, the data related to wind speed and its direction was obtained from SMHI for a hub-height of 100 m. Based on this data, a wind rose was plotted with the help of Enviroware® Software and is shown in **Figure 2**. Based on wind roses from 2012, as shown in **Figure 2**, t the wind speed in this region is observed to span over an interval, with a dominant direction around 180° even if all wind directions are represented during a year. Enercon E−101 wind turbines are used in this project to provide the required amount of electric power (a maximum of 60 MW).

The diameter of this turbine is 101 m; an optimal distance between each of the 20 wind turbines is chosen in such a way that low wake, low noise and smaller interference effects should be observed. Based on these criteria, an optimal spacing is chosen amongst the turbines so that even at low wind speeds, sufficient power is obtained. In this project, it is assumed that a 7-rotor diameter distance is useful between the turbines located towards the main wind direction, while a 6-rotor diameter distance is optimum for the wind turbines facing an orthogonal wind direction [18], *i.e.*

- Main wind direction distance: 707 m.
- Orthogonal wind direction distance: 606 m.

The power plants are located in two areas and based on the aforementioned distance between the installed wind turbines. Both fields are assumed to be identical in size, with total area equal to 3.427 km^2. Moreover, the length of each field is found to be almost three times its width.

3.3.2. Wind Power Generation

Besides selecting a suitable location for the wind turbines, another important factor is their connection with the regional grid. This is a complex issue due to the unstable nature of wind power. The following issues from a general perspective must first be addressed properly [20]:

- Power quality problems in terms of voltage dips and frequency variations.
- Due to difficulties in predicting the wind power production, operation of power system can be significantly affected.
- The considerable impact of wind inconsistency on system operating costs.
- Power imbalance issues, *i.e.* more wind power generation than required or vice versa.
- The need for reliable transmission planning to allow the electric power to flow from generation sources to load points.

Figure 3 illustrates the wind speed and the generation during 2012. It is evident that the maximum wind speed at 100 - 110 meters of hub-height was obtained around 14 m/sec with the lowest around 1.5 m/sec. Another important observation from the graph is the trend of wind speeds with respect to variations in the

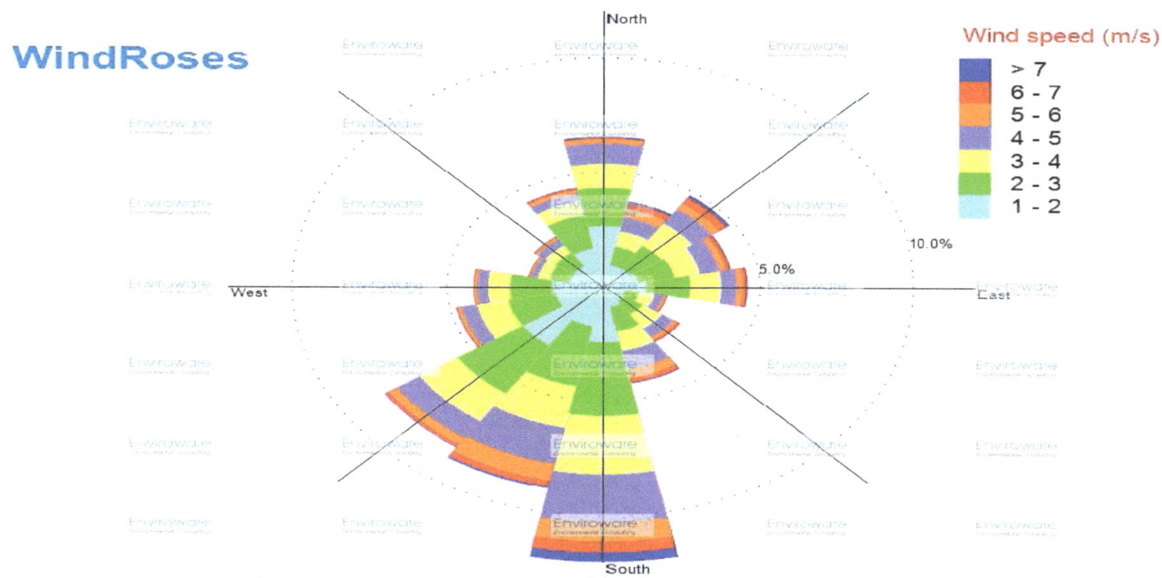

Figure 2. Wind Roses showing the wind direction at Stöpsjön.

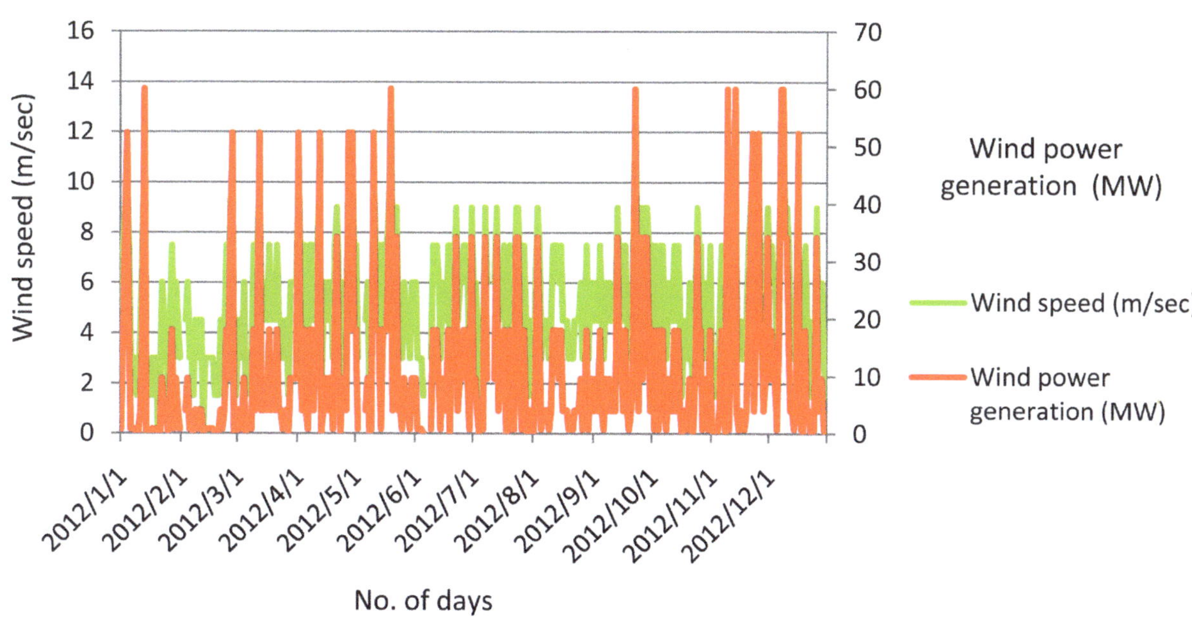

Figure 3. Wind speed versus power generation.

seasons. Similarly, the average wind speed during the whole year is observed to be around 6 m/sec.

3.3.3. Wind Energy Output

Two important inputs from the projected wind farm site are required for calculating the annual electricity output: 1) the power curve of the wind turbine and 2) the annual frequency distribution of the wind speeds. Both of these factors may help in providing a sufficient prognosis of electricity that a wind turbine can generate when installed at the given site [18]. A power curve gives an a rough estimate of the electric power that a wind turbine is expected to produce at different wind speeds [18], whereas the frequency distribution provides information regarding wind speeds at a site and information on how many hours a year the wind will blow at a certain wind speed [18].

From the power curve shown in **Figure 4**, it is evident that the cut-in speed of an E−101 wind turbine is around 3 m/sec, *i.e.* when the turbine starts to generate electricity. Moreover, it indicates that the turbine is able to generate maximum (nominal) output power when it experiences a wind speed of around 12 m/sec. **Table 2** indicates the wind speed intervals and the annual frequency distribution with calculated power for the year of 2012 for the aforementioned type of wind turbines located in the specified wind farm area.

Based on data from **Table 2**, the annual electricity production is calculated to be around 68.19 GWh. Furthermore, in **Table 3** and **Table 4**, wind speed values across VL3 conductor and in the wind farm are mentioned for the whole year of 2012 (the data obtained from SMHI [16]).

3.4. Dynamic Line Rating Approach Applied in the Case Study

3.4.1. Weather Based Dynamic Rating

The ampacity of a power system conductor is its ability to carry the maximum RMS (root mean square) current continuously without exceeding the temperature limit [21]. Hence, it limits the actual capability of transmitting electric power [22]. If the real-time weather-based information is used in place of fixed assumptions, then actual capacity of overhead lines can be obtained. This study is developed on the basis of equations provided in IEEE-738-2006 standard (see section0) [14] and the real-time weather data [22]. Mainly, DLR is used for two

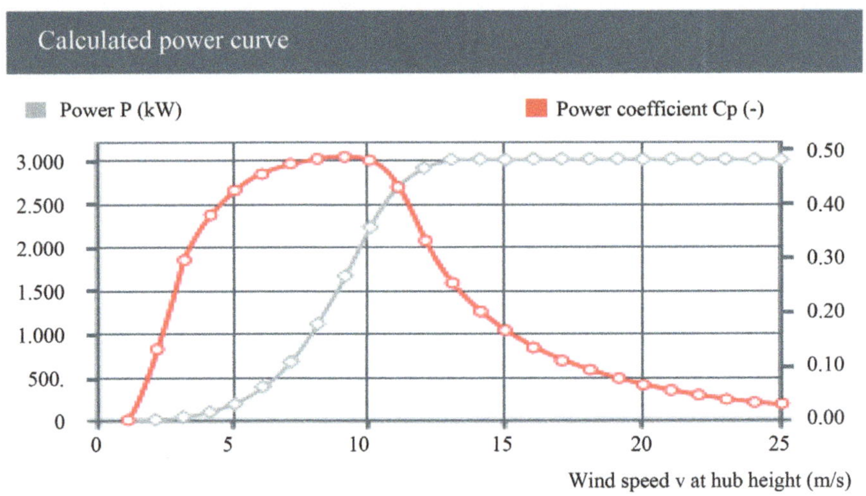

Figure 4. Power curve of enercon e−101 wind turbine [14].

Table 2. Annual frequency distribution of wind speeds.

Wind speed intervals [m/sec]	Frequency distribution [%]	No. of hours/year	Calculated power [MW]
0 - 0.75	8.70	762	0.00
0.75 - 2.25	14.75	1296	0.03
2.25 - 3.75	22.48	1975	0.74
3.75 - 5.25	21.07	1851	4.00
5.25 - 6.75	14.98	1316	9.58
6.75 - 8.25	10.35	909	19.90
8.25 - 9.75	4.57	402	34.20
9.75 - 11.25	2.27	200	52.30
11.25 - 12.75	0.57	50	60.00
12.75 - 14.25	0.24	21	60.00
>14.25	0.02	2	60.00

Table 3. Wind Speed [m/s] across Line VL3 and the wind farm from January to June 2013.

Day	January		February		March		April		May		June	
	*	**	*	**	*	**	*	**	*	**	*	**
1	2	3	3	4.5	2	3	3	4.5	3	4.5	2	3
2	5	7.5	4	6	4	6	5	7.5	4	6	5	7.5
3	6	9	2	3	1	1.5	3	4.5	1	1.5	4	6
4	7	10.5	2	3	2	3	2	3	2	3	5	7.5
5	5	7.5	1	1.5	2	3	5	7.5	7	10.5	4	6
6	2	3	3	4.5	2	3	5	7.5	5	7.5	2	3
7	2	3	2	3	5	7.5	3	4.5	5	7.5	4	6
8	2	3	3	4.5	3	4.5	5	7.5	5	7.5	3	4.5
9	1	1.5	3	4.5	5	7.5	5	7.5	2	3	3	4.5
10	1	1.5	0	0	7	10.5	7	10.5	5	7.5	5	7.5
11	3	4.5	2	3	3	4.5	1	1.5	5	7.5	1	1.5
12	5	7.5	2	3	3	4.5	3	4.5	5	7.5	5	7.5
13	8	12	2	3	4	6	4	6	6	9	4	6
14	1	1.5	2	3	3	4.5	3	4.5	9	13.5	6	9
15	1	1.5	2	3	5	7.5	4	6	3	4.5	3	4.5
16	1	1.5	1	1.5	3	4.5	4	6	4	6	4	6
17	2	3	1	1.5	4	6	2	3	6	9	4	6
18	1	1.5	1	1.5	3	4.5	5	7.5	3	4.5	5	7.5
19	2	3	3	4.5	5	7.5	6	9	3	4.5	5	7.5
20	0	0	2	3	3	4.5	4	6	2	3	4	6
21	1	1.5	3	4.5	3	4.5	1	1.5	4	6	2	3
22	4	6	5	7.5	2	3	4	6	3	4.5	6	9
23	3	4.5	4	6	2	3	4	6	3	4.5	4	6
24	3	4.5	6	9	3	4.5	3	4.5	2	3	4	6
25	1	1.5	7	10.5	1	1.5	7	10.5	4	6	3	4.5
26	3	4.5	1	1.5	4	6	5	7.5	4	6	1	1.5
27	5	7.5	1	1.5	4	6	7	10.5	4	6	2	3
28	2	3	3	4.5	4	6	5	7.5	2	3	2	3
29	4	6			4	6	5	7.5	2	3	6	9
30	3	4.5			7	10.5	2	3	2	3	4	6
31	2	3			5	7.5			1	1.5		

*Wind Speed across line (VL3) [m/s]. **Wind Speed across wind farm [m/s].

purposes: first, to increase the capacity of overhead conductor in terms of transmitting the maximum electric current and second, to help in the transfer of electricity during peak load and emergency states [22].

To gain an overall picture of the changes in conductor (VL3) ampacity based on SLR assumptions, see **Table 5**. The results shown in **Table 5** are calculated based on typical fixed weather assumptions, *i.e.* the highest ambient temperature, the lowest wind speed and the maximum solar radiation (during daytime) for a 30-kilometer long ACSR overhead conductor. Furthermore, it is observed from **Table 5** that wind speed and its direction has

Table 4. Wind Speed [m/s] across Line VL3 and the wind farm from July to December 2013.

Day	July		August		September		October		November		December	
	*	**	*	**	*	**	*	**	*	**	*	**
1	4	6	5	7.5	3	4.5	5	7.5	6	9	3	4.5
2	6	9	5	7.5	6	9	3	4.5	8	12	3	4.5
3	3	4.5	4	6	4	6	5	7.5	3	4.5	2	3
4	4	6	5	7.5	5	7.5	3	4.5	2	3	3	4.5
5	4	6	3	4.5	3	4.5	1	1.5	3	4.5	7	10.5
6	5	7.5	3	4.5	5	7.5	3	4.5	2	3	3	4.5
7	1	1.5	3	4.5	3	4.5	3	4.5	2	3	2	3
8	2	3	2	3	2	3	2	3	3	4.5	1	1.5
9	5	7.5	2	3	3	4.5	4	6	5	7.5	5	7.5
10	2	3	2	3	4	6	2	3	6	9	3	4.5
11	2	3	3	4.5	9	13.5	1	1.5	7	10.5	3	4.5
12	6	9	3	4.5	5	7.5	4	6	3	4.5	1	1.5
13	6	9	3	4.5	3	4.5	6	9	5	7.5	1	1.5
14	4	6	4	6	6	9	5	7.5	7	10.5	2	3
15	2	3	2	3	4	6	5	7.5	5	7.5	6	9
16	5	7.5	3	4.5	6	9	2	3	5	7.5	3	4.5
17	3	4.5	5	7.5	6	9	4	6	3	4.5	4	6
18	1	1.5	2	3	5	7.5	4	6	4	6	4	6
19	3	4.5	4	6	3	4.5	2	3	6	9	2	3
20	2	3	4	6	5	7.5	5	7.5	4	6	1	1.5
21	1	1.5	3	4.5	2	3	2	3	5	7.5	3	4.5
22	4	6	4	6	5	7.5	1	1.5	4	6	3	4.5
23	6	9	3	4.5	2	3	1	1.5	4	6	6	9
24	5	7.5	5	7.5	5	7.5	1	1.5	2	3	4	6
25	2	3	3	4.5	5	7.5	3	4.5	5	7.5	1	1.5
26	3	4.5	2	3	3	4.5	5	7.5	8	12	2	3
27	3	4.5	4	6	2	3	4	6	8	12	3	4.5
28	3	4.5	3	4.5	4	6	1	1.5	6	9	3	4.5
29	2	3	4	6	3	4.5	8	12	6	9	8	12
30	2	3	3	4.5	3	4.5	2	3	5	7.5	3	4.5
31	3	4.5	3	4.5			5	7.5			5	7.5

*Wind Speed across line (VL3) [m/s]. **Wind Speed across wind farm [m/s].

a huge impact on increasing or decreasing the conductor ampacity compared with changes in ambient temperature.

Low ambient temperature combined with high wind speed and its direction (perpendicular to the position of the conductor) increases the conductor ampacity significantly.

3.4.2. Line Current

Line current represents the flow of current through a conductor during normal state in the power system. In this

Table 5. A typical example of weather effect on lineampacity, VL3 OH-line (30 km) with SLR assumptions: 30˚C, 0.6 m/sec and Day-time.

Variation in Weather Parameter (s)	Change in Conductor ampacity
Ambient Temperature (˚C)	
+5˚C Variation	21.7% Decrease in Capacity
−5˚C Variation	17.7% Increase in Capacity
Wind Speed (m/sec) at line corridor	
1 m/sec Increase	
45˚ angle	24.8% Increase in Capacity
90˚ angle	36.9% Increase in Capacity

project, line current was changed by varying the loads in the region around VL3. The load situation was adjusted according to the philosophy discussed in section 3.2. After variations in the system loads and by addition of wind power generation, the current and temperature across "VL3" overhead line were analyzed. Line currents were calculated with PSS/E® (power system simulation) software.

The 60 MW wind farm was connected at 10 km distance from VL3 overhead line conductor in Fortum's 130 kV regional network. The network was simulated in PSS/e® software in the presence of wind farm. Besides the effect of wind power on line loading, the effect of loads connected in 130 kV regional network was also investigated in this research work. The rating of these loads was observed for both summer and winter case scenarios. Furthermore, the critical span of VL3 was considered around 6 m. Hence, loading of VL3 line conductor as well as its dynamic ampacity was calculated based on its critical span.

3.4.3. Static and Dynamic Ampacities

The dynamic and/or static ampacity is based upon two main factors, *i.e.* physical characteristics of the conductor and environmental parameters [9], with sub categories such as conductor diameter, conductor temperature, ambient temperature, wind speed, angle between wind speed and conductor, and solar radiation. The dynamic and static line ampacities for VL3 overhead line are calculated in this section in addition to line current (based on the monthly load changes and the hourly wind power generation), which is illustrated in **Figure 5**. Note that the static rating is not totally static since it is divided into one static "winter" and one static "summer" rating.

Regarding network operations, the analysis shows that during winter, the dynamic and even the static line ampacity is sufficient in terms of electricity flow through the overhead conductor as required based on the load demand. However, in summer, the situation is different, due to excessive ambient temperature: the line cannot be allowed to transmit the required flow of electricity due to the risk of conductor deterioration or sag problems.

As **Figure 5** shows, in winter, both SLR and DLR are higher than the line current and hence the line can be loaded up to its full capacity without any risk during this period. However, during summer timings, the line cannot be allowed to carry the required current (as is clear from the graph, both SLR and DLR are seen to be lower than the line current). The significance of this graph is that with the traditional approach, network operators need to consider corrective actions or curtailments during all summer months. With DLR, corrective actions or curtailments are only necessary during a few summer days.

4. Economic Analysis of Ampacity Upgrading Methods

4.1. Introduction

Based on technological advancements in the wind energy sector, the cost of electricity produced from wind power is significantly reduced: one seventh of the electricity cost in the early eighties [23]. When carrying out an economic analysis of the wind power system, it should be checked whether this project will be profitable (even on a short-term basis) or not from an economic perspective. The decision -making plays a vital role in this regard and needs to take into account the interests of all the stakeholders involved. From a wind farm owner's perspective, the economic analysis of wind power generation involves three main categories of economic costs [20]:

- Capital costs related to construction of the wind farm.
- Fixed operation costs.
- External costs of the wind power generation (termed the leakage costs).

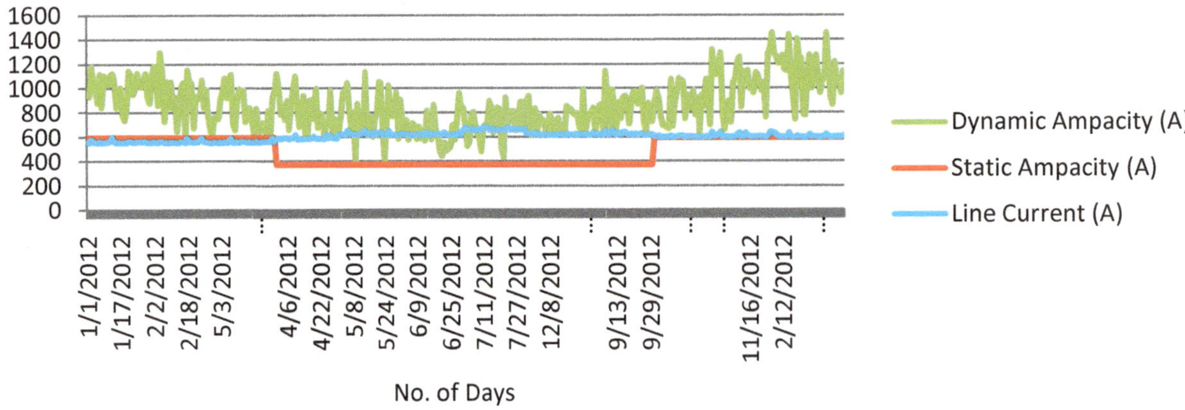

Figure 5. Static and Dynamic Ampacities (A) versus the Line Current (A) in 2012.

In this project, besides wind power analysis, the DLR approach is also taken into account. Based on DLR, better information on the actual ampacity of an overhead line may allow an increased flow of energy. A direct economic advantage for a DSO can be to save money in the short or long term by postponing or avoiding upgrade of existing line (s) or construction of new line (s). Moreover, the delay in investments may help in enhancing the effectiveness of spent money [11]. However, since power system infrastructure is a regulated natural monopoly, it can be difficult to estimate exactly the share of the benefits each stakeholder receives. From a DLR perspective, many financial benefits can be directly or indirectly achieved (depending on the type of regulation) [11].

- Cheaper electricity for consumers (a DLR advantage for society).
- Better prices for the wind power owners (in terms of lower connection fee).
- Improvement in economic use of transmission lines (useful for the DSO).
- Making asset utilization cost effective (valuable from a DSO perspective).

4.2. Economic Analysis of Ampacity Upgrading Methods

4.2.1. Introduction

By adopting the DLR approach, the capacity of an overhead conductor is increased to a significant extent by allowing the transfer of more electric power [12]. Other options if the capacity needs to be increased are to replace the existing conductor with a high temperature-sustaining conductor or to construct a new line in parallel with the existing conductor. The hypothesis is that these kinds of solutions are less appropriate from economic, environmental and social perspectives [11]. DLR allows operators opportunities for delivering more power during its peak demand and avoiding unnecessary load shedding under contingency conditions [24]. Similarly, the power capacity limitations for any DSO can also be expensive, while a small increase in the capacity of the overhead lines may have significant economic effects [19]. In the following sections, annual revenue based on increment in capacity is calculated, both for a DLR approach to an existing conductor (VL3) and for traditional solutions, including investing in a new overhead conductor (area 593 mm^2). Finally, the economic value of postponing the power system investments will also be taken into account.

4.2.2. Net Annual Income Based on the DLR Approach

The dynamic ampacity in this project was calculated for the whole year of 2012 and was based on a particular hour of the day. The selection of that hour was assumed as random and was chosen within a 13-hour range (7 am to 7 pm). Based on this ampacity rating, the allowed energy flow E_{DLR} during the selected hour was calculated for the whole year of 2012 and is given as:

$$E_{\text{DLR}} = 38.51\ \text{GW} \cdot \text{h/year}$$

Similarly, based on the static ampacity rating, the allowed energy flow E_{SLR} (for a single hour in a day) through a "VL3" overhead line for the whole year of 2012 was calculated as:

$$E_{\text{SLR}} = 21.68\ \text{GW} \cdot \text{h/year}$$

From the calculations of E_{DLR} and E_{SLR}, the theoretical maximum energy flow through the "VL3" overhead line on the basis of dynamic ampacity is around 1.8 times higher than if static ampacity is used. Hence, the energy that is curtailed by the SLR approach can be transferred with the help of the DLR technique across the same overhead conductor. Hence, the increased possible transfer of energy through VL3 overhead line by replacing SLR with DLR is equal to:

$$\Delta E_{DLR} = E_{DLR} - E_{SLR} = 16.83 \text{ GW}\cdot\text{h/year}$$

Furthermore, applying the real-time dynamic rating technique across an existing overhead conductor requires equipping the line with useful communication and computer tools for transferring the real-time data from the line corridor to the control room. The different monitoring and communication tools are available in this regard; CAT-1 transmission line monitoring system is one of them [25]. If these tools are installed across an existing overhead conductor for continuous real-time monitoring and transmission of data to the DSO's control center then it costs around \$200,000 (one time product cost estimation) for a single existing transmission line conductor (excluding the shipment, installation and O&M costs) [20]. Now converting it to SEK (1 SEK ≈ \$0.15), the first time expenditure D_{DLR} will be equivalent to:

$$D_{\text{DLR}} = 1.317 \text{ MSEK}$$

With the help of the DLR technique, the transfer of electricity through the existing overhead conductor is significantly higher than its flow based on the traditional SLR approach. The net annual income I_{DLR} based on this allowed energy flow at an assumed electricity price (λ) of $0.369\ \text{SEK/kW}\cdot\text{h}$ [26] for the year of 2012 will be:

$$I_{\text{DLR}} = \Delta E_{\text{DLR}} * \lambda - D_{\text{DLR}} = 4.89 \text{ MSEK}$$

Furthermore, due to the involvement of different stakeholders in the electricity market, it is difficult and complex to estimate the income or profit for each stakeholder involved, *i.e.* how much share should be allocated to the wind power owners, the utilities and the DSOs when the energy is transferred based on the DLR approach (particularly due to its dynamic nature). Hence, to avoid such assumption-based profits or incomes for each stakeholder, it is better to focus on the economic analysis based on the benefit of capacity increase across an overhead line. This financial benefit after implementation of the DLR approach across the existing overhead line (VL3) is now calculated as:

$$B_{\text{DLR}} = \frac{I_{\text{DLR}}}{\Delta E_{\text{DLR}}} = 0.29 \text{ MSEK/GW}\cdot\text{h}$$

B_{DLR}, indicates that an increment of 1 GW·h energy flow through an existing overhead line (VL3) may yield a *theoretical* benefit of 0.29 MSEK during the first year (as it assumes 1.317 MSEK/year in operating costs, *i.e.* valid for first year only). The economic benefit in reality is more complex and will be discussed later in section 4.3.

4.2.3. Net Annual Income in Case of Upgrading the Line

The allowed energy flow E_u through the new planned overhead conductor of area 593 mm^2 based on its static ampacity (63% of that of the existing conductor) during the whole year of 2012 is found to be around 35 GW·h/year:

$$E_u = 35.37 \text{ GW}\cdot\text{h/year}$$

Replacement of the existing line "VL3" with a new conductor that has an area of 593 mm^2 and the length of 30 km will require an approximate total capital cost $C_{u,t}$ of:

$$C_{u,t} = 32.1 \text{ MSEK}$$

Assuming that the cost related to upgrading the conductor is financed by a loan at a nominal interest rate of 7.5% (real interest rate + expected inflation) and the loan is required to be paid in a period of 20 years, then, based on this data, the capital cost of upgrading the conductor on annual basis $C_{u,a}$ can be calculated with the help of the annuity method. Moreover, by this method, the value of the annuity (based on pay-back period and the interest rate) can be calculated with the help of Equation (4.1) [18]:

$$a = \frac{r \times (1+r)^n}{(1+r)^n - 1}, \tag{4.1}$$

where a is the annuity, r is the nominal interest rate, and n is the pay-back period in years. Based on Equation (4.1), the annuity will be:

$$a = \frac{0.075 \times (1+0.075)^{20}}{(1+0.075)^{20} - 1} = 0.097$$

After calculation of the annuity, the annual capital cost of replacing the existing overhead conductor (VL3) with a new overhead conductor (of area 593 mm^2) can be calculated with the help of Equation (4.2) [18], *i.e.*

$$C_{u,a} = a \times C_{u,t} \tag{4.2}$$

$$C_{u,a} = 0.097 \times 32.1 \times 10^6 = 3.11 \text{ MSEK}$$

After replacement of the existing conductor with a new overhead line that has 1.1 times larger cross-sectional area, the transfer of electricity (based on SLR approach) through this new overhead line is increased significantly compared to its flow through the existing smaller cross-sectional area overhead line. Moreover, the energy (based on static ampacity) that is curtailed by existing conductor $\Delta E_{u,a}$ can be transferred through the new overhead line and is given as:

$$\Delta E_{u,a} = 13.69 \text{ GW} \cdot \text{h/year}$$

$$\lambda = 0.369 \text{ SEK/kW} \cdot \text{h}$$

$$I_{u,a} = \Delta E_{u,a} \times \lambda - C_{u,a}$$

or,

$$I_{u,a} = 13.69 \times 0.369 - 3.11 = 1.94 \text{ MSEK/year}$$

where λ is the current electricity price and $I_{u,a}$ is the net annual income after the line upgrading. The economic analysis of the increase in conductor capacity when the existing overhead line is upgraded with a new overhead conductor with the aforementioned specifications will be:

$$B_u = \frac{I_{u,a}}{\Delta E_{u,a}} = 0.14 \text{ MSEK/GW} \cdot \text{h}$$

Based on this value, it is observed that, after the upgrading of the existing conductor, an increment of 1 GW·h energy flow through the new overhead line may yield a theoretic maximal benefit of 0.14 MSEK during a single year.

4.2.4. Net Annual Income in Case of Building a New Line

An approximate total capital cost $C_{n,t}$ associated with building a new overhead line of about 30 km in length with an area of 593 mm^2, designed for 130 kV operating voltage is:

$$C_{n,t} = 39.3 \text{ MSEK}$$

Similar to upgrading of the line, if the construction of a new overhead conductor is financed in the form of a bank loan at a nominal interest rate of 7.5% with a pay-back period of 20 years, then the annual capital cost $C_{n,a}$ calculated with the help of annuity method from Equations (4.1) and (4.2) will be:

$$C_{n,a} = a \times C_{n,t}$$

where,

$$a = \frac{0.075 \times (1+0.075)^{20}}{(1+0.075)^{20} - 1} = 0.097$$

Hence, the annual capital cost for the new line construction will be:

$$C_{n,a} = 0.097 \times 39.3 \times 10^6 = 3.81 \text{ MSEK}$$

After construction of a new overhead line with 1.1 times larger cross-sectional area, the transfer of electricity (based on the SLR approach) through this new overhead line is significantly increased compared with its flow through the existing smaller cross-sectional area overhead line. Furthermore, the energy (based on static ampacity) that is curtailed by existing conductor $\Delta E_{n,a}$ can be transferred through the new overhead conductor and is given as:

$$\Delta E_{n,a} = \Delta E_{u,a} = 13.69 \text{ GW} \cdot \text{h/year}$$

$$\lambda = 0.369 \text{ SEK/kW} \cdot \text{h}$$

$$I_{n,a} = \Delta E_{n,a} \times \lambda - C_{n,a}$$

or

$$I_{n,a} = 13.69 \times 0.369 - 3.81 = 1.24 \text{ MSEK}$$

where λ is the current electricity price and $I_{n,a}$ is the net annual income after the new line construction. The economic analysis related to increase in the conductor capacity when a new overhead line (with aforementioned specifications) is constructed in place of existing conductor will be around:

$$B_n = \frac{I_{n,a}}{\Delta E_{n,a}} = 0.09 \text{ MSEK/GW} \cdot \text{h}$$

Based on this value, it is observed that an increment of 1 GW·h energy flow through the new overhead line may yield a theoretical benefit of 0.09 MSEK during a single year.

4.2.5. Comparison amongst Ampacity Upgrading Methods

In this section, a comparison amongst different ampacity upgrading techniques is taken into account. The ampacity upgrading methods for replacement of a smaller overhead conductor with a new larger overhead conductor and new line construction are considered as conventional approaches for the transmission of electricity that otherwise cannot be transferred through the old smaller overhead conductor. From a technical perspective, any of these conventional ampacity-upgrading methods can be considered useful in terms of required electricity transmission but from an economic perspective, these methods are less desirable. In comparison with conventional ampacity upgrading methods, it is observed that the dynamic line rating approach is not only useful from a technical perspective but also feasible from an economic perspective. Furthermore, **Table 6** gives an overall picture of the annual benefit from the increase in conductor capacity with the help of different ampacity upgrading methods.

From **Table 6**, it is observed that the DLR approach is significantly profitable in comparison to conventional ampacity upgrading techniques. However, on the basis of limited information, it is difficult to find and compare the exact turnovers from the aforementioned ampacity upgrading techniques. For example, the energy transfer through conductor replacement or through a new line construction is based on a static ampacity approach which may yield the predictable revenue, whereas the revenue based on dynamic line ampacity cannot be predicted due to significant variations in the conductor capacity.

Therefore, the revenue-based comparison cannot be estimated amongst static and dynamic line ratings. Moreover, the cost of dynamic rating equipment and their control room issues are uncertainties when estimating the profits obtained from the dynamic line ratings compared with the profits based on conventional static ampacity techniques. Therefore, these figures cannot be regarded as reliable and may differ in reality.

Table 6. Annual benefit from ampacity upgrading solutions.

Ampacity Upgrading Solution (s)	MSEK/GW·h
Dynamic Line Rating	0.29
Conductor Upgrading	0.14
New Line construction	0.09

5. Conclusions

This paper presented analysis results from an application study where dynamic line rating was implemented across an overhead line located in a 130 kV sub-transmission system. The static as well as dynamic ampacities has been calculated to estimate the accurate range of loading an overhead line throughout the whole year. The application study also investigated the integration of a 60 MW wind power park. Based on wind power production and power demand in the system, the line current and conductor temperature has been calculated. Thereafter, a comparison amongst conductor ampacities and the line current was carried out to investigate how much further the overhead conductor can be rated (loaded).

The results of this paper indicate that dynamic line rating significantly has the potential to improve the capacity of a power component (here exemplified by an overhead conductor) and to facilitate wind power integration. During winter, due to low ambient temperature and minimum solar radiation, the dynamic ampacity of an overhead conductor is found to be many times higher than its static ampacity. However, during peak summer days, both ampacity ratings are almost identical and lower than actual line current.

From an economic study, it is observed that the ampacity upgrading of an overhead conductor on the basis of dynamic rating is significantly profitable in comparison with conductor replacement or new line construction techniques. The main contribution of the paper is to combine theoretical calculations and modeling with real application. Besides received results and conclusions, data from this paper can be valuable because of reference material within other studies of wind power and/or dynamic rating.

References

[1] Ipakchi, A. and Albuyeh, F. (2009) Grid of the Future. *IEEE Power and Energy Magazine*, **7**, 52-62. http://dx.doi.org/10.1109/MPE.2008.931384

[2] Lopes, J.A., Hatziargyriou, N., Mutale, J. *et al.* (2007) Integrating Distributed Generation into Electric Power Systems: A Review of Drivers, Challenges and Opportunities. *Electric Power Systems Research*, **77**, 1189-1203. http://dx.doi.org/10.1016/j.epsr.2006.08.016

[3] Wallnerström, C.J. (2011) On Incentives affecting Risk and Asset Management of Power Distribution. PhD Thesis, KTH, Stockholm.

[4] Shaker, H., Fotuhi-Firuzabad, M. and Aminifar, F. (2012) Fuzzy Dynamic Thermal Rating of Transmission Lines. *IEEE Transactions on Power Delivery*, **27**, 1885-1892.

[5] Simms, M. and Meegahapola, L. (2013) Comparative Analysis of Dynamic Line Rating Models and Feasibility to Minimise Energy Losses in Wind Rich Power Networks. *Energy Conversion and Management*, **75**, 11-20. http://dx.doi.org/10.1016/j.enconman.2013.06.003

[6] Michiorri, A., Currie, R. and Taylor, P. (2011) Dynamic Line Ratings Deployment on the Orkney Smart Grid. *Proceedings of the* 21*st International Conference on Electricity Distribution*, Frankfurt, 6-9 June 2011.

[7] Wallnerström, C.J., Huang, Y. and Söder, L. (2014) Impact from Dynamic Line Rating on Wind Power Integration. *IEEE Transactions on Smart Grid*, 6, 343-350.

[8] Wallnerström, C.J., Hilber, P., Söderström, P., Saers, R. and Hansson, O. (2014) Potential of Dynamic Rating in Sweden. *Proceedings of the* 2014 *International Conference on Probabilistic Methods Applied to Power Systems* (*PMAPS*), Durham, 7-10 July 2014, 1-6. http://dx.doi.org/10.1109/pmaps.2014.6960605

[9] Holbert, K.E. and Heydt, G.T. (2001) Prospects for Dynamic Transmission Circuit Ratings. *Proceedings of the* 2001 *IEEE International Symposium on Circuits and Systems*, Sydney, 6-9 May 2001, 205-208. http://dx.doi.org/10.1109/iscas.2001.921283

[10] Douglass, D.A. (1988) Weather-Dependent versus Static Thermal Line Ratings (Power Overhead Lines). *IEEE Transactions on Power Delivery*, **3**, 742-753.

[11] Sandy, K.A. (2010) Dynamic Line Ratings for Optimal and Reliable Power Flow, Enhanced Power Flow for the Smart Grid. *Proceedings of the FERC Technical Conference*, Washington DC, 24 June 2010.

[12] Kim, S.D. and Morcos, M.M. (2013) An Application of Dynamic Thermal Line Rating Control System to Up-Rate the Ampacity of Overhead Transmission Lines. *IEEE Transactions on power delivery*, **28**, 1231-1232.

[13] Merrell, J., Mike P.D. and Jared R.J. (2008) Dynamic Line Ratings for the Cowlitz-LaGrande Transmission Lines. Washington State University, Washington DC.

[14] IEEE (1993) IEEE Standard for Calculating the Current-Temperature of Bare Overhead Conductors. IEEE, New York, c1-59.

[15] Talpur, S. (2013) Dynamic Line Rating Implementation as an Approach to Handle Wind Power Integration: A Feasibility Analysis in a Sub-Transmission System Owned by Fortum Distribution AB.

[16] SMH (2012) Swedish Metrological and Hydrological Institute. http://www.smhi.se

[17] Blaabjerg, F., *et al.* (2006) Power Electronics in Wind Turbine Systems. *The Proceedings of the IEEE* 5*th International Power Electronics and Motion Control Conference*, Shanghai, 14-16 August 2006, 1-11.

[18] Wizelius, T (2007) Developing Wind Power Projects: Theory and Practice. Routledge, London.

[19] Enercon Product Overview. E-101 Wind Turbine. http://www.enercon.de/p/downloads/ENERCON_PU_en.pdf

[20] Deutsche Gesellschaft für Technische Zusammenarbeit (GTZ) GmbH (2006) Feasibility Study for Wind Park Development in Ethiopia and Capacity Building. Mesobo-Harena Wind Park Site.

[21] Williams, N. and Jeffrey, S. (2007) NEC Q & A: Questions & Answers on the National Electrical Code. Jones & Bartlett Learning, Burlington.

[22] Kim, D.-M., *et al.* (2006) Prediction of Dynamic Line Rating Based on Assessment Risk by Time Series Weather Model. Proceedings of the *PMAPS* 2006 *International Conference on Probabilistic Methods Applied to Power System*, Stockholm, 11-15 June 2006, 1-7. http://dx.doi.org/10.1109/pmaps.2006.360329

[23] Georgilakis, P.S (2008) Technical Challenges Associated with the Integration of Wind Power into Power Systems. *Renewable and Sustainable Energy Reviews*, **12**, 852-863. http://dx.doi.org/10.1016/j.rser.2006.10.007

[24] Douglass, D.A., *et al.* (2000) Dynamic Thermal Ratings Realize Circuit Load Limits. *Computer Applications in Power, IEEE*, **13**, 38-44. http://dx.doi.org/10.1109/67.814665

[25] Seppa, T.O. (2007) Reliability and Real Time Transmission Line Ratings. Nexans, Ridgefield.

[26] (2013) Nord Pool Spot, Elspot Prices (EUR/MWh). http://www.nordpoolspot.com/Market-data1/Elspot/Area-Prices/ALL1/Hourly/

22

Integrating Roundabouts with Freight Roadway Networks

Ranjit Prasad Godavarthy[1], Eugene R. Russell[2]

[1]Upper Great Plains Transportation Institute, North Dakota State University, Fargo, USA
[2]Department of Civil Engineering, Kansas State University, Manhattan, USA
Email: ranjit228@gmail.com, geno@ksu.edu

Abstract

The demand for freight transportation in the United States has grown rapidly in the past couple of decades; the rate at which the highway capacity is growing is not able to meet the freight-transportation needs, resulting in congestion and delay, ultimately affecting the users and the public with increased prices for the goods delivered, unreliable delivery times, and air-pollution concerns. Freight bottlenecks are a major cause of recurring congestion which accounts for about 40% of total vehicle hours of delay in the United States. Intersections for urban freight-roadway networks are one of the major freight bottlenecks and are considered to be a significant contributor for congestion and delay. Improving the efficiency at urban intersections with high truck traffic can address the freight-traffic congestion, leading to optimized goods movement as well as decreased delays, congestion, and emissions, thus enhancing the air quality in and around the communities. With the roundabout intersection control being proven as a safe, operationally efficient, and environment-friendly control treatment, a greater use of roundabouts with urban freight-roadway networks and their feasibility is analyzed in this study. The control for most urban intersections is a signalized treatment; a performance analysis is conducted for selected signalized intersections in urban freight-roadway networks, comparing the intersections by adapting a roundabout control theoretically using SIDRA (signalized and un-signalized intersection design and research aid) Intersection software. Various parameters, such as the intersection's level of service, the effective intersection capacity, the average control delay, vehicular emissions [carbon dioxide (CO_2), hydrocarbons (HC), carbon monoxide (CO), and nitrogen oxides (NO_x)], the intersection's annual delay, and the intersection's annual cost, are selected for comparison.

Keywords

Congestion on Roads and Highways, Modern Roundabouts, Truck and Vehicular Emissions

1. Background and Literature Review

A recent study conducted by Federal Highway Administration (FHWA) has estimated that about 40% of traffic congestion in general, as opposed to freight congestion specifically, is caused by bottlenecks, resulting in stop-and-go traffic flow and long backups [1]. The study found that freight bottlenecks caused upwards of 243 million truck hours of delay and that the direct user cost for this delay was about $7.8 billion per year. Signalized, arterial intersections account for a total of 18% of the delay—about 43 million hours of delay—for different freight routes that are comprised of urban freight corridors, intercity freight corridors, truck-access routes, and intermodal connectors[1].

There are also air-pollution concerns as a result of heavy congestion in urban areas. According to Freight Facts and Figures from the Office of Freight Management and Operations [2], diesel-fueled heavy trucks emit small amounts of carbon monoxide (CO) but large amounts of nitrogen oxides (NO_x) when compared to gasoline-fueled cars. Freight transportation contributes to 27% of the total NO_x emissions and one-third of the particulate-matter emissions that are 10 microns in diameter (PM-10) from mobile sources in the United States. Trucks have a two-third share of the NO_x emissions from the freight sector. Apart from the above emissions, the transportation sector releases large quantities of greenhouse gases (GHGs), such as carbon dioxide, methane, nitrous oxide, and hydrofluorocarbons, and these gases trap heat in the atmosphere which affects the earth's temperature. Therefore, the increased truck congestion at urban intersections can affect the air quality by the emissions which can be mitigated with better traffic-flow techniques such as less delay at urban intersections [2].

The above challenges clearly show that there is a need to improve traffic flow at interchanges, intersections, and other transportation facilities in order to better accommodate vehicles and trucks with less congestion and delay time, thereby saving many dollars and improving the environment. This study's scope focuses on improving performance and traffic flow while decreasing congestion at intersections with high truck volumes. One of the three strategies to decrease traffic congestion is to operate the existing traffic capacity more efficiently [3]. While most intersections for urban freight-roadway networks (designated roads that are more frequently used by trucks for goods movement) have signalized control, roundabout control can also be considered as a possible solution to improve efficiency as well as to reduce delay and emissions, considering that roundabouts' safety and operational benefits are well documented [4]-[6]. Roundabouts are also conventionally designed for effectively accommodating trucks. Some recent studies have also addressed oversize-overweight vehicles' accommodation concerns at roundabouts by maintaining roundabout features such as an external truck apron, center-island truck apron, a custom center-island design, and traversable splitter islands [4] [7]-[10]. Russell *et al.* [11]-[13] have conducted surveys with trucking agencies that haul heavy trucks in the United States to understand freight bottlenecks on state highways; Russell *et al.* concluded that intersections with signs or signals were considered as freight bottlenecks by most trucking agencies when compared to a roundabout control. Mandavilli *et al.* [14] have studied the environmental impact of single-lane and double-lane modern roundabouts in Kansas; the authors found that roundabouts operated more effectively than the previous intersection control (all-way stop control or two-way stop control) to reduce vehicular emissions.

While the environmental impact of roundabouts has been proven with prior research, the environmental impacts of roundabouts in urban freight-roadway networks expecting high truck traffic are not known. According to a 2008 United States Environmental Protection Agency report [15], the average idle emission for light-duty gasoline-fueled vehicles includes 71.22 grams/hr of CO emissions, 3.51 grams/hr of NO_x emissions, and 3.16 grams/hr of total hydrocarbon (HC) emissions. In contrast, the average idle emissions for heavy-duty, gasoline-fueled vehicles include 151.90 grams/hr (113% more) of CO emissions, 5.33 grams/hr (52% more) of NO_x emissions, and 7.26 grams/hr (129% more) of total hydrocarbon emissions; the average idle emissions for heavy-duty diesel vehicles include 25.63 grams/hr (64% less) of CO emissions, 33.76 grams/hr (861% more) of NO_x emissions, and 3.50 grams/hr (11% more) of total hydrocarbon emissions [15]. It can be understood from the above numbers that emissions at roundabouts with a high volume of trucks can be substantially high; therefore, the environmental impacts for such roundabouts need to be validated to understand if roundabouts can be included with urban freight-roadway networks.

2. Study Methodology

The study's objective was to investigate integrating a greater use of roundabouts for freight roadway networks in and around urban areas in order to optimize goods movements as well as to decrease delay, congestion, and emissions, thus enhancing air quality in and around the communities. Most intersections for urban freight-roadway networks were signalized intersections; therefore, the delay and emissions for vehicles at signalized intersections were analyzed and theoretically compared to a roundabout to understand if roundabouts can be designed and built at these intersections for better delay and emission reductions.

This comparative analysis was achieved by analyzing various performance measures for selected signalized intersections in urban freight-roadway networks, comparing signalized intersections by adapting a roundabout control theoretically using the SIDRA Intersection software. The analyzed performance measures included intersection level of service (LOS), effective intersection capacity (the ratio of the total intersection demand flow to the intersection's degree of saturation), average control delay (the delay to a vehicle which decelerates from the approach cruise speed to a full stop, waits, and then accelerates to the exiting cruise speed), vehicular emissions [CO_2, HC, CO, and NO_x], the intersection's annual delay, and the intersection's annual cost (the vehicles' operating cost along with the time cost for drivers and passengers). Average control delays and vehicular emissions (CO, HC, CO_2, and NO_x) were used as key performance measures to make conclusions about the most effective intersection control. SIDRA Intersection is an advanced micro-analytical, traffic-evaluation tool that is used worldwide for intersection capacity, level of service, and performance analysis, this software is the only program known to the authors that is capable of making these comparisons for traditional intersection control vs. roundabouts [16]. SIDRA Intersection version 5.1 was used for this study's performance analysis.

This study was conducted by selecting cities of different sizes, such as a small city, a medium city, and a big city in Kansas, United States based on the area population. The different cities considered were as follows: Manhattan as a small-sized city, Topeka as a medium city, and Overland Park as a big city. City officials for each locale were contacted to obtain the freight roadway networks (when available) or routes that had a high trucking activity. Based on the obtained routes with high trucking activity, intersections which had higher truck usage and reliable traffic-count data were selected to conduct a performance analysis with the SIDRA Intersection software.

2.1. Small City: Manhattan, Kansas

The City of Manhattan did not have a developed freight roadway network to identify the roads that were heavily used by trucks. However, staff from the city traffic engineer's office helped to identify six signalized intersections (listed in **Table 1**) that expect high truck traffic. These six signalized intersections had traffic counts available from the city traffic engineer which were used to calculate the peak-hour factor; however, the data were not useful for SIDRA Intersection analysis because directional traffic counts (the number of left-turning, through-movement, and right-turning vehicles) were not available for each approach at an intersection. Therefore, manual traffic counts were conducted at all six intersections to obtain directional traffic counts, the percentage of trucks using the intersection, signal phasing, and signal timing.

The performance of the six signalized intersections was analyzed using the SIDRA Intersection software and was theoretically compared with a roundabout control for the same traffic-count data and percentage of trucks by using the SIDRA Intersection software; the results are presented in **Table 1** and **Table 2**. The procedure for conducting the performance analysis is explained and demonstrated in the following section. The results for the performance analysis are summarized in **Table 1** and **Table 2**. For all six intersections that were studied in Manhattan, the roundabout control proved to have better LOS, more effective intersection capacity, less control delay, fewer emissions, and a better annual intersection performance when compared to the signalized control.

Performance Comparison of Signalized Intersection Control with Roundabout Control Using the Sidra Intersection Software

The theoretical performance-comparison procedure for a signalized intersection with a roundabout control that utilizes the SIDRA Intersection software is demonstrated by using the McCall Road and Hays Drive signalized intersection in Manhattan, Kansas. Initially, the geometry of the signalized intersection was studied (on site and/or using Google maps), and an intersection was designed in the SIDRA Intersection software with details such as the number of lanes, the lane length, and the lane configuration for each intersection approach.

Part a of **Figure 1** shows the Google map for the McCall Road and Hays Drive intersection, and part b of **Figure 1** illustrates the intersection geometry developed with the SIDRA Intersection software. Similarly, signal

Table 1. Performance measures for the signalized control vs. roundabout control at intersections in small, medium, and large cities.

Intersection, truck %	Intersection control	LR*	Intersection LOS	EIC** (veh/hr)	Intersection's annual performance	
					Delay (veh-h/y)	Cost ($/y)
City of Manhattan, Kansas						
Fort Riley Blvd & Richards Rd, 5.7%	Signalized		C	2,916	9,040	432,088
	Roundabout	Yes	A	5,727	2,597	338,010
McCall Rd & Hays Drive, 8.5%	Signalized		C	1,711	7,548	319,484
	Roundabout	No	A	4,895	1,802	234,757
McCall Rd & Carlson Rd, 4.6%	Signalized		B	3,229	1,741	134,433
	Roundabout	No	A	6,531	829	117,662
Seth Childs Rd & Southwind, 4.3%	Signalized		D	3,948	19,891	713,632
	Roundabout	Yes	A	6,347	4,662	482,394
Tutle Creek Blvd & Kimball, 6.6%	Signalized		B	3,062	4,858	333,376
	Roundabout	Yes	B	2,919	3,061	292,332
Tutle Creek Blvd & McCall Rd, 8.7%	Signalized		B	3,088	4,563	327,388
	Roundabout	Yes	A	5,487	1,807	279,811
City of Topeka, Kansas						
SW Wanamaker & Huntoon, 5%	Signalized		C	4,084	15,919	716,747
	Roundabout	No	B	5,120	6,420	560,171
SW Wanamaker & SW Winding, 5%	Signalized		C	3,443	15,830	630,998
	Roundabout	No	B	4,659	6,122	474,474
SW Wanamaker & Westridge Mall, 5%	Signalized		B	3,918	3,853	349,141
	Roundabout	No	A	6,084	2,515	322,484
Wanamaker & 17th St, 5%	Signalized		C	3,966	12,654	560,257
	Roundabout	Yes	B	4,713	4,926	435,638
Wanamaker & 19th St, 5%	Signalized		C	2,671	11,234	492,154
	Roundabout	No	A	6,119	3,265	364,537
City of Overland Park, Kansas						
Antioch & 135, 2%	Signalized		E	4,814	51,766	1,391,678
	Roundabout	Yes	D	4,767	41,629	123,731
Metcalf & 75, 2%	Signalized		F	5,040	140,362	2,732,601
	Roundabout	No	F	4,272	115,808	2,361,954
Nall& 119, 2%	Signalized		F	4,256	123,712	2,500,024
	Roundabout	Yes	F	4,546	103,624	2,209,390
Nall& College Blvd, 2%	Signalized		E	6,308	71,590	1,846,954
	Roundabout	Yes	F	3,634	197,777	3,586,945
Roe & 119, 2%	Signalized		E	4,046	45,299	1,146,175
	Roundabout	Yes	B	4,745	9,795	639,893
Metcalf & 119, 2%	Signalized		E	4,976	48,330	1,338,656
	Roundabout	Yes	D	5,348	30,898	1,100,765

*LR = Approach number of lanes reduced for a roundabout, **EIC = Effective Intersection Capacity.

Table 2. Average control delay and emissions for a signalized control vs. roundabout control for small, medium, and large cities.

Intersection, truck %	Intersection control	Average control delay (seconds/vehicle)	Emissions (kg/h)			
			CO_2	Hydro carbons	CO	NO_x
City of Manhattan, Kansas						
Fort Riley Blvd & Richards Rd, 5.7%	Signalized	25.7	577.7	0.93	43.15	1.36
	Roundabout	7.4	509.8	0.78	39.39	1.25
McCall Rd & Hays Drive, 8.5%	Signalized	32	408.0	0.65	29.19	0.91
	Roundabout	7.6	351.1	0.52	26.46	0.84
McCall Rd & Carlson Rd, 4.6%	Signalized	13.7	191.5	0.31	14.56	0.46
	Roundabout	6.5	176.3	0.27	13.28	0.43
Seth Childs Rd & Southwind, 4.3%	Signalized	40.2	862.1	1.48	64.52	1.95
	Roundabout	9.4	704.6	1.12	55.76	1.73
Tutle Creek Blvd & Kimball, 6.6%	Signalized	17.2	479.6	0.78	38.73	1.18
	Roundabout	10.8	456.0	0.71	37.84	1.17
Tutle Creek Blvd & McCall Rd, 8.7%	Signalized	15.8	404.1	0.63	27.58	0.83
	Roundabout	6.3	367.3	0.54	24.22	0.75
City of Topeka, Kansas						
SW Wanamaker & Huntoon, 5%	Signalized	29.2	953.5	1.60	75.67	2.29
	Roundabout	11.8	847.3	1.35	69.39	2.14
SW Wanamaker & SW Winding, 5%	Signalized	33.7	786.6	1.38	61.54	1.84
	Roundabout	13	682.7	1.13	55.67	1.70
SW Wanamaker & Westridge Mall, 5%	Signalized	11.3	518.6	0.82	40.62	1.28
	Roundabout	7.4	492.3	0.76	38.38	1.23
Wanamaker & 17th St, 5%	Signalized	29.8	745.2	1.25	58.95	1.79
	Roundabout	11.6	662.1	1.06	54.66	1.68
Wanamaker & 19th St, 5%	Signalized	30.3	657.2	1.10	52.44	1.59
	Roundabout	8.8	556.5	0.87	44.65	1.40
City of Overland Park, Kansas						
Antioch & 135, 2%	Signalized	66.4	1,518.5	2.78	111.44	3.30
	Roundabout	53.4	1,440.4	2.58	111.82	3.29
Metcalf & 75, 2%	Signalized	158.5	2,398.3	4.71	150.61	4.38
	Roundabout	130.8	2,222.5	4.29	152.64	4.38
Nall& 119, 2%	Signalized	136.9	2,296.0	4.46	150.74	4.38
	Roundabout	114.7	2,149.1	4.10	151.47	4.44
Nall& College Blvd, 2%	Signalized	73.2	2,003.9	3.70	150.72	4.38
	Roundabout	202.4	3,053.8	6.12	195.26	5.49
Roe & 119, 2%	Signalized	74.7	1,209.5	2.23	85.55	2.56
	Roundabout	16.2	907.7	1.52	74.45	2.26
Metcalf & 119, 2%	Signalized	60.7	1,459.0	2.62	101.32	3.12
	Roundabout	38.8	1,370.6	2.41	109.33	3.24

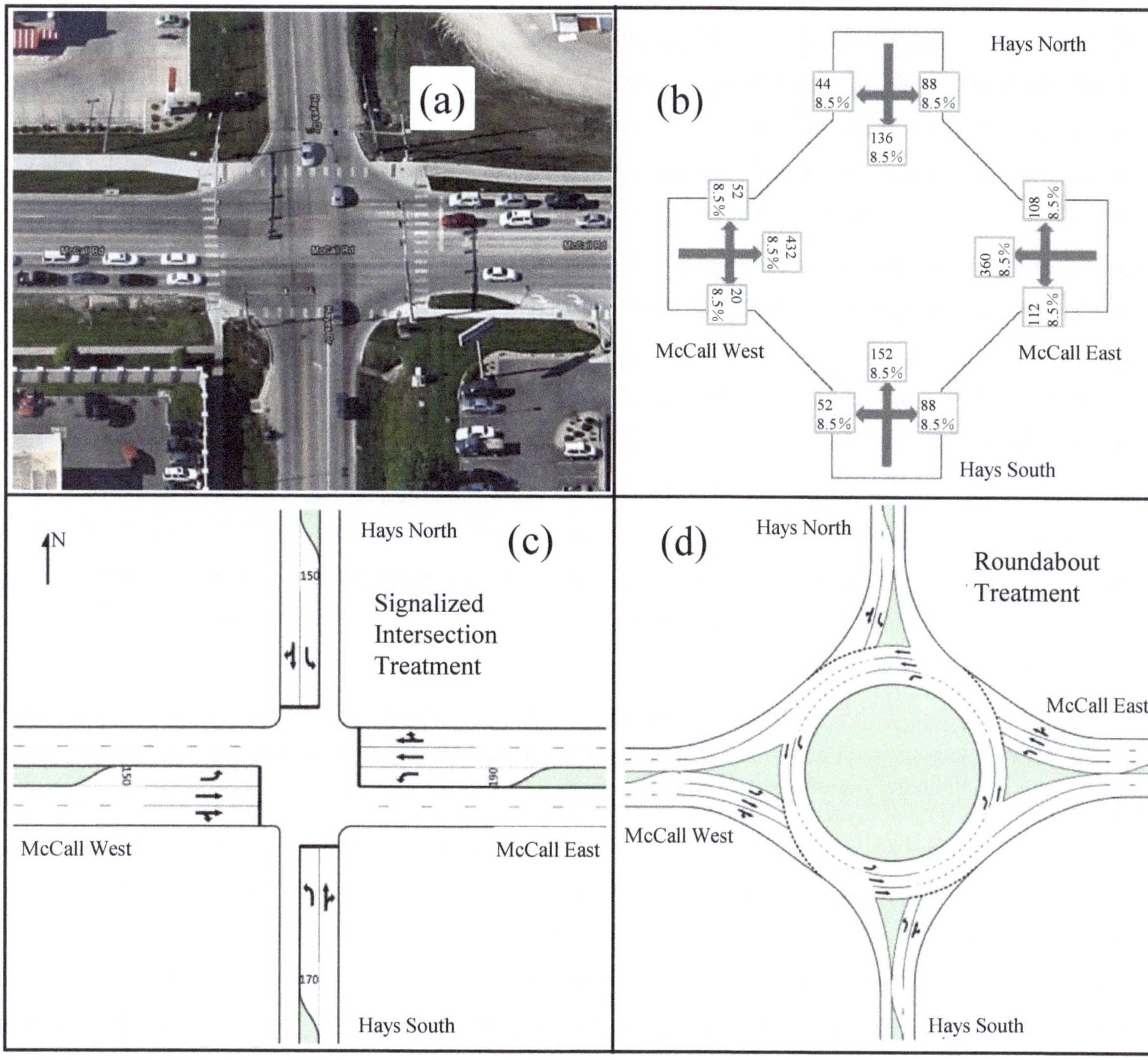

Figure 1. Analysis of the McCall road and Hays drive intersection using the SIDRA Intersection 5.1 software.

phasing and signal timing were designed for the signalized intersection with the SIDRA Intersection software. While the city traffic engineer had traffic-count data available for this intersection, the data were not useful (except for the peak-hour factor data) for the study due to lack of directional traffic counts. Therefore, manual data-counting techniques were used to determine the directional traffic counts and the percentage of trucks for all approaches. Intersection traffic counts and the percentage of trucks counted during one hour were used in the SIDRA Intersection software (illustrated in part c of **Figure 1**). Other input data that were used for the SIDRA Intersection software included peak-hour factor, the approach cruise speed (speed limit on site), the exiting cruise speed (speed limit on site), queue space for a normal vehicle (a standard of 25 ft was assumed), queue space for a heavy vehicle (truck; a standard of 73 ft was assumed), vehicle length for a normal vehicle (a standard of 17 ft was assumed), and vehicle length for a heavy vehicle (a standard of 73 ft was assumed). Using the above input data, the SIDRA Intersection software was used to analyze the intersection's LOS, the effective intersection capacity, the intersection's annual performance, the average control delay, and the vehicular emissions (CO, HC, CO_2, and NO_x) for the signalized intersection control at McCall Road and Hays Drive in Manhattan, Kansas. The findings are summarized in **Table 1** and **Table 2**.

The same intersection traffic-count data, percentage of trucks, and all other parameters (except the signal phasing and timing) mentioned above for the signalized-intersection analysis were used to theoretically construct a roundabout at the McCall Road and Hays Drive intersection using the SIDRA Intersection software. A standard center-island diameter of 100 ft, an entry radius of 100 ft, and an entry angle of 30 degrees were used to construct the roundabout in the SIDRA Intersection software. The number of circulating lanes was determined based on the number of left-turn and through-movement lanes for each approach. Also, roundabouts with more than three circulating lanes were not considered in this study because four-lane roundabouts had not been constructed in the United States and were not compatible with the SIDRA Intersection software. Therefore, even when one or more approaches for a signalized intersection had a total number of left-turn and through lanes that was more than three, a three-lane roundabout was designed theoretically, and the signalized intersection was compared to it. A three-lane roundabout (illustrated in part d of **Figure 1**) was generated using the SIDRA Intersection software for the McCall Road and Hays Drive intersection. The performance measures were calculated using the SIDRA Intersection software and are summarized in **Table 1** and **Table 2**.

It can be observed from **Table 1** that the LOS of the McCall Road and Hays Drive signalized intersection (LOS C) was improved by adapting a roundabout control (LOS A). Further, by adapting a roundabout control for the McCall Road and Hays Drive intersection, the average control delay was decreased by 76.3% (**Table 2**), the intersection's annual delay was decreased by 76.1% (**Table 1**), and the intersection's annual cost decreased by 26.5% (**Table 1**) when compared to a signalized control. Vehicle emissions, such as CO, HC, CO_2, and NO_x, were decreased by 13.9%, 20%, 9.4%, and 7.7%, respectively. Therefore, it can be understood that, at the McCall Road and Hays Drive intersection, which has significant truck traffic (8.5% of total traffic), a roundabout control instead of a signalized control can improve traffic flow (better LOS), reduce congestion (observed reduction in the delay), and enhance air quality in and around the communities (observed reduction in emissions) which can then facilitate optimum goods movement in Manhattan's urban freight-roadway network.

2.2. Medium City: Topeka, Kansas

Officials from the city of Topeka did not respond to the request for acquiring data about city freight-roadway networks, intersections with a high truck volume, or intersection traffic counts. Therefore, traffic counts for five of Topeka's major intersections (listed in **Table 1**) in were obtained from Mr. Dean Landman and Dr. Eugene Russell's [17] previous studies, and the intersection control at these five intersections was a signalized control. These five signalized intersections, with their traffic counts, were used for the study assuming that 5% of the total vehicles were trucks.

Intersection analysis was performed using the SIDRA Intersection software for all five selected intersections by testing the effectiveness of the signalized control vs. roundabout control, and the results are presented in **Table 1** and **Table 2**. It can be summarized from **Table 1** and **Table 2** that, for all the five intersections that were studied in Topeka, the roundabout control proved to have a better LOS, more effective intersection capacity, less control delay, fewer emissions, and better annual intersection performance when compared to the signalized control.

2.3. Big City: Overland Park, Kansas

The City of Overland Park had a freight roadway network that was already developed; therefore, the task of selecting intersections for this city was simple. Based on discussions with the city traffic engineer for Overland Park, Mr. Brian Shields, and based on the traffic data availability from the city for this study, six signalized intersections (listed in **Table 1**) were selected as optimum for studying the performance comparison between the signalized intersection control that was present and a roundabout control that was theoretically created using the SIDRA Intersection software. Directional traffic counts, signal phasing, and signal timing data were provided by the city traffic engineer's office. A 2% truck usage among the total vehicles utilizing these six intersections was suggested by the city traffic engineer for the performance-comparison computations.

Intersection analysis was performed using the SIDRA Intersection software for all six selected intersections in order to test the effectiveness of signalized intersection control and roundabout control; the results were presented in **Table 1** and **Table 2**. From the six selected intersections in Overland Park, it was observed that the current intersection geometry would require designing four-lane or five-lane roundabouts, which the authors considered to not be practical at this time in the United States. Therefore, for such situations, the roundabout design was downgraded to a three-lane roundabout, and the performance was measured and compared with signalized intersections. Among the six intersections, roundabout designs were downgraded for five intersections by designing a three-lane roundabout which might bring down the roundabout's benefit trend. While there have been intersection annual-cost benefits, and interaction annual-delay benefits observed for all six intersections with a roundabout control, average control delay benefits, emission-reduction benefits, and effective intersection-capacity benefits observed for less than six intersections with a roundabout control, the benefits were not significant and consistent among all the intersections considered. Further, the LOS for a roundabout control in five of the six intersections considered was either equal to or lower than the LOS observed for a signalized intersection control. One reason for the poor performance of roundabout control in this city can be attributed to the fact that five of the six selected intersections were designed with a smaller number of lanes than were originally present on site with the signalized intersection control. Control delays and vehicular emissions, such as CO, HC, CO_2, and NO_x, were used as performance measures to make conclusions about the most effective intersection control.

2.4. Statistical Analysis for Testing the Performance Measures for Signalized and Roundabout Intersection Control in Urban Freight-Roadway Networks

Average control delay and vehicular emissions, such as CO, HC, CO_2, and NO_x, were considered to be critical in the study to determine the optimum intersection control. A randomized complete-block design statistical analysis was conducted for the selected intersections in Manhattan, Topeka, and Overland Park in order to determine the desirable intersection control that can be used with urban freight-roadway networks. For this randomized complete-block design analysis, the experimental units are stratified into homogeneous blocks. Each intersection control type is randomly assigned to an equal number of experimental units for each block such that a precise comparison between the intersection control types can be made within the experimental units. **Figure 2** illustrates the design layout for intersections in Manhattan, Kansas. Each city's selected intersections were used as blocks, such that any difference in the response (average control delay or emissions) caused by a specific intersection control can be associated with the blocks. It was assumed that there is no interaction between the intersection control types and the intersections. The randomized complete-block design experiment was conducted multiple times with various response variables, such as average control delay, CO emissions, HC emissions, CO_2 emissions, and NO_x emissions.

Statistical Model for the Randomized Complete Blocks [18]: The linear model for this study can be written as

$$Y_{ij} = \mu + \tau_i + b_j + e_{ij}, \text{where } i = 1, \cdots, t \text{ and } j = 1, 2, \cdots, b$$

1) Y_{ij} is the observed response [for example: observed control delay (in seconds/vehicle)] for the ith intersection type with respect to the jth intersection.

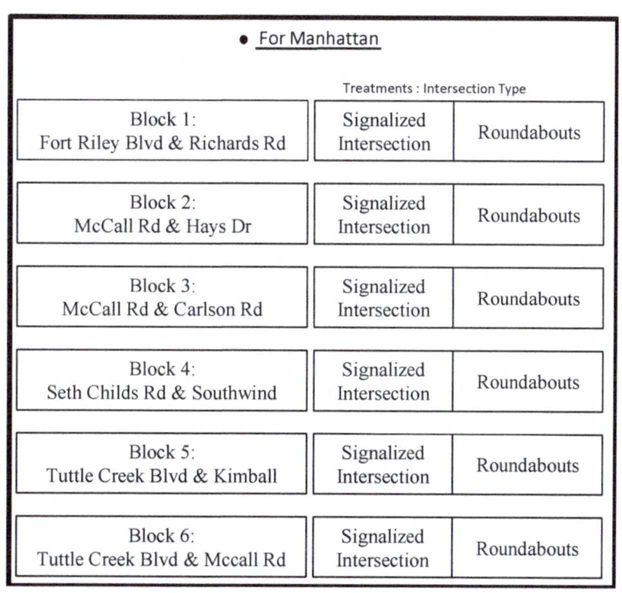

Figure 2. Randomized complete-block design layout for Manhattan intersections.

2) μ is the overall mean response (for example: overall mean control delay measured in seconds/vehicle).

3) τ_i is the fixed effect of ith intersection type on the response.

4) The random block effect, b_j, represents the effect of intersection j on the response, with mean 0 and variance σ^2_b.

5) E_{ij} is the experimental error which is independently and identically distributed (IID) as a normal distribution, with mean 0 and variance σ^2.

The results from the hypothesis testing are summarized in **Table 3**, and it was concluded that statistically significant results were observed for a reduction in the average control delay and vehicle emissions (CO, HC, CO_2, and NO_x) when theoretically implementing a roundabout control at intersections for freight roadway networks instead of having a signalized control for small and medium cities. Although of limited universal application, using a series of roundabouts on freight routes, or routes with a large number of trucks, is something the authors believe has never been studied and/or published; the local results show that the roundabouts in Kansas' small and medium cities' freight roadway networks can reduce the drivers' delay, thereby decreasing congestion and improving the traffic flow. It can also be concluded that the emissions were decreased by implementing a roundabout control for small and medium cities' urban, freight intersections, leading towards a better environment. While statistically significant results were observed for the average control delay reduction with a roundabout control instead of a signalized control in a big city, statistically significant results were not observed for vehicular emissions. It has to be understood that, as the city's size increases, the intersections' complexity increases for urban freight-roadway networks, necessitating the usage of multiple lanes for improved traffic management which is when roundabout installation might get challenging and might not be possible.

3. Conclusions and Recommendations

This study has analyzed integrating the greater use of roundabouts for freight roadway networks in and around urban areas in order to optimize goods movements as well as to decrease vehicle delays, congestion, and emissions, thus enhancing air quality in and around the communities. For this purpose, three different cities (Manhattan as a small city, Topeka as a medium-sized city, and Overland Park as a big city) in Kansas were considered to analyze the effectiveness of roundabout control for freight roadway networks when compared to a signalized control.

It was concluded that the roundabouts in Kansas' small and medium cities' freight roadway networks could reduce the delay for the trucks and other vehicles, could decrease congestion, and could contribute to improved traffic flow. It could also be concluded that emissions were reduced, decreasing air pollution, by implementing a roundabout control for small and medium cities' urban freight-route intersections. Statistically significant results were also observed for decreased average control delay and reduced vehicular emissions (CO, HC, CO_2, and NO_x) by implementing a roundabout control with freight roadway networks for selected intersections in small and medium-sized cities.

For selected intersections in Overland Park, the roundabout control did not yield a statistically significant result to conclude that they decreased emissions when compared to a signalized control. As discussed in the body of this paper, these results could be due to a small sample, the limitations with the number of lanes that are practical at this time in the United States, and the limitations for the software version used. Due to the design limitations for the large-city example used in this study, the results about the benefits of integrating roundabouts into freight roadway networks in large cities were inconclusive. It is recommended that additional, more comprehensive studies be conducted to verify the potential benefits.

Table 3. Results from the randomized complete-block design analysis.

Performance measure	Statistically significant reduction observed with roundabout control instead of a signalized control?		
	Manhattan	Topeka	Overland Park
Average control delay	Yes	Yes	Yes
CO_2 emissions	Yes	Yes	No
HC emissions	Yes	Yes	No
CO emissions	Yes	Yes	No
NO_x	Yes	Yes	No

It has to be understood that all intersections on an urban freight-roadway network may not be suitable for roundabout control; by implementing roundabout control at a few locations, the overall freight movement and emissions can be optimized. In this study, the city engineers have guided the selection of intersections which are suitable locations for a performance comparison of the current intersection control with a theoretical roundabout control. While the results obtained from the SIDRA Intersection software are reliable, it is recommended that the emission results obtained from the software output should be checked with a U.S. field study to validate the results.

Acknowledgements

The authors would like to acknowledge Mr. Brian Shields and Mr. Dean Landman for providing useful data for this study.

References

[1] Cambridge Systematics and Battelle Memorial Institute (2005) An Initial Assessment of Freight Bottlenecks on Highways, Federal Highway Administration.

[2] Office of Freight Management and Operations (2009) Freight Facts and Figures, Federal Highway Administration.

[3] Cambridge Systematics and Texas Transportation Institute (2004) Traffic Congestion and Reliability: Linking Solutions to Problems, Federal Highway Administration.

[4] Rodegerdts, L., Bansen, J., Tiesler, C., Knudsen, J., Myers, E., Johnson, M., Moule, M., Persaud, B., Lyon, C., Hallmark, S., Isebrands, H., Crown, R.B., Guichet, B. and O'Brien, A. (2010) Roundabouts: An Informational Guide—Second Edition. NCHRP Report 672, Transportation Research Board of the National Academics, Washington DC.

[5] Russell, E.R., Luttrell, G. and Rys, M. (2002) Roundabout Studies in Kansas. *4th Transportation Specialty Conference of the Canadian Society for Civil Engineering*, Canada.

[6] Russell, E.R., Mandavilli, S. and Rys, M. (2005) Operational Performance of Kansas Roundabouts: Phase II. Report No: K-TRAN: KSU-02-4.

[7] Gingrich, M. and Waddell, E. (2008) Accommodating Trucks in Single and Multilane Roundabouts. *Transportation Research Board, National Roundabout Conference*, Kansas City, 12-18 May 2008.

[8] Godavarthy, R.P. (2012) Network and Design Concepts for Accommodating Large Trucks at Roundabouts. Ph.D. Dissertation, Kansas State University.

[9] Godavarthy, R.P., Russell, E.R. and Landman, D. (2012) Accommodating Oversize/Overweight Vehicles at Roundabouts: Survey Results and Preliminary Designs. Research in Agricultural & Applied Economics.

[10] Godavarthy, R. and Russell, E. (2015) Low-Clearance Truck's Vertical Requirements at Roundabouts. *Journal of Transportation Technologies*, **5**, 214-222. http://dx.doi.org/10.4236/jtts.2015.54020

[11] Russell, E.R., Landman, D. and Godavarthy, R.P. (2013) A Study of Accommodating Oversize Overweight Vehicles (OSOW) at Roundabouts. *Transportation Research Board Annual Meeting*: *Compendium of Papers*, Washington DC.

[12] Russell, E.R., Landman, D. and Godavarthy, R.P. (2013) Accommodating Oversize Overweight Vehicles at Roundabouts. Report No. K-TRAN: KSU-10-1, Kansas Department of Transportation.

[13] Russell, E.R., Landman, D. and Godavarthy, R.P. (2013) Key Findings and Conclusions from the Study: Accommodating Oversize/Overweight Vehicles at Roundabouts. 2013 *Conference and Exhibition of the Transportation Association of Canada—Transportation*: *Better-Faster-Safer*.

[14] Mandavalli, S., Rys, M.J. and Russell, E.R. (2008) Environmental Impact of Modern Roundabouts. *International Journal of Industrial Ergonomics*, **38**, 135-142. http://dx.doi.org/10.1016/j.ergon.2006.11.003

[15] United States Environmental Protection Agency (2008) Idling Vehicle Emissions for Passenger Cars, Light-Duty Trucks, and Heavy Duty Trucks, EPA420-F-08-025.

[16] SIDRA INTERSECTION 5.1 Version Software (2012) SIDRA INTERSECTION User Guide. http://www.sidrasolutions.com/downloads/SIDRAINTERSECTION_UserGuide.pdf

[17] Russell, E.R., Landman, D. and Godavarthy, R.P. (2012) A Study of the Impact of Roundabouts on Traffic Flows and Business. Report No. K-TRAN: KSU-09-10, Kansas Department of Transportation.

[18] Kuehl, R.O. (1994) Statistical Principles of Research Design and Analysis. Duxbury Press.

23

Performance of Solar-Powered Air Conditioning System under AlMadinah AlMunawwarah Climatic Conditions

Khaled S. Al Qdah

Mechanical Engineering Department, Faculty of Engineering, Taibah University, Medina, KSA
Email: dr.khaled_qdah@yahoo.com

Abstract

Nowadays, solar-powered air conditioning has witnessed an increased progress because air conditioning system is almost a must in every building in Saudi Arabia where the outside temperature in summer is higher than 42°C. AlMadinah AlMunawwarah is one of the important holy places in the world, and therefore, this paper investigates the design and performance of solar-powered air conditioning system integrated with photovoltaic (PV) system which consists of PV panels, solar charger, inverter and batteries. This air conditioning system can be used in non-electrified areas near AlMadinah where the cost of electricity for this area is very high. The first step in this project is the cooling load calculations for the selected space that will be cooled and it was found about 1 ton refrigeration (3.52 kW). Based on this cooling load, the photovoltaic (PV) system has been estimated and built with the necessary connections. Data logging system has been used to measure the temperatures at the main components in the cycle like compressor inlet and outlet as well as evaporator and condenser. The input powers for the system as well as the coefficient of performance (COP) for the system under AlMadinah climatic conditions were measured along the day. The coefficient of performance varies from 2.16 to 4.22 for the system and these results compared with the conventional system performance with a good agreement.

Keywords

Performance, Solar Energy, COP, Remote Areas, Climatic Conditions

1. Introduction

Ten years ago energy consumption for cooling has increased dramatically in Saudi Arabia. The main reasons for this increasing energy demand for summer air-conditioning are the population growth, increased living standards and comfort demands as well as type of architecture; this increasing demand is one of the most significant costs to Saudi Arabia households during summer. AlMadinah AlMunawwarah one of the most holy places has a very sunny climate, with high demand for air conditioning on hot summer's days; the electricity grid increasingly faces the danger of overload due to air conditioner use, which would cause essential service disruption and severe economic impact. Solar air conditioning is one of the few solutions provided to overcome the summer high demand and addresses peak loading, and does so with reduced environmental impact at AlMadinah AlMunawwarah. Air conditioning systems in use are most often built around a vapor compression system driven by grid-electricity. In remote Saudi's villages, far from the grids, electric energy is usually supplied by diesel generators. In most of these cases, the supply with diesel fuel becomes highly expensive. Solar air conditioning might be a way to reduce the demand for electricity. Therefore, this work focuses in the design and construction of solar-powered air conditioning system integrated with photovoltaic (PV) system. According to the solar radiation map shown in **Figure 1** [1], it can be seen that solar energy is the best source to generate electricity in Saudi Arabia and specifically AlMadinah city where the annual intensity of solar radiation varies from 2200 to 2400 kWh/m^2, making it potential to consider the solar energy as the best option for power generation.

A lot of research is conducted for this purpose especially in countries where there is high availability of solar energy just like Saudi Arabia and the most previous works that conducted solar cooling and solar air conditioning are as follows.

Sayigh (1981) [2] performed that comparisons for a hot, arid climate like that of Riyadh, Saudi Arabia are made for four solar air conditioning systems. Mattarolo (1982) [3] conducted a general survey of solar-powered air conditioning systems, including descriptions of a continuous open cycle absorption cooler, and dehumidification-humidification systems. Solar-powered air conditioning system was designed, installed and operated in Singapore (Bong, 1987 [4]). Aly and Fathalah (1988) [5] presented a combined absorption-desiccant high performance air conditioning system suitable for hot humid areas. Feasibility of utilizing solar power for comfort cooling in Hong Kong has been studied; a solar-powered absorption air conditioning system was designed and successfully constructed on the campus of the University of Hong Kong (Yeung, 1992 [6]). Tsilingiris (1993) [7]

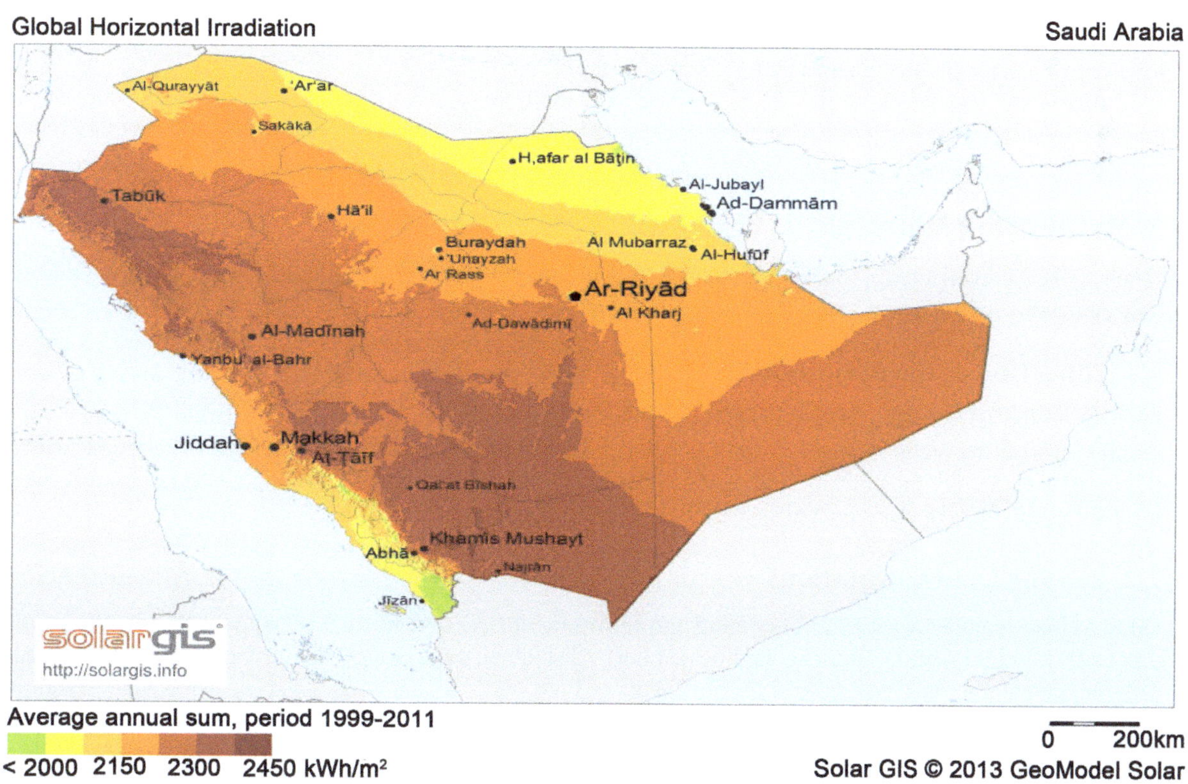

Figure 1. Annual solar radiation map for Saudi Arabia.

conducted the development of a theoretical microcomputer model suitable for performance predictions and investigation of the operational behavior of a simple solar cooling system for small residential applications. A hybrid open-cycle vapor absorption and liquid desiccant system using LiBr for the process of absorption and dehumidification has been simulated and the COP so obtained is about 50% higher than that of conventional vapor absorption (Khalid Ahmed *et al.*, 1997 [8]). Li and Sumathy (2000) [9] reviewed solar-powered air conditioning systems with the absorption air of lithium bromide and water. A number of attempts have been made by researchers to improve the performance of the solar applied air conditioning (chiller) subsystems. The feasibility of solar-powered absorption cooling technology under Tunisian conditions has been carried out (Moncef Balghouthi, 2005 [10]). Mittal *et al.* (2005) [11] presented a work that will help many researchers work in the area of air conditioning and provide them with fundamental knowledge on absorption systems. Sayigh (2007) [12] conducted the theoretical behavior of thermal parameters and their interaction in absorption cooling systems powered with solar energy. Constantinos *et al.* (2007) [13] described the main results of the EU project SACE (Solar Air Conditioning in Europe), aimed to assess the state-of-the-art, future needs and overall prospects of solar cooling in Europe. Solar-powered adsorption air conditioning system was designed and installed in the green building of Shanghai Research Institute of Building Science (Zhai *et al.*, 2008 [14]). Umberto *et al.* (2009) [15] analyzed the technical and economic feasibility of solar absorption cooling systems, designed for two different application fields: industrial refrigeration and air conditioning. Paula Naukkarinen (2009) [16] discussed the solar air conditioning and its role in alleviating the energy crisis of the Mediterranean hotels. Koroneos.*et al.* (2010) [17] investigated the use of thermal solar system in the medical centre applications. A solar hybrid desiccant air conditioning system has been configured, experimentally investigated and theoretically analyzed (Dong La *et al.*, 2011 [18]). The potential applications and advantages of powering solar air-conditioning systems using concentrator augmented solar collector (Dan *et al.*, 2012 [19]). Lavinia and Mario (2012) [20] investigated the solar heating and air-conditioning by GSHP coupled to PV system for a cost-effective high-energy performance building.

It can be seen that many researchers conducted solar cooling and solar-powered air conditioning systems but none of these conducted the potential of solar power air conditioner in Saudi Arabia especially AlMadinah region in spite of its climatic conditions, which is the motivation of this work. This project will focus on the design and construction of solar-powered air conditioning system integrated with photovoltaic (PV) system which is easy to use in AlMadinah AlMunawwarah and the non-electrified remote areas outside AlMadinah where the intensity of solar radiation reaches to 7.5 kWh/m^2 a day.

In this work, the design of the solar-powered air conditioning system is described in detail and its performance over an entire cooling season will be investigated by utilizing the PV panels to generate necessary power for small-scale air conditioning system about 1 TR and compare the system performance with the conventional system.

2. Method and Materials

Many tools are available for sizing the PV system with battery storage. The steps below summarize the methodology for this investigation.

2.1. Collection of the Required Meteorological Data

Such as design temperatures, humidity, tilt angle, daily and annual values of solar radiation for the site and the number of working hours. The space that has been chosen to be cooled in this work is the office room with 4 × 4 × 3 m used by faculty member located inside AlMadinah and the first step to size the air conditioner is the cooling load estimation for this selected space.

2.2. Cooling Load Calculations

Cooling load includes what kind of cooling and how much of cooling needed. The space that will be cooled in this work for eight working hours a day located near AlMadinah city. Hourly analysis program (HAP) used, this program calculating the cooling load for any location. First of all the site must be specified as well design inside and outside design temperature and relative humidity, thickness and type of insulation materials used and the building structure. The design temperature used are 45˚C as a maximum temperature during summer and the comfort temperature or the inside temperature is 23˚C and the relative humidity of 50% [21]. Finally the output of the program will be displayed for each month. **Figure 2** displays the maximum and minimum cooling load for

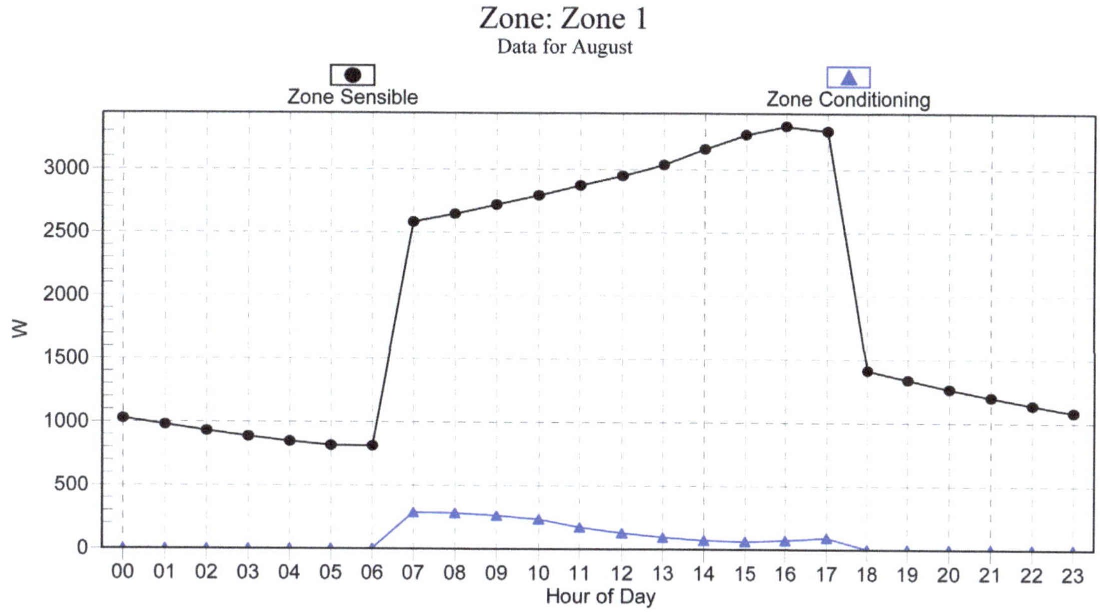

Figure 2. Maximum cooling load during August.

this space. It was found that the maximum cooling load during August it is about 3.4 kW where as the minimum cooling load during January 2.4 kW.

2.3. Design and Sizing of the Air Conditioning System

Based on cooling load calculations, the air conditioning unit has been selected with the following specification: unit capacity 3.52 kW or 1 ton refrigeration (1 TR) split unit, 220 V, input power 1.250 - 1.374 kW to provide the suitable COP.

2.4. PV System Sizing and Material Selection

Based on air conditioning unit capacity that has been selected under AlMadinah climatic conditions and the data collected, where the intensity of solar radiation about 7.5 kWh/m^2, the PV solar-powered system specifications can be selected and design.

A photovoltaic system is an array of components designed to supply usable electric power for a variety of purposes [22]. The sun delvers its energy to us in two main forms, heat and light. There are two main types of solar power systems, namely, solar thermal systems that convert heat to electricity, and solar PV systems that convert sunlight directly into usable direct current (DC) electricity. One or more DC to alternating current (AC) power converters which called inverters. PV cells are made from layers of semi-conducting material, usually silicon. When light shines on the cell it creates an electric field across the layers. The stronger the sunshine, the more electricity is produced. Groups of cells are mounted together in panels or modules that can be mounted on your roof [22]. The peak sun hour is essential in order to know the number of PV modules to be installed. Before doing so, the power that can be assumed generated by the PV modules must be determined based on solar irradiance of the location.

The function of charger is to regulate the voltage and current coming from the solar panel going to the battery. The battery is the key components in PV-SA systems as it act as energy back-up for the renewable energy systems. It also functions as storage devices for storing PV generated electricity during cloudy days and at night. In order to apply this system in AC load, the inverter is needed to convert the DC electricity generated by the PV panel into AC. The AC load is a common type of load and easily available with cheaper in price.

As we mentioned previously, the purpose of charge controller is to regulate the current from the PV module to prevent the batteries from overcharging. A charge controller is used to sense when the batteries are fully charged and to stop, or decrease, the amount of current flowing to the battery [23]. The solar energy is received by the PV module and transform into electrical energy. The electrical energy is then being regulated by charge controller either by supplies it directly into the load or charges the batteries. As the electrical energy coming from the PV module is in DC, inverter will convert it into AC as the compressor needs AC to operate. The electricity provided by the panel array and battery is DC at a fixed voltage. The voltage provided might not match what is required by the load. A direct/alternating (DC/AC) converter, known as inverter, converts the DC current from batteries into AC.

The charge controller specification are, 12 V, rate charging control 24 Amp.

A voltage regulator is designed to automatically maintain a constant voltage level

$$COP_{cooling} = \frac{\text{Desiredoutput}}{\text{InputPower}} \quad [24]$$

$$W = \frac{3520\text{ W}}{2.52} = 1397\text{ W}$$

where the acceptable COP for the system recommended by the supplier should be 2.52.

Input power range: 1250 to 1397 W.

We need 1397 W with 8 working hour per day

$$\text{work} \times \text{working hours in day} = \text{work hours perday}$$

$$13978 \times 8 = 11176\text{ W} \cdot \text{hr/day}$$

Minimum hour for sunlight per day = 6 hr/day

$$\text{Total load capacity } = \frac{\text{work per day}}{\text{minimum hour for sunlight per day}}$$

$$\text{Total load capacity } = \frac{11176\text{ W} \cdot \text{hr/day}}{6\text{ hr/day}} = 1863\text{ W} \; [24]$$

We need photovoltaic solar system with capacity 1863 W.

The available size of panel with the standard 310 W output.

$$\text{Number of panels} = \frac{\text{Total Capacity}}{\text{The output for 1 panel}} = \frac{1863\text{ W}}{310\text{ W}} = 6\text{ Panels}$$

therefore we need 6 panel at 310 W each. Assuming drop voltage 30% [23]

$$150 \times 30\% = 45\text{ A} \cdot \text{h}$$

The minimum efficiency of battery 90%.

The actual amperes of battery is $(90\% \times 150) - 45 = 90\text{ A} \cdot \text{h}$.

The actual capacity of one battery: 90 A·h, working hours: 8 h.

The actual Amp. $\frac{90\text{ A} \cdot \text{h}}{8\text{ h}} = 11.25\text{ A}$.

The running current of air conditioning unit is 5.68 A - 7.6 A with starting current of 22.8 Amp.

Number of battery $\frac{22.8\text{ A}}{11.25\text{ A}} = 2\text{ batterys : for 8 hour.}$

Number of battery's for two days $2 \times \frac{48\text{ h}}{8\text{ h}} = 12\text{ batteries.}$

With total power stored 1863 W, 12 Volt and 150 Ah.

The inverter specification are: DC input 48 V, AC output 220 - 240 V, output power 5 kW and the charge controller 12 Volts each and charge rating 24 Ampers with over load and short circuit protection.

2.5. PV Installation and Tilt Angle Selection

In order to make the system work efficiently, the inclination angle for the panels (Tilt angle) should be carefully selected. The optimum tilt varied from month to another and the collected solar energy depends on the optimum panel tilt for each month. It was found that the average optimum tilt angle at Madinah for the winter months is 37° and for the summer months is 12°. So, the yearly average tilt panel is 23.5° which nearly corresponding to the latitude of Madinah site 24.5° [24].

2.6. PV System Installation and Connections

After the material selection for PV panels, batteries, charge controller and inverter then using suitable basement, cabinets to protect the system, the frame for the panels designed at an optimum tilt angle, the system installed with the help of technical as shown in **Figure 3** which explain the actual final cycle that includes the air conditioning unit and power supply system that can be used at any time; the system has been tested for 8 hours per day with a full capacity.

3. Results and Discussion

The most common type of air conditioning is technically referred to the vapor-compression refrigeration system. The operation of the air conditioning system starts when the refrigerant flows across the evaporator inside the space to absorb heat. The refrigerant that went into the evaporator leaves as vapor. Then, the low pressure and cool vapor is taken outside and compressed by the compressor to become a hot, high pressure gas. Compressor is electrically operated can be described as the heart of air conditioning system as it pump refrigerant throughout the system. The main function of a compressor is to compress refrigerant vapor to a high pressure, making it hot for the circulation process. Next, the hot vapor pass through the condenser and gives off some of its heat as outdoor air is blown across the condenser coil. The warm liquid is carried back to the evaporator to repeat the cycle again.

The system performance measured by measuring the temperature at each previous component in the cycle by using data logging system connected with thermocouples, these data transfer through the USB to the computer for analysis where a lot of readings carried out through a specific time. The system worked between 9 am to 17 pm usually for 8 hours a day, Extensive measurement were taken through the two months ago.

To find the coefficient of performance which represents the system efficiency, the main properties for the working fluid which are; temperature, pressure, enthalpy and entropy must be reported, the refrigerant or the working fluid in this cycle is R22. Suction and discharge pressure for the system measured as well as mass flow rate in the cycle. **Figure 4** represents the refrigeration cycle on the temperature-entropy (T-S) diagram [25].

From this figure the points from 1 - 4 can be described as follow:

Point4: T_4: Temperature at the inlet of the compressor where the liquid enters as dry saturated vapor.

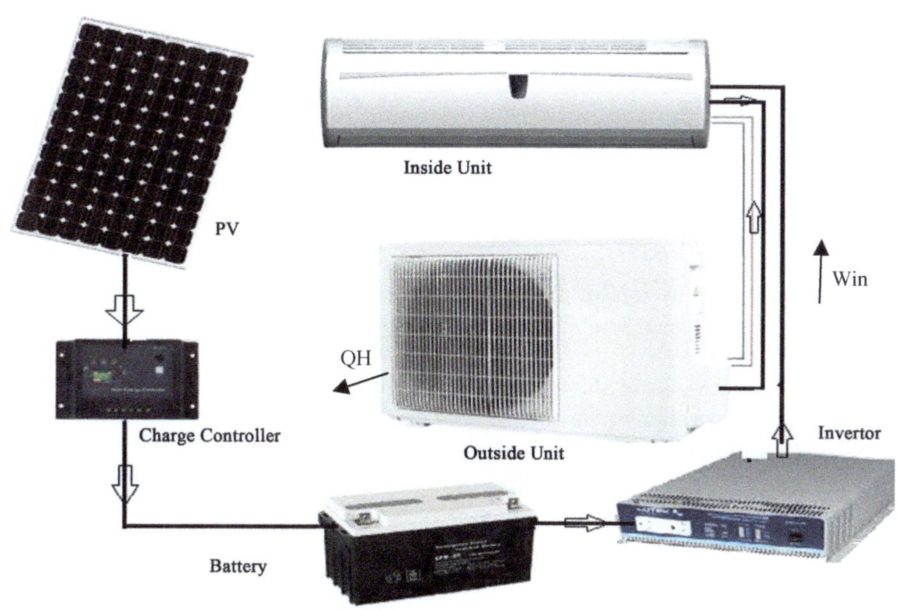

Figure 3. Final actual system cycle.

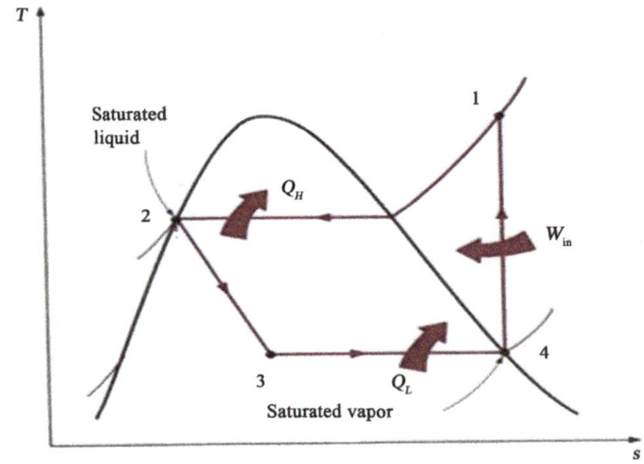

Figure 4. Refrigeration cycle.

T_1: Temperature at compressor outlet where the refrigerant leaves the compressor as superheated vapor.

T_2: Temperature of condenser where the refrigerant leaves as saturated liquid.

T_3: Temperature of evaporator where evaporation takes place at constant temperature.

The quality x for point 4 is one and for point 2 is zero

$$x_4 = 1 \text{ and } x_2 = 0 \quad \dot{m} \approx 0.03 \text{ Kg/s}$$

$h_2 = h_3$ where h is the enthalpy (kJ/kg) and the pressure $P_2 = P_1$.

From the tables of refrigerant (R22) properties available online [26], the enthalpy can be found directly at specified temperature and pressure. The cooling load or refrigeration capacity $QL = \dot{m}(h_4 - h_3)$.

The input power or work input is W or $h_1 - h_4$ and the direction shown in **Figure 5**.

$$\text{COP} = \frac{h_4 - h_3}{h_1 - h_4} [24], \text{COP} = \frac{\dot{m}(h_4 - h_3)}{p} [24]$$

Figure 5 represents the variation of COP with time during last May, it can be seen that the higher value of COP measured in the morning where the outside temperature still low and the system is fully charged. The same trend can be seen in **Figure 6** for the system performance through June where the summer takes place and the outside temperatures higher than 40˚C. The COP is a function of compressor input work and the desired output, the higher values indicated that the cooling capacity is at its optimum values during the first hours then the compressor needs more work which means lower COP. From these figures the COP can be considered with its normal ranges for these types of air conditioning system powered by conventional power systems but the PV system is clean, safe and reliable system in spite of the high cost.

During the last two months, the system performance examined under variable intervals of time and the temperatures at each point in the cycle were measured by the thermo couples KT-type connected to Omega data logging system. **Figure 7** display the variation of temperature for each component during last may in the morning where the temperature outside temperature in this day reaches 39˚C. The compressor outlet temperature which is the highest temperature in the cycle varied between 40˚C to 64˚C at constant value of condenser temperature 23˚C. Generally, the inlet of the compressor and exit of evaporator temperature usually closed to each other which mean normal operation to the air conditioning system.

Figure 8 shows the variation of temperatures with time in the afternoon from 1 pm to 5 pm at the beginning of June where the outside temperature 42˚C, the compressor usually switch off by the control system when in the temperature of the space 23˚C, small deviation between the inlet of compressor and evaporator temperature during this test which is due to outside temperature variation and the continuous door opening for the room from the occupants.

Figure 9 display the variation of temperature during June from 12 to 3:00 pm, Normal operation for the system but the system COP seemed to be lower than that for the optimum value because during that day the

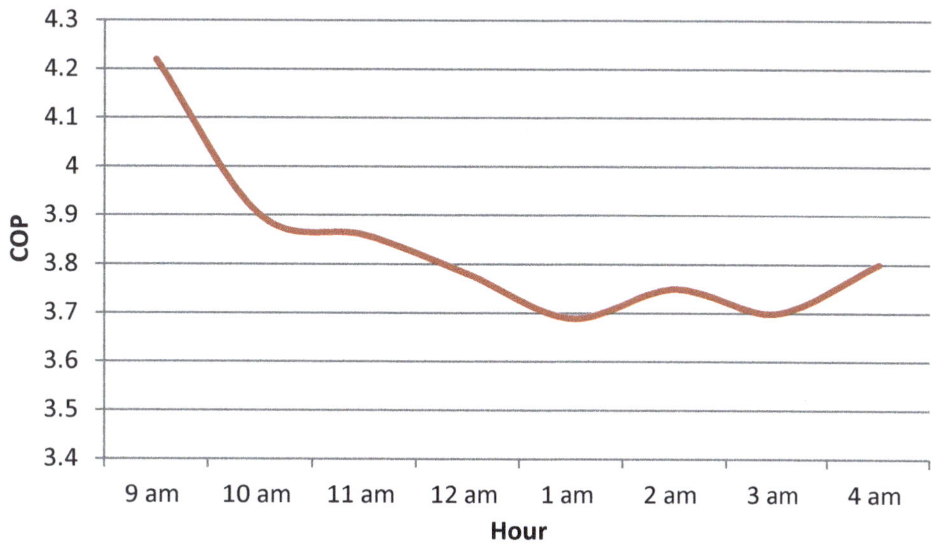

Figure 5. Coefficient of performance during May.

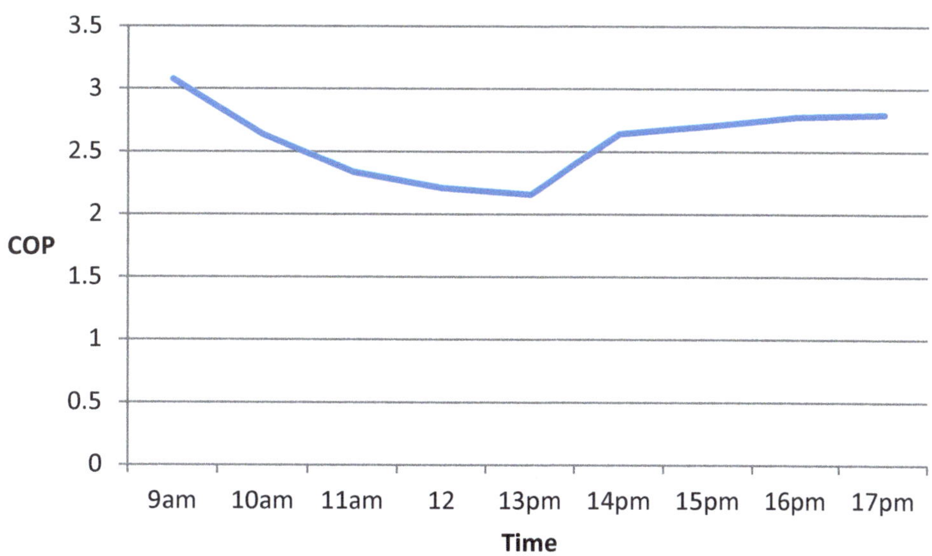

Figure 6. Coefficient of performance during June.

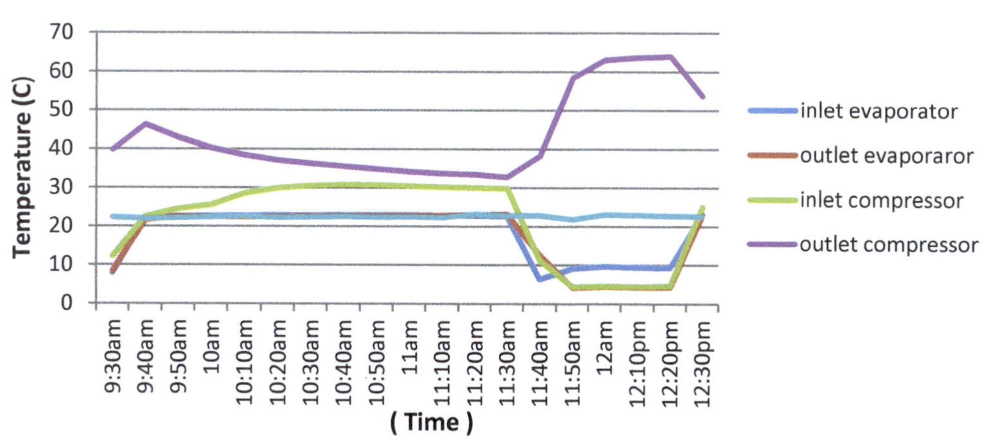

Figure 7. Variation of cycle temperature from 9:30 am to 12:30 pm during May.

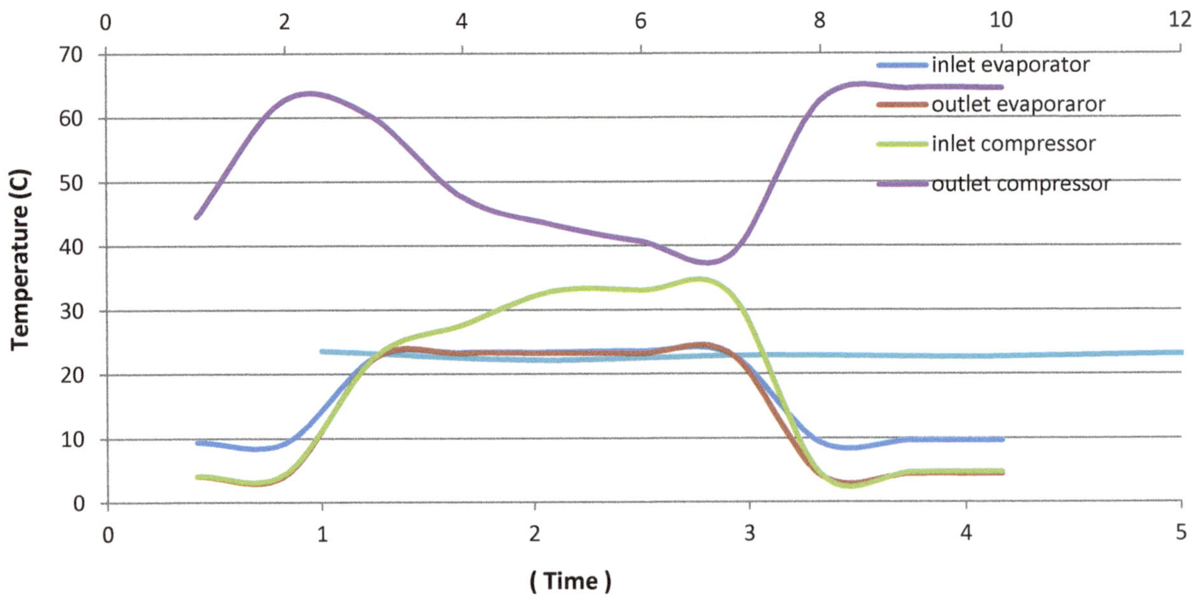

Figure 8. Temperatutre variations from 1:00 pm to 5:00 pm.

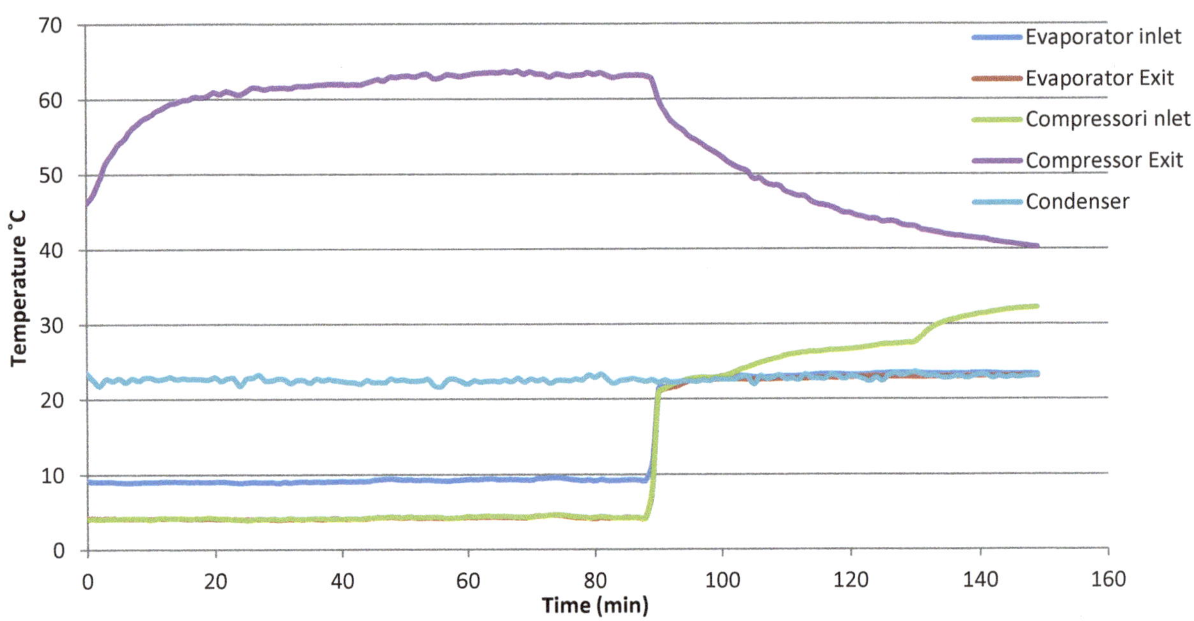

Figure 9. Temperature variations from 12:30 pm to 3:00 pm.

weather was stormy and the dust cover the pV panels and the input power affected by this action. It is necessary to clean and wash the PV panels if we want to get a higher output from these pannels. From the results obtained using solar energy as a power source it can be said that solar air-conditioning might be a way to reduce the demand for electricity. Therefore, this work focuses in the design and construction of solar-powered air conditioning system integrated with photovoltaic (PV) system.

Solar air conditioning at an economically competitive level could reduce electricity costs for residential and small buildings and this would cut the growth of peak electric demand.

About 70% of the power used in Saudi Arabia in the summertime goes into air conditioning (according to Saudi Electricity Company). The approximate cost for the system including installation about 46,000 SR and this cost can be considered high when we compare it with electricity price in Saudi Arabia. Recent price drops of PV panels however have changed this system can be utilized efficiently in remote areas where the electricity generation becomes more cost. In spite of the cost, solar energy as a power source to the air conditioning system is one of the few solutions that provides to overcome the summer high demand and addresses peak loading. On hot summer's days, the electricity grid increasingly faces the danger of overload due to air conditioner use in addition to its advantages as a clean and safe energy. The air conditioning system can be operated on solar and can be used in non-electrified areas. As we all known, solar energy is cost effective, renewable and environmentally friendly.

4. Conclusions

This work focuses on design, construction and testing the performance of solar-powered air conditioning system integrated with photovoltaic (PV) system and applying it under AlMadinah AlMunawwarah climatic conditions. The project comes as a solution to reduce the electric demand especially during the summer and can be used efficiently in remote area where the electricity is not available. It was designed by two different methods: one of them is preferred upon the other because of the acceptable range of the voltage and the easy connection. The COP for the system is found to be with a good agreement with conventional system

Also, it is extremely difficult to keep the performance of the system components to be constant over entire experimental period as the components deteriorate with time.

As the rate of air-conditioned demand is expected to grow, the exploitation of solar energy, especially in Gulf countries, like Saudi Arabia, during the hot season, seems to be a valuable option to mitigate the consumption of conventional fuels due to the cooling requirements. Solar air conditioning systems can be a reasonable alternative to conventional air conditioning systems. There are several characteristics that must be considered to know either on the PV system or on the air conditioning system such as electrical equivalent, characteristic curve, and factors affect the output of PV cells.

Solar energy as a power source can reduce peak energy demand and increase the use of renewable energy, while providing energy savings to the end user.

Acknowledgements

The author would like to express deep thanks and gratitude to the Deanship of Scientific Research at Taibah University for the technical and financial supports and encouragement.

References

[1] http://solargis.info/doc/free-solar-radiation-maps-GHI

[2] Sayigh, A.A.M. (1981) Solar Air Conditioning in a Hot Arid Climate. *Proceedings of the Second Miami International Conference*, Miami Beach, 10-13 December 1979, 761-773.

[3] Mattarolo, L. (1982) Solar Powered Air Conditioning Systems: A General Survey. *International Journal of Refrigeration*, **5**, 371-379. http://dx.doi.org/10.1016/0140-7007(82)90059-7

[4] Bong, T.Y, Ng, K.C. and Tay, A.O. (1987) Performance Study of a Solar-Powered Air-Conditioning System. *Solar Energy*, **39**, 173-182. http://dx.doi.org/10.1016/S0038-092X(87)80025-7

[5] Aly, S.E. and Fathalah, K.A. (1988) Combined Absorption-Desiccant Solar Powered Air Conditioning System. *Wärme- und Stoffübertragung*, **23**, 111-121. http://dx.doi.org/10.1007/BF01637133

[6] Yeung, M.R., Yuen, P.K., Dunn, A. and Cornish, L.S. (1992) Performance of a Solar-Powered Air Conditioning System in Hong Kong. *Solar Energy*, **48**, 309-319. http://dx.doi.org/10.1016/0038-092X(92)90059-J

[7] Tsilingiris, P.T. (1993) Theoretical Modelling of a Solar Air Conditioning System for Domestic Applications. *Energy Conversion and Management*, **34**, 523-531. http://dx.doi.org/10.1016/0196-8904(93)90143-X

[8] Khalid Ahmed, C.S., Gandhidasan, P. and Al-Farayedhi, A.A. (1997) Simulation of a Hybrid Liquid Desiccant Based Air-Conditioning System. *Applied Thermal Engineering*, **17**, 125-134. http://dx.doi.org/10.1016/S1359-4311(96)00025-7

[9] Li, Z.F. and Sumathy, K. (2000) Technology Development in the Solar Absorption Air-Conditioning Systems. *Renewable and Sustainable Energy Reviews*, **4**, 267-293. http://dx.doi.org/10.1016/S1364-0321(99)00016-7

[10] Moncef, B., Chahbani, M.H. and Amenallah, G. (2005) Solar Powered Air Conditioning as a Solution to Reduce Environmental Pollution in Tunisia. *Desalination*, **185**, 105-110. http://dx.doi.org/10.1016/j.desal.2005.03.073

[11] Mittal, V., Kasana, K.S. and Thakur, N.S. (2005) The Study of Solar Absorption Air-Conditioning Systems. *Journal of*

Energy in Southern Africa, **16**, 59-66.

[12] Sayegh, M.A. (2007) The Solar Contribution to Air Conditioning Systems for Residential Buildings. *Desalination*, **209**, 171-176. http://dx.doi.org/10.1016/j.desal.2007.04.038

[13] Constantinos, A., Gershon, G., Hans-Martin, H., Carlos, A., Infante, F., Erich, P., Wang, L. and Edo, W. (2007) Solar Air Conditioning in Europe—An Overview. *Renewable and Sustainable Energy Reviews*, **11**, 299-314. http://dx.doi.org/10.1016/j.rser.2005.02.003

[14] Zhai, X.Q., Wang, R.Z., Wu, J.Y., Dai, Y.J. and Ma, Q. (2008) Design and Performance of a Solar-Powered Air-Conditioning System in a Green Building. *Applied Energy*, **85**, 297-311. http://dx.doi.org/10.1016/j.apenergy.2007.07.016

[15] Umberto, D., Stefania, P. and Paolo, S. (2009) Solar-Powered Cooling Systems: Technical and Economic Analysis on Industrial Refrigeration and Air-Conditioning Applications. *Applied Energy*, **86**, 1376-1386. http://dx.doi.org/10.1016/j.apenergy.2009.01.011

[16] Naukkarinen, P. (2009) Solar Air Conditioning and Its Role in Alleviating the Energy Crisis of the Mediterranean Hotels. *International Journal of Sustainable Energy*, **28**, 93-100.

[17] Koroneos, C., Nanaki, E. and Xydis, G. (2015) Solar Air Conditioning Systems and Their Applicability: An Exergy Approach. *Resources, Conservation and Recycling*, **55**, 74-82. http://dx.doi.org/10.1016/j.resconrec.2010.07.005

[18] La, D., Dai, Y.J., Li, Y., Ge, T.S. and Wang, R.Z. (2011) Case Study and Theoretical Analysis of a Solar Driven Two-Stage Rotary Desiccant Cooling System Assisted by Vapor Compression Air-Conditioning. *Solar Energy*, **85**, 2997-3009. http://dx.doi.org/10.1016/j.solener.2011.08.039

[19] Nkwetta, D.N. and Smyth, M. (2012) The Potential Applications and Advantages of Powering Solar Air-Conditioning Systems Using Concentrator Augmented Solar Collectors. *Applied Energy*, **89**, 380-386. http://dx.doi.org/10.1016/j.apenergy.2011.07.050

[20] Lavinia, C.T., Mario, M. and Claudio, D.P. (2012) Solar Heating and Air-Conditioning by GSHP Coupled to PV System for a Cost Effective High Energy Performance Building. *Energy Procedia*, **30**, 683-692. http://dx.doi.org/10.1016/j.egypro.2012.11.078

[21] McDowall, R. (2007) Fundamentals of HVAC Systems. American Society of Heating, Refrigerating and Air-Conditioning Engineers, Inc. and Elsevier Inc., Atlanta.

[22] Saad, S.S., Daut, I., Misrun, M.I., Champakeow, S. and Ahmad, N.S. (2010) Study of Photovoltaic and Inverter Characteristics. Universiti Malaysia Perlis (UniMAP), Malaysia.

[23] Shaari, S., Omar, A.M., Haris, A.H., Sulaiman, S.I. and Muhammad, K.S. (2009) Solar Photovoltaic Power: Design and Installation of Stand-Alone Systems. Pusat Tenaga Malaysia, 46-52, 119-129.

[24] Benghanem, M. (2011) Optimization of Tilt Angle for Solar Panel: Case Study for Madinah, Saudi Arabia. *Applied Energy*, **88**, 1427-1433.

[25] Cengel, Y.A. and Boles, M.A. (2015) Thermodynamics: An Engineering Approach. 8th Edition, McGraw Hill, New York.

[26] http://www.engineeringtoolbox.com/r22-properties-d_365.html

Permissions

All chapters in this book were first published in SGRE, by Scientific Research Publishing; hereby published with permission under the Creative Commons Attribution License or equivalent. Every chapter published in this book has been scrutinized by our experts. Their significance has been extensively debated. The topics covered herein carry significant findings which will fuel the growth of the discipline. They may even be implemented as practical applications or may be referred to as a beginning point for another development.

The contributors of this book come from diverse backgrounds, making this book a truly international effort. This book will bring forth new frontiers with its revolutionizing research information and detailed analysis of the nascent developments around the world.

We would like to thank all the contributing authors for lending their expertise to make the book truly unique. They have played a crucial role in the development of this book. Without their invaluable contributions this book wouldn't have been possible. They have made vital efforts to compile up to date information on the varied aspects of this subject to make this book a valuable addition to the collection of many professionals and students.

This book was conceptualized with the vision of imparting up-to-date information and advanced data in this field. To ensure the same, a matchless editorial board was set up. Every individual on the board went through rigorous rounds of assessment to prove their worth. After which they invested a large part of their time researching and compiling the most relevant data for our readers.

The editorial board has been involved in producing this book since its inception. They have spent rigorous hours researching and exploring the diverse topics which have resulted in the successful publishing of this book. They have passed on their knowledge of decades through this book. To expedite this challenging task, the publisher supported the team at every step. A small team of assistant editors was also appointed to further simplify the editing procedure and attain best results for the readers.

Apart from the editorial board, the designing team has also invested a significant amount of their time in understanding the subject and creating the most relevant covers. They scrutinized every image to scout for the most suitable representation of the subject and create an appropriate cover for the book.

The publishing team has been an ardent support to the editorial, designing and production team. Their endless efforts to recruit the best for this project, has resulted in the accomplishment of this book. They are a veteran in the field of academics and their pool of knowledge is as vast as their experience in printing. Their expertise and guidance has proved useful at every step. Their uncompromising quality standards have made this book an exceptional effort. Their encouragement from time to time has been an inspiration for everyone.

The publisher and the editorial board hope that this book will prove to be a valuable piece of knowledge for researchers, students, practitioners and scholars across the globe.

List of Contributors

June M. Lukuyu and Judith B. Cardell
Picker Engineering Program, Smith College, Northampton, USA

T. Harinarayana
Gujarat Energy Research & Management Institute (GERMI), Gandhinagar, India

K. Jaya Kashyap
Malla Reddy Engineering College, Hyderabad, India

Jianxiong Qin and Pei Zhang
College of Tourism Historical Culture, Southwest University for Nationalities, Chengdu, China

Guiping Deng
Science Department, Jiuzhaigou National Nature Reserve Administration, Jiuzhai, China

Lu Chen
Institude of Geographic Sciences and Natural Resources Research, CAS, Beijing, China

Emodi Nnaemeka Vincent and Samson D. Yusuf
Department of Technology Management Economics and Policy Program, College of Engineering, Seoul National University, Seoul, South Korea

Kshitij Tiwari
Department of Electrical Engineering, Indian Institute of Technology, Kanpur, India
Gujarat Energy Research and Management Institute, Gandhinagar, India

Tirumalachetty Harinarayana
Gujarat Energy Research and Management Institute, Gandhinagar, India

Yasushi Iwasaki and Kayoko Yamamoto
Graduate School of Information Systems, University of Electro-Communications, Tokyo, Japan

Razzaqul Ahshan, Mohammad Tariq Iqbal, George K. I. Mann and John E. Quaicoe
Faculty of Engineering and Applied Science, Memorial University of Newfoundland, St. John's, Canada

Razzaqul Ahshan, Mohammad Tariq Iqbal, George K. I. Mann and John E. Quaicoe
Faculty of Engineering and Applied Science, Memorial University of Newfoundland, St. John's, Canada

Mohammed El Alami and Mohamed Habibi
Laboratory of Systems and Telecommunications Engineering Decision, Sciences Faculty, Ibn tofail University, Kenitra, Morocco

Seddik Bri
Electrical Engineering Department, High School of Technology, Moulay Ismail University, Meknes, Morocco

Woo-Kyu Chae, Hak-Ju Lee, Sung-Wook Hwang and Il-Keun Song
Research Institute, Korea Electric Power Corporation, Daejeon-si, Korea

Jae-Eon Kim
Chungbuk National University, Chungbuk, Korea

Alo Mikola, Teet-Andrus Koiv and Hendrik Voll
Department of Environmental Engineering, Tallinn University of Technology, Tallinn, Estonia

Meshack Hawi and Hiram Ndiritu
Department of Mechanical Engineering, JKUAT, Nairobi, Kenya

Robert Kiplimo
Department of Marine Engineering, JKUAT, Nairobi, Kenya

M. L. Elhafyani, S. Zouggar and A. Aziz
Laboratory of Electrical Engineering and Maintenance (LEEM), University Mohammed 1, Oujda, Morocco

M. Benkaddour
Department of Physics, Faculty of Science, University Mohammed 1, Oujda, Morocco

Ghullam Mustafa Bhutto
Department of Energy Technology, Aalborg University, Copenhagen, Denmark
Department of Electrical Engineering, Quaid-e-Awam University of Engineering, Science & Technology, Nawabshah, Pakistan

Ehsan Ali
Department of Electrical Engineering, Quaid-e-Awam University of Engineering, Science & Technology, Nawabshah, Pakistan
IM2NP, Aix-Marseille University, Marseille, France

Jagdesh Kumar and Muhammad Akram Bhayo
Department of Electrical Engineering, Quaid-e-Awam University of Engineering, Science & Technology, Nawabshah, Pakistan

Vincent Nnaemeka Emodi, Samson D. Yusuf and Kyun-Jin Boo
Department of Technology Management, Economics and Policy Program, College of Engineering, Seoul National University, Seoul, South Korea

Sameh Alsaqoor
Mechanical Engineering Department, Tafila Technical University, Tafila, Jordan

Khaled S. AlQdah
Mechanical Engineering Department, Faculty of Engineering, Taibah University, Medina, KSA

Sheeraz Kirmani
Department of Electrical Engineering, Jamia Millia Islamia, New Delhi, India

Samia Tabassum, Mashudur Rahaman, Muhammad Shahriar Bashar, Saidul Islam, Afrina Sharmin, Abdullah Yousuf Imam, Azizul Hoque, Nahid Mahbub, Sayeda Khatun and Mahfuza Khanam
Institute of Fuel Research and Development, Bangladesh Council of Scientific and Industrial Research, Dhaka, Bangladesh

Akira Nishimura, Masanobu Kakita, Junsuke Murata, Toshitake Ando,
Yasunari Kamada and Masafumi Hirota
Division of Mechanical Engineering, Graduate School of Engineering, Mie University, Tsu, Japan

Mohan Lal Kolhe
Faculty of Engineering and Science, University of Agder, Grimstad, Norway

Hermes José Loschi, Julio Leon, Yuzo Iano, Ernesto Ruppert Filho, Fabrizzio Daibert Conte, Telmo Cardoso Lustosa and Priscila Oliveira Freitas
Department of Communications, Faculty of Electrical and Computer Engineering, University of Campinas (UNICAMP), Campinas/SP, Brazil

Saifal Talpur, Carl Johan Wallnerstrom, Patrik Hilber and Christer Flood
School of Electrical Engineering, KTH Royal Institute of Technology, Fortum Distribution AB, Stockholm, Sweden

Ranjit Prasad Godavarthy
Upper Great Plains Transportation Institute, North Dakota State University, Fargo, USA

Eugene R. Russell
Department of Civil Engineering, Kansas State University, Manhattan, USA

Khaled S. Al Qdah
Mechanical Engineering Department, Faculty of Engineering, Taibah University, Medina, KSA